WISDOM ON LIVING A SUSTAINABLE, BACK-TO-BASICS LIFE

INWOOD '74

WISDOM ON LIVING A SUSTAINABLE, BACK-TO-BASICS LIFE

By **Robert Inwood & Christian Bruyère**

Foreword by **Ashley English**

New York / London
www.sterlingpublishing.com

STERLING and the distinctive Sterling logo are registered
trademarks of Sterling Publishing Co., Inc.

Library of Congress Cataloging-in-Publication Data Available

10 9 8 7 6 5 4 3 2 1

Published by Sterling Publishing Co., Inc.
387 Park Avenue South, New York, NY 10016

This book comprises material from the following Sterling titles:
In Harmony with Nature © 1981 by Christian Bruyère
Country Comforts © 1976 by Christian Bruyère

Distributed in Canada by Sterling Publishing
c/o Canadian Manda Group, 165 Dufferin Street
Toronto, Ontario, Canada M6K 3H6
Distributed in the United Kingdom by GMC Distribution Services
Castle Place, 166 High Street, Lewes, East Sussex, England BN7 1XU
Distributed in Australia by Capricorn Link (Australia) Pty. Ltd.
P.O. Box 704, Windsor, NSW 2756, Australia

Design by Yeon Kim

Printed in China

Sterling ISBN 978-1-4027-5774-7

For information about custom editions, special sales, premium and
corporate purchases, please contact Sterling Special Sales
Department at 800-805-5489 or specialsales@sterlingpublishing.com.

Contents

Section Three

ANIMAL SHELTERS 156

Section Four

STONE STRUCTURES & BLACKSMITHING 224

Section Five

HEATING STRUCTURES & ROOT CELLARS 290

Foreword

Ashley English

It's interesting how memories of things made by hand have a way of staying close to the surface. I'll never forget the first pie I baked, a pecan-laden beauty, and eyeing it anxiously as it cooled on my mother's back porch. The same goes for the ramshackle tree forts and scrap-wood skateboard ramps my brother and I cobbled together, not to mention the (in retrospect hideous) sweatshirt I sewed in high school. For as long as I can remember, my hands have been in constant motion: stirring, whisking, chopping, lifting, hammering, gathering, and otherwise feeling, creating, and crafting my way through life. In many senses, what I've made by hand has defined who I am. Within the pages of *Building Country Comforts* it becomes quickly evident that the handmade has also similarly defined authors Christian Bruyère and Robert Inwood.

I find few things as rewarding as what I'll collectively refer to as the "slow life." Whether that means rendering apples gathered from my mother's abundant tree into apple butter, transforming milk into yogurt, building raised beds in my kitchen garden, or simply gathering around our oak dining table for a home-cooked meal with friends and family, if it takes time, patience, and perseverance to do it, count me in. There is something so deeply satisfying, in perhaps even a primal sense, in creating something by one's own hands, for one's own purposes. Witnessing, from concept to completion, the genesis, fruition, growth, and harvest of goods is one of life's greatest rewards.

As a culture, an increasing number of individuals are recognizing and embracing the wonder, pleasure, and abiding satisfaction that stems from the handcrafted, homemade, slow life. While some never strayed far from it (the authors of this book, are themselves dyed-in-the-wool "slow life" denizens), a new collective is forming—both domestically and internationally—and learning to build their own greenhouses, generate renewable sources of energy, and knit their own sweaters. Together, they're discovering means of meeting their own food needs, taking up hammers to fashion their own living quarters, and connecting with landscapes in new, and yet familiar, ways. Collectively, they're redirecting the toll they exact on this planet, seeking solace, balance, and, dare I say, sanity in the pace afforded by living the slow life.

The even greater, less obvious reward offered from engaging in this inherently tactile way of living—greater than a pantry lined with home-canned provisions, a hand-dug root cellar, or even an entire living structure built stone-by-stone—is found in the community it creates. The slow life invites inclusion, celebrates camaraderie, and, arguably, at least from personal experience, necessitates forging networks with others. Whether it's sharing skills, wisdom, tools, or sweat equity, the slow life teaches us that the greatest pleasures in life are made possible, and more meaningful, when they are experienced together. In this book, that message resonates implicitly, coursing quietly through each chapter.

When I first began my own "slow life," I learned that no matter how many books I read or digital resources I perused, it wasn't until I interacted directly with members of my community that my learning curve truly gained momentum. Rubbing shoulders with beekeepers, farmers, cheese-makers, and long-time food preservationists expedited the scope of my knowledge beyond compare. Within the pages of *Building Country Comforts*, a similar theme prevails: both the authors, as well as the homesteaders profiled, found their buildings were raised faster, their trenches dug deeper, and their lives made richer when brought into hand-shaking and meal-sharing distance with the citizens inhabiting their stomping grounds. I've discovered all too often that, in the spirit of doing-it-yourself, individuals find themselves ultimately going-it-alone. I'd argue, and this book showcases, that creating goods ourselves needn't take place in isolation.

If you're looking to further develop and accentuate your own handcrafted, homemade life, *Building Country Comforts* is an ideal read. Christian Bruyère and Robert Inwood have penned a manual for incorporating slow-life structures into your homestead. From blacksmithing to stacking an outdoor stone stove, from building raised garden beds to erecting a solar greenhouse, this book guides the reader in an informed, personable, and experienced manner. It's an indispensable tool for showing you not only how the authors handcrafted these items, but also what they learned along the way. You'll read about the mistakes they made, and uncover solutions for bettering your own projects. The authors are intimately acquainted with the best approach and offer seasoned how-to advice for each building idea presented. Originally written over four decades ago, Bruyère and Inwood have been honing their chops (both literally and figuratively!) for some time. Why reinvent the wheel when they've already greased it so thoroughly for you?

For me, *Building Country Comforts* is an obvious, vital, and requisite addition to the slow-life library. I've got books on preserving food, keeping bees, maintaining a flock of chickens, crafting furniture, making soap, growing fruits, and concocting my own cleaning and beauty products, so it only stands to reason that I'd have a go-to manual on building structures for my homestead. Like those trusted resources, this book offers clear, seasoned, thorough instruction—with details on materials and techniques—for successfully achieving the project at hand. The authors know their subjects inside and out, in large part because they've been living with these structures and "comforts" themselves for so long, having had the time to witness their staying power and to see where improvements could be made.

Furthermore, as the how-to books mentioned above illustrate and as *Building Country Comforts* reinforces, abundant living space isn't a prerequisite to engage in slow-life activities. It isn't necessary to have vast tracts of land in order to build your own saunas, chicken shelters, or cold frames. In fact, the "country" part isn't even remotely necessary. If you live in and urban center and have a small plot in your backyard, you will still find troves of valuable projects here. Of course, if you do have a larger patch of green, then the sky is truly the limit for the numerous inspirational ideas on display. The point is that crafting a homemade, hand-built, slow life needn't wait to begin until you've secured that bit of acreage you pine for daily. It can begin now, right where you are, with whatever vista *you* wake to each morning.

For some folks, the stumbling block preventing them from truly embracing a slow life involves a variation of the question "why take the time to do these activities?" Why build your own greenhouse when you could order a pre-made one, delivered wholly intact right to your door? Why whip your own butter, when the packaged sticks lining grocery store coolers have always served you well? Why put in a root cellar at all, when modern refrigeration, cross-country highways, and overnight planes enable nearly year-round access to any type of produce under the sun? Here's why: because it puts you back in charge. As so many aspects of modern life become mechanized, routinized, and otherwise removed from the domestic sphere, precious knowledge for basic self-sufficiency skills becomes lost. Knowing how to provide for and meet the most essential needs of ourselves, our families, and our communities (in other words, the creation of food, clothing, energy, and shelter) far surpasses the level of "hobby"-type interests. Being able to—on our own—meet our basic human needs allows us to resume our position in the driver's seat, wherein we navigate the course of our lives, no longer subject to the whims of economic swings, weather patterns, or political volatility. Aside from creating structures, products, and goods and services, we create invaluable lifelines when we engage in a hands-on approach to existence. If we need a scarf, we can knit it ourselves. If we desire fresh eggs from a reliable, nutritious source, we can raise our own flock of laying hens. If we want fresh produce, even in winter, we can erect winter-hardy cold frames to help us achieve that goal. In short, we can have what we want, when we want it, and how we want it if we're able to meet our own needs.

When asked to define what constitutes "homemade living," the over-arching title of my series of books on small-scale homesteading, I typically reply that, for me, the homemade living approach is anything involving a tactile, hands-on orientation. That might manifest in a number of ways, but the consistent thread lies in the goal of achieving a purposeful, mindful, directed, and inherently involved approach to living one's life. Part and parcel of that orientation is a personal relationship with the land, and in working with one's hands to create a rich home life and, perhaps even a livelihood. To one person that might mean growing a market garden and selling the produce for a small profit; for another that could translate to creating a fully self-sufficient homestead—providing shelter, food, basic energy needs and more. Whichever direction you're moving in, *Building Country Comforts* can help.

Building Country Comforts will become a vital tool in satisfying your desire for achieving greater self-sufficiency. The authors, who have devoted their lives to creating, living in, and advancing the design of these helpful "comforts," are primed and ready to help you craft your own handmade, homemade, slow life. Get ready to start hammering, sawing, stacking, and forging a whole new set of close-to-the-surface memories!

Ashley English
Candler, North Carolina

Introduction

Robert Inwood

Building Country Comforts is the fourth edition of two books originally published in the mid 1970s: *In Harmony with Nature* and *Country Comforts*. This release marks the first time that all of the material in both of the classic homesteading manuals has been reprinted—combined under one cover—since the books' initial publication. Both books were born of necessity, as both Christian Bruyère and I were contemplating the design and construction of our own rural homesteads. While investigating various building concepts and techniques, and in discussing these matters with local friends and acquaintances, who were further along in this same endeavor, a wealth of valuable information began to emerge—inspiration struck! At that point in time, there were several coffee table–style books that showcased unique owner built rural houses, but relatively few books available that described how the structures were actually conceived and constructed. Christian and I set out to fill that void, presenting the material in ways that would be easily understandable to people who might not have a personal background in design or construction. We researched local, exemplary examples of sustainably-made structures and asked the builders to share their personal experiences in creating them. In almost every instance, the homesteaders gladly shared their knowledge and the stories behind their buildings. Christian and I would typically visit the homestead in person to speak with the builders and view their works. Many of the exact words and phrases of the builders appear in the text today—capturing the personalities behind this practical wisdom. While Christian recorded the interviews and took photographs, I would sketch—sometimes completing drawings, sometimes making notes for technical illustrations. Together we created a narrative that would make it easy for the average reader to understand the construction principles being described.

In the ensuing forty-odd years since these books were originally published, much has changed in the world. One might question their relevance in the increasingly high-tech existence that many live today. However, the continuing appeal of these volumes lies not only in the timeless techniques they offer but also in their themes that transcend trends. Perhaps the strongest of these is the underlying sub-text in all of the homesteaders' tales: self-sufficiency! In a world where events seem to be, more and more out of one's control, and where basic human needs (such as providing shelter and food for oneself and one's family) are often left to "specialists" and corporate entities, the message that YOU can "do it yourself" resonates more strongly than ever. Current buzz words—like sustainable, regional, or green evoke a collective desire to return to a lifestyle that is more conscionable and personally controlled. The voices and images contained within this book reach across the decades and continue to lend sound advice and inspiration for those who do seek self-fulfillment through responsibility, through an attempt to benefit the earth and environment—in the present and for the future. Christian and I express our continuing gratitude to the many individuals (some, sadly, now deceased) who generously shared their experiences in the creation this book. And we send our best wishes to those just now making their own way for a happy and self-actualized life "in harmony with nature."

Robert Inwood
Illustrator & co-author
British Columbia, 2010

Section 1

PLANT SHELTERS, GARDENING & OUTBUILDINGS

CHAPTER 1

Ground Type or Sunken Greenhouse

Life's too short to be trading it for money. I did just that for twenty-five years. But for the last thirteen years, I've spent a lot of energy working hard on a worthwhile project—our homestead. I used to work on long jobs like constructing the Alaska Highway, back in 1947. I saved up my money, and by the time I got around to spending it, it had devalued so much that it just didn't seem worth the trouble of accumulating it. Nevertheless, I spent a few more years traveling around, working where work was available, until one day I found a place I wanted to call home. It was in a beautiful spot, secluded, at the end of a road, with an old house on it. We bought the land very reasonably and immediately began fixing up the place. Soon we got the feeling we didn't want to move again.

We were inspired to make lasting improvements and to develop the land to accommodate our needs. We became compulsive homesteaders, building and planting and raising and hunting. The resources were all ours to make do with as we wanted. We could waste them through ignorance and idleness or learn about what we had to work with, producing what we needed right on our own homestead.

Now we have nearly everything we need in case there comes a time when we can't just go out and buy it. We have good soil, with water coming down the mountainside to irrigate our place. We have fruit trees, and pasture to feed our livestock. There are many wild animals that we hunt, and there is timber for constructing the necessary buildings. And we have some good neighbors who are also conscientious homesteaders. We do a lot of trading with them. Some years they have crops that are excellent when some of ours fail, so we trade with them. You can't go it alone—there's just too much that needs to be done. You have to cooperate with your neighbors, each concentrating on particular tasks.

We decided we needed a greenhouse so we could have some control over our crops. Depending on good weather may work in some milder areas, but in our region, we are constantly threatened with late frosts and long stretches of foul weather. Some years, we lose a lot of crops. Several times I watched my wife's frustration after she labored over her little plants, transplanting and weeding them, just to see them destroyed by a late frost or flooded by a long, heavy spring rain, which interrupted their growing cycle. It frustrated me, too, knowing we wouldn't have enough of a harvest to last us through a long winter without having to go out and buy a lot of produce.

We figured that the best place to build a greenhouse would be in a clearing with plenty of southern exposure, preferably with some protection from the cold north wind. With good exposure to the sun, the greenhouse could capture the solar heat and retain it, providing a nourishing environment for the plants. Protection on the north side would help keep in the warmth and cut down on the energy needed to heat the structure.

Another way we could economize on energy and hold in the heat would be to build the greenhouse into the ground. So, we dug a 2' hole for the structure so that it would be protected by the ground's moisture and warmth. Even when the temperature above ground level is -20°, it remains only 40° below the frost level. Everything in the ground keeps warmer, so less fuel is needed to protect the starting plants through the early spring. We would have gone even deeper into the ground, but we encountered huge boulders (**FIG. 1-1**). As an added precaution, to prevent periodic cold spells from destroying our early efforts, we also heated the greenhouse with an efficient wood-heating system.

FIG. 1-1

FIG. 1-2

FIG. 1-3

Such a heating system should be foolproof. It must be able to give off continuous heat through the late winter and into the early spring danger period, when a momentary frost could wipe out all your preparation work. It should be located somewhere along the north wall, so that plants to the north—farthest from the sun—can get the most heat.

Our heater burns dry cottonwood very efficiently. You put a stick in and it burns to the last ash, going twenty-four hours or more without the need for reloading. I designed it using an old wood heater door, welded onto an old sawdust burner firebox. It has a 12" galvanized "smoke pipe" which is connected to the firebox, extending the length of the greenhouse (**FIG. 1-2**). This smoke pipe acts as a heat exchanger unit, radiating the heat throughout the greenhouse with the help of an aluminum reflective shield that reflects it into the room. This unit is on a slight incline pitch from the firebox to the east wall, 4" in 14", so the

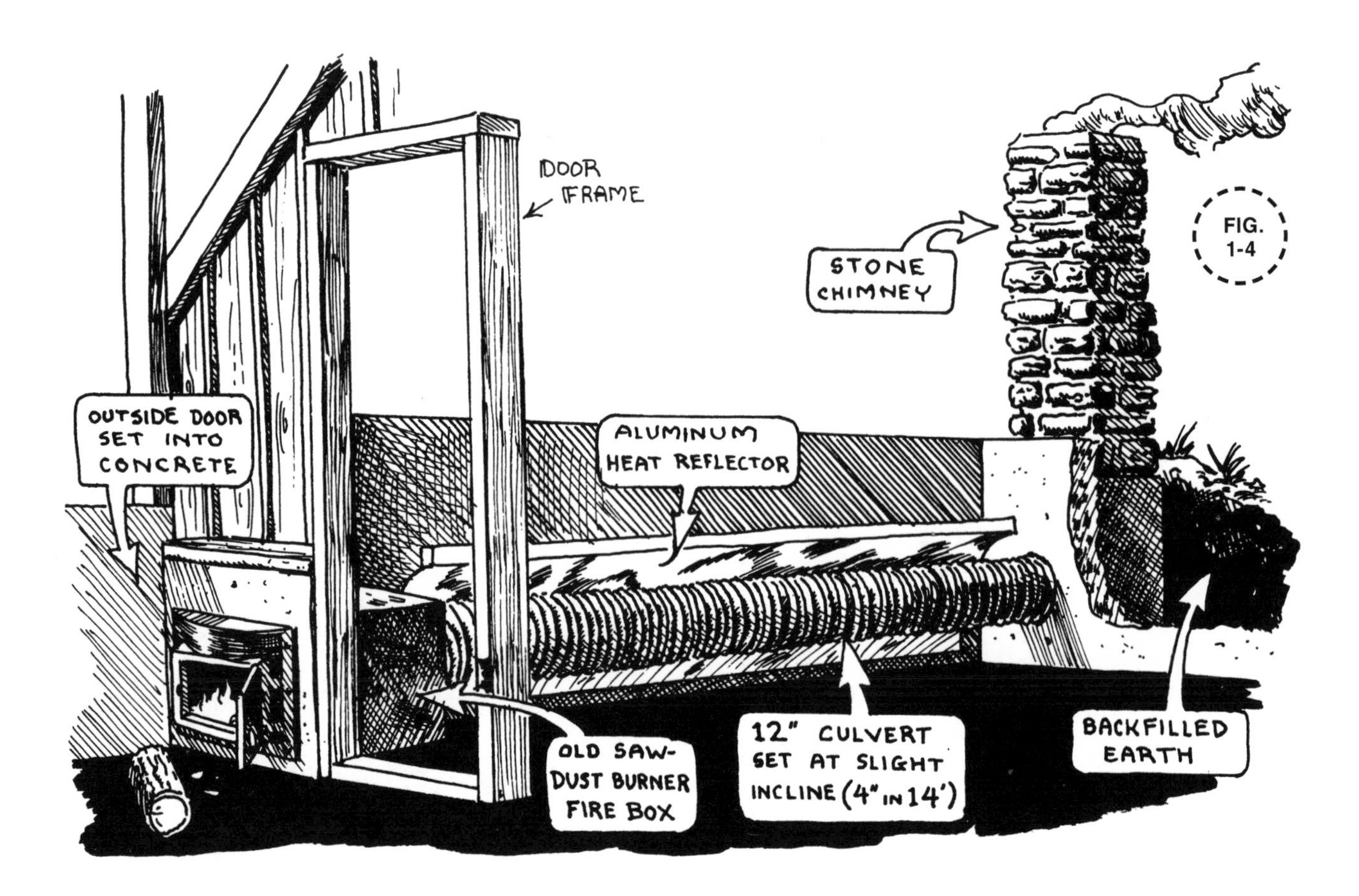

smoke can properly travel through it and out the stonework flue chamber and stovepipe chimney located outside the east wall. This incline also allows the creosote to run back into the firebox instead of accumulating in the smoke pipe; but an even greater incline would have been better. The loading door is outside to prevent drastic temperature changes when refueling, and also to provide the oxygen from outside instead of depleting the moisture content from the greenhouse (**FIG. 1-3 and 1-4**).

The outside dimensions of the greenhouse proper are 7' in width by 14' ⅜" in length. The length was determined by the width of the glass panes, plus space between the glass (tongue of rafters), plus ⅛" allowance for each pane (to be explained in detail later). It was important to me that the building be laid out so the glass would come out evenly—because the glass had been precut.

This greenhouse has an outside porch area, which extends to the west. The outer dimensions of the porch are 7 × 7 feet, making it a 6½ × 6½ foot enclosure for protection from the weather when stoking the fire. It is also a shelter for the firewood, keeping it dry throughout the year (**FIG. 1-5**). This shelter was included in the design plans and was allowed for when we set up the forms for the concrete foundation.

The actual construction did not begin until most of the designing had been completed. I even figured out the braces for interior benches and shelves so that iron pegs could be sunk into the setting concrete, instead of drilling holes into it later to accommodate them. All openings and interruptions in the walls had to be figured

FIG. 1-5

out exactly when building the forms, otherwise changes might have forced us into altering these spaces after the concrete was poured.

On a small job like this, where the pouring of concrete was done by one person alone, the pour was very slow and did not require a solid form, especially since a lot of stones were used. Stone is harder than any man-made material. You can save money by using all the stones the walls will take, providing you put concrete between every stone to cement them together. The forms were made of one-by-ten horizontal boards, braced by two-by-four uprights, which were held at the proper width by spacers. They were wired together near the top and bottom to secure them in place. The wires, of course, were left in the concrete after the forms were removed. The south and west walls were poured first. Then the same form panels were used for the other walls, so as to conserve on materials. The whole job was more than I wanted to pour at one time, working by myself.

I began the concrete work by pouring a footing around the greenhouse. This footing was the height of the one-by-tens used for the inside form. The outside form was a dirt bank, 12" away. A footing should be at least twice the width of the walls. My walls were to be 6" in width. The height of each wall depends on the slope of the land. The front, south wall needed to be 24", so a form was built to that height. This form extended from the east wall to 3' beyond the greenhouse proper to allow for a stairway at the center of the south porch wall. The west wall was the same height and the north porch wall was stepped up a few more inches, making it level with the ground. It had to be stepped up again to 48" at the beginning of the greenhouse to compensate for the sloping ground. It remained this height until after the northern half of the east wall. It was then stepped down to 24" to allow the southern light in. Here, the form was built around the smoke pipe as the inner west wall form was built around the firebox, both already being in place. Two pilasters were added at equal distances of 7' along the back wall to protect it against eventual back pressure from the expanding ground in winter. These pilasters were 10" wide, stretching diagonally from the top of the north wall to 3' beyond it at the base. Each pilaster was reinforced with cross-welded metal rods for added support. Holes were then drilled into the forms to accommodate the bolts and pegs for the shelves and tray before the cement was poured (**FIG. 1-6**).

I used a mix of Portland cement and sand (1 part cement to 1.5 parts sand), putting in as many stones as possible. Cement was used instead of concrete blocks because cement allows moisture through the walls, while concrete blocks have air spaces, which cause the moisture to evaporate within them. Moisture is necessary to create humidity and should be allowed in whenever possible. I tamped the freshly poured concrete slightly with a stick to press it down and then hit

the bracing studs really hard several times with a hammer to make it settle. You can see the air bubbles come up and the top of the concrete level right off as you rap on the studs. After the concrete was tamped and leveled, bolts were sunk into the setting concrete to accommodate the two-by-six sills that would be resting above these walls (**FIG. 1-7**).

I used rough-cut material throughout the greenhouse to get the full 2" out of the board instead of losing ½" with the planing. The front sills overlap the concrete by about 2", shielding it from direct runoff. These sills are tapered 12 percent on the outside edge to prevent runoff from entering into the greenhouse. The taper enables the runoff to drain down into the dirt outside (**FIG. 1-8**).

Instead of working with the usual 16" to 24" center spacing with the wall studs and the roof

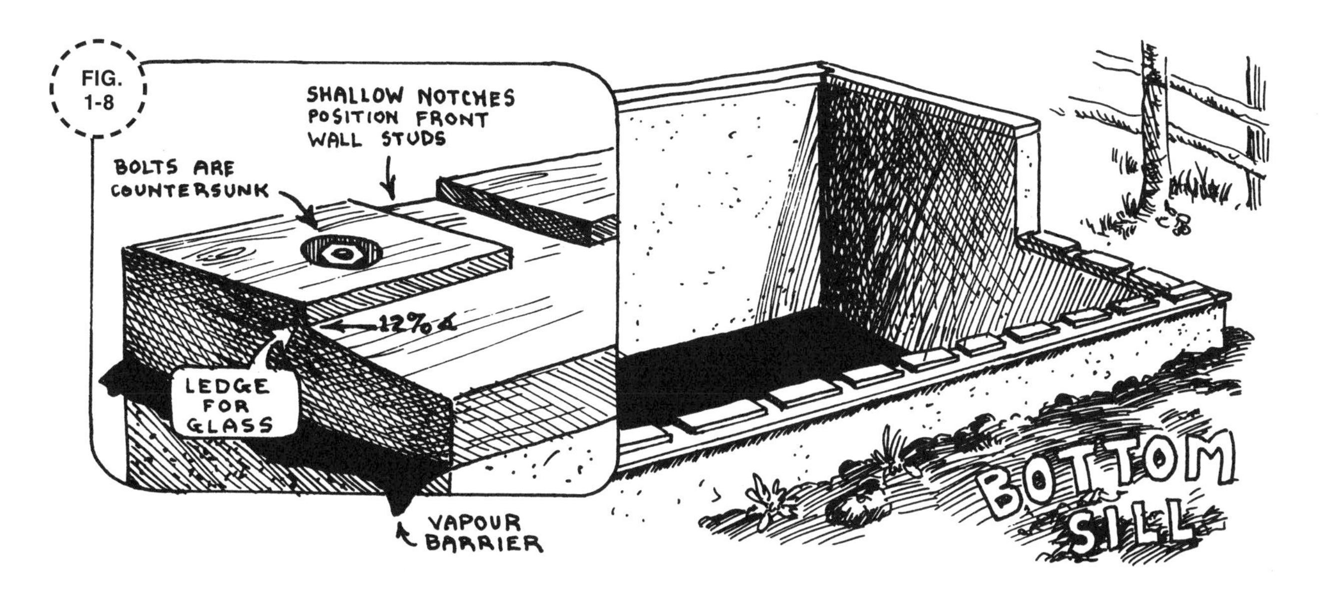

rafters, I calculated the spaces so the distances between all the studs and rafters would be equal. I had to do this because all the glass I used was salvaged ¼" glass that was already precut at 12" widths. The glass was salvaged from a plate glass company. I got a great price on it—50 cents per foot—because most people can't use small ¼" pieces, since sashes are not made for them. I bought all the plate they had from which 12"-wide pieces could be cut—length did not matter. I had to have the company do all the cutting because ¼" is too difficult to attempt to cut without the proper facilities. I planned on 13 panels each with 12"-wide panes. This meant that each panel had to be 12", plus the space between the glass taken by the 1" stud or rafter tongue, plus ⅛" allowance for clearance on either side of the glass, plus 2" on each wall for the width of the corner two-by-four, making the total wall length 14' 4⅜". Each of the studs and rafters were cut out of the rough-cut two-by-six stock, making them plenty strong, at such narrow centers, to hold any snow load, even if single-weight glass was used. I wanted these rafters to be strong because within ten years, the rot will weaken them. It's best to be prepared in the beginning for eventual rotting, instead of having to do the job over again in a few years.

The pitch of a roof that is composed of only rafters and glass should be quite steep, having at least a ½' or 3½' in each 7' pitch, so as to allow the accumulating snow to slip down off the panes.

The studs between the short wall's windows were cut slightly over 24" in length to accommodate the 24" windows. They were rabbeted on the outside end so that all the protruding tongues between the window laps would be centered 1" high and 1" across. The widths of the laps were each approximately ½" depending on the exact width of the stud. The studs were placed at slightly over 12" centers and were capped with a two-by-six wall plate, which was lap-notched into the crossing side-wall plates at either end. The glass was then set in and framed with the ¾" strips left over from rabbeting the studs (**FIG. 1-9 AND 1-10**).

Next came the rafters. I cut them in such a way that I got two rafters out of every 10' board; using the same angled top cut for both sides, they provided the necessary pitch. These rafters were rabbeted in the same manner as were the front studs and were cut at angles, top and bottom, to allow them to seat as much as possible at the

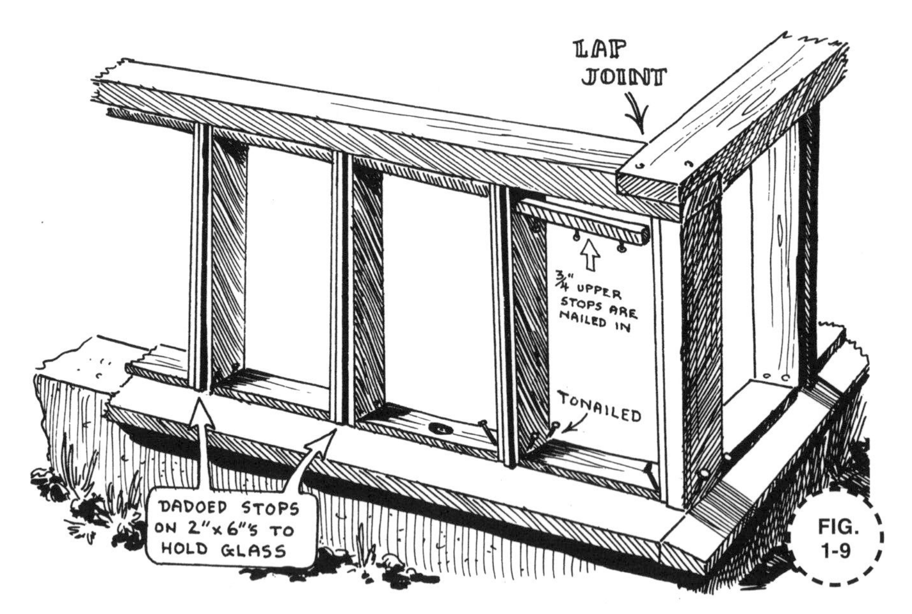

two-by-eight ridge beam and the top wall plates. Since the rafter tongues between the panels were ¾" higher than the glass, planks could be put across them, resting on the wood instead of coming in contact with the glass and possibly shattering it (**FIG. 1-11**).

The glass was now ready to be set in as shingles would be, starting from rafter bottom up. Before laying in these lapped panes, I first set a bed of putty down. I used black caulking compound for this job, because it does not dry and crack as readily as regular putty does. Small metal S-clips were then nailed into the rafter bottoms on either side of the glass. The length of the bottom pane was then measured and another set of clips were nailed in where the top of this pane meets the rafters. The bottom pane was then seated into the putty between the clips and the above piece was lapped about ½" to ¾" over it. At this overlapping, a tapered wedge was set in above the rafter seats to compensate for the lap, allowing the glass to set snug into the putty. This wedge was feathered with a hand plane until it was the proper size.

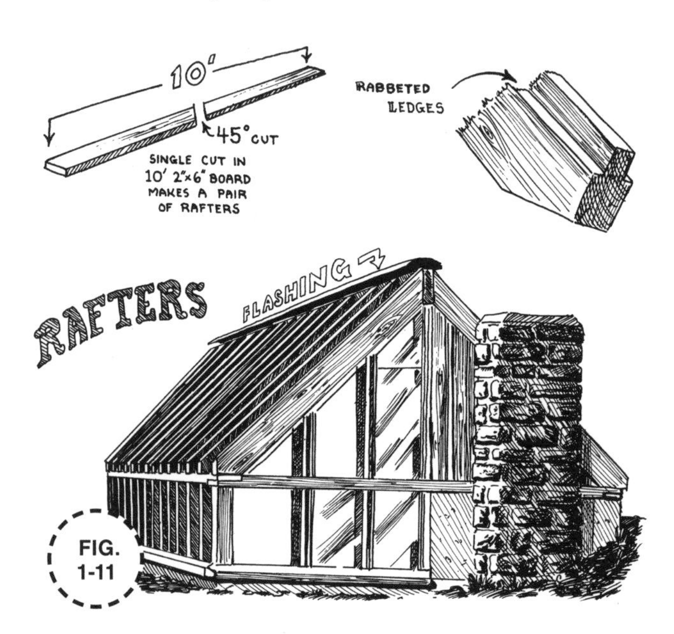

After the bottom panes in each panel were seated in place, a measurement was taken of the space between the lower end of the glass and the top plate. This measurement designated the size of the "bird stops," called this because they do actually stop birds from coming in (**FIG. 1-12**).

It is important to have a roof that allows the source of light to come from straight overhead, because plants will reach toward the light. So, if light comes in from overhead, plants will grow straight.

The north side of the east wall and the west wall, where the front door is located, were sheathed in on both sides with 1" cedar, because these areas get very little direct sunlight. The front door has 6' 4" of vertical clearance, and was constructed of rough-cut vertical one-by-fours held together with a Z-brace. Thin baton strips were nailed on over the board junctions to seal them off. Above this door, I installed a vent. The vent cover was hinged at the bottom with a strip of rubber, allowing it to open to the inside (**FIG. 1-13**). Greenhouses should have good vents high up where the plants won't get the direct draft. High vents also help circulate the air to prevent hot air from collecting at the top, which would eventually get the greenhouse warm enough to burn the plants on warm days.

Just outside the vent area, above the doorway, a two-by-eight crosspiece was nailed in place to accommodate the porch rafter bottoms. Uprights were set up above the outer west wall and a wall plate was nailed in across them for the outer rafter bottoms. The opposing rafter tops then met at the ridge beam. Stringers for the shakes were nailed in place and an inverted V-shaped plate was attached where this roof separated

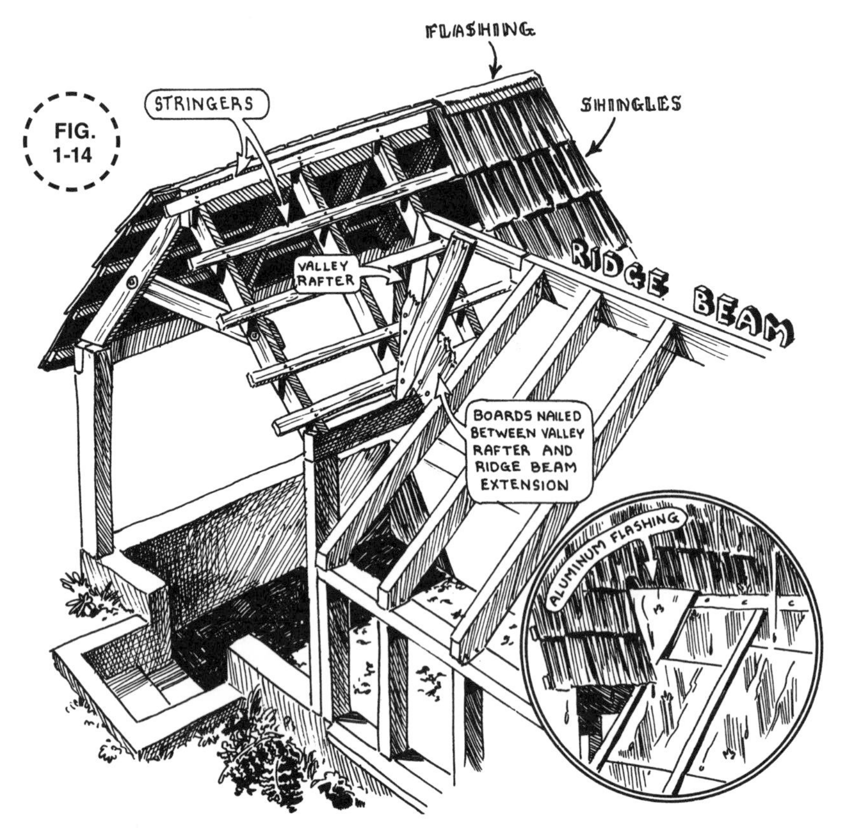

from the greenhouse roof. A small ridge cap was extended from the apex of that inverted V to the greenhouse roof peak. This area between the roofs was sheathed in and capped with aluminum flashing to prevent leakage into the porch area. Shakes were then split and put on the porch roof. Another piece of aluminum flashing was nailed in above the shakes at the peak, to seal the roof (**FIG. 1-14, 1-15, 1-16**).

The wood heating system described works excellently in heating and producing humidity for maximum growth in the late winter and early spring. In fact, the plants closest to it—and farthest away from the southern sun—grew much faster and healthier then those closest to the direct sunlight (**FIG. 1-17**). This could be attributed to the hazy, overcast early spring weather during which time the direct sun seldom comes out. For humidity, we simply set a kettle of water on the firebox and let it steam into the greenhouse. And if we wanted more moisture content in the air, we just hosed down the dirt floor and let the remaining moisture circulate throughout. Plants thrive on humidity and should always be supplied with plenty of it.

The bottom shelf along the back wall for the plant flats is supported over two pipe rails bracketed to the east wall and held up with a wire support at the west wall. The pipes are also braced in the center by upright pipes resting

on concrete footings. Along these pipes, two-by-four blocks are spaced and notched over them to hold the crossing shelf boards (FIG. 1-18). Another narrower shelf, also for the starter flats, was suspended above this lower shelf. It is held with blocks nailed to the side walls at either end and supported in the center by a wire bracket hanging from the above rafter (FIG. 1-19).

The flats were made from the cedar shake scraps. They are 13" long, 4¾" wide, and 3" deep. The plants are transferred to these flats shortly after they are started in the front tray, then they are transplanted into the garden. To avoid crowding, there are never more than two rows of plants in each flat—(FIG. 1-20).

The front tray is used for starting the plants, mixing the soil, and growing plants that will not be transplanted into the garden until well after the last frost threat, if at all. This tray extends from one end of the south wall to the other, held on brackets attached to peg bolts (**FIG. 1-21**). It is supported at the center by a crossing pipe rail, which is encircling a peg bolt at the south wall end and is held up on an upright welded to it on the inside end. The tray bottom is of corregated galvanized roofing, which lines the boards on either side and is turned up 1" at the ends to hold water. I had to split the ends every few inches on the ribs with a pair of shears to make the bend, otherwise it would just widen itself and go out of shape (**FIG. 1-22**). I didn't want to use wood for the tray bottom because it would just rot out after a couple of years and need replacing. I was lucky to have an old piece of galvanized roofing. It is good, heavy stuff, not like what's sold today. Everything is now made so thin and cheap—except the prices—there's nothing "cheap" about the prices.

We added only cow manure to the soil. Even without the use of chemical fertilizers, all the plants grew very well in the greenhouse. We definitely wanted to stay away from using any such inorganic conditioners, because they quickly deplete the soil by speeding up the growing process beyond natural limitations, exhausting it prematurely.

The leaf lettuce just shot right up, tasting as good as if it were grown in the garden. The peppers were enormous. They really benefitted from the heating system. Even the peppers closest to the sunlight were dwarfed compared to the ones nearest the heater. We started them in late February and by May, they were already flowering. Some plants didn't do as well, but I attribute this to the soil. It was a bit too hard, which does not allow for good drainage. It also seemed to be in need of lime, which is a deficiency in most of the

soil in this area. Lime deficiency also makes sour weeds grow well, and we had plenty of them. Ashes are a cheap, excellent source of lime. They are also a good mulch and help keep worms and pests out of the soil. Squash was by far our fastest growing greenhouse crop. It grew so huge even by early spring that we had to take out a few panes and add a temporary plastic shelter to the greeenhouse to contain it long before it was safe enough weather-wise to transplant it to the garden (**FIG. 1-23 AND 1-24**).

Because the greenhouse is a closed shelter, and bees can't come in naturally to pollinate the plants, we help the process along by rubbing the blossoms from one plant with the end of a feather and transferring the pollen to another unfertilized plant (**FIG. 1-25**).

All and all we are extremely pleased with the greenhouse and feel that our efforts were well worth it. Even with the slightest fire from the wood heating system, we can keep the tomato plants going until December, getting several quarts of ripe tomatoes. We even have enough to give to neighbors. Outside, most years, the frost kills the tomatoes before the first ones get a chance to really ripen.

CHAPTER 2

Solar Greenhouse

It had become pretty obvious to me that purchasing food in the near future might be an issue. The quality of food that would be available for purchasing would soon be substandard, anyway, because wholesalers and distributors had been milking the food dollar with increasing ruthlessness. Growers were getting less and less percentage-wise for their efforts, so the production of high-quality produce was getting neglected, since monetary greed leaves no room for the incentive of quality to flourish. The "syndrome" of city living is such that people in large urban centers spend all their work energy making money to purchase their necessary food items. Consequently, city dwellers have no training in developing their own crops in case the food is not shipped in—a potentially helpless situation to be in. If food shipments were cut off, all the money in the world probably couldn't purchase those simple little items we've learned to so take for granted.

And so we made the inevitable move from the city, built a temporary shelter to house our large family, and began developing a piece of raw land. Of course, one of our first major concerns was to prepare an area for a garden that would feed us—we were ten to twelve adults. This wasn't an easy task since the ground we were working with was 50 percent gravel and 50 percent light sand, with not enough soil to stain your hands when wetted down. We had a shredder, so we went to all the neighbors' yards and collected their leaves and shredded them. The leaf shreddings decomposed well. Soon we didn't even notice them in our garden, except for the presence of a new fiber in the ground. We also searched and scrounged the nearby hillsides for topsoil, and within a few months, we developed the garden plot into workable soil.

An indispensable item that we have had a great deal of success with is the compost bin. We built a square frame out of poles, about 4' in height, 8' on each side, to hold leaves, manure, garbage, and other organic substances that quickly decompose. In such a small square pile, the decomposing process works extremely fast from the bottom center, outward. Within a year,

most of the organic substances in this pile break down into rich humus, which we spread over the garden and till in to build up the soil (**FIG. 2-1**).

The first year on our land was one of unpredictable weather. It was rainy and cold all spring, and even in summer there wasn't enough direct sunlight to sustain a lot of our crops—especially since the soil needed so much attention. Because of this, and also to avoid having to deal with the weather, we decided to expend our energies on developing a controlled-environment greenhouse. Here, we could bring up most of our vegetables in flats and then transplant them after they'd had a proper start. This way we'd also have more than a month jump on the other people in the area. It is risky to start most crops in the ground before June in our area, because of late spring frosts, which annually take their toll. We just start them in the greenhouse, thin them out right in the flats, and then transplant them into the ground later. Everything took off right away in the greenhouse. This was very encouraging. If it grew half as well outside, we'd have excellent crops. We didn't lose any of the greenhouse-started plants. It was unlike having hit and miss rows of plants—we controlled exactly what we wanted. We put in 137 tomato plants without losing a single plant. And that was true of all our transplants—they went right on growing rapidly. It was definitely worth the effort that we put into this type of controlled environment. I wouldn't do it any other way, except maybe to add a greenhouse to an already existing dwelling house, so that the greenhouse could get the heat exposure from the dwelling all year-round.

The site for the greenhouse was an electric power line clearing with good southerly exposure. It gets the morning sun as well as uninterrupted exposure throughout the day. We leveled off the top of the clearing and dug corner holes outlining a 12 × 16 foot area. Other holes were dug at 4' intervals along these periphery lines. We creosoted cedar posts, which were then put into the ground in these holes. We chose a size that would allow us to keep fragile plants, such as peppers and cucumbers, throughout the entire growing season. Fortunately, I had a transit to make sure these posts were really level. Cedar sills were then cut and placed above these posts, just above ground level. I figured that creosoted cedar on porous soil such as this is good for at least forty years, so why worry about a concrete foundation for such a lightweight structure? (**FIG. 2-2**)

FIG. 2-2

FIG. 2-3

A conventional stud wall was built over the sill in front and on the sides up to about 30", to the bench height, leaving a space on the west wall for a door. We're short people, so we prefer that height. The windows and glass began at this level. Cedar was also used for the sheathing inside and out, with 3" batons between each of the vertical boards. Next, 3½" of fiberglass insulation were put in between the studs with an 18-pound building paper backing on either side. A 4-mil plastic vapor barrier was then stapled in over the building paper (**FIG. 2-3**).

The high 12' back wall was constructed in the same manner. This tall back wall provided us with the necessary pitch and gave us total exposure to the whole greenhouse by making it possible for us to use a shed-roofed design rather than a hip design. If the roof went up and back down as it would in the hip-roofed design, the storage space on the walls would have been cut down considerably. You need a lot of storage space in a greenhouse for fertilizer and soil preparations, such as bonemeal and peat moss. Also, the high north wall provides protection against the cold northern winds and weather (**FIG. 2-4**).

FIG. 2-4

Even with the sheltering north wall and the shed-roofed design, there has been no problem with plants leaning toward the south, not even in early spring, when the arch of the sun is furthest away. The deflection seems to be pretty even in the greenhouse throughout the growing season. If this problem should arise, it can be easily remedied just by turning the flats around. This will force the plants to grow back the other way.

The spaces between the side wall studding depended upon the width of the glass used. After we searched around, we finally found a man who wanted to sell all his used glass pieces. We made him an offer for the whole lot and brought the assortment home. We found the greatest width we could use, saving the most glass, and cut all the glass to that 21" width. The height didn't matter. It could be as long as possible. Our pieces ran between 2" to 20". The longest pieces were saved for the roof because of the greater spans.

We were dealing with double-pane and thicker glass, which was really a hassle to cut. But before long, we became quite proficient in glass cutting. Some of the glass had extreme tensile strength and didn't cut properly. To cut such miscast glass at absolute right angles was nearly impossible because it has a tendency to break away on your line, undercutting itself (**FIG. 2-5 AND 2-6**). If you stack glass with this kind of break, it leaves a small opening between panes, allowing

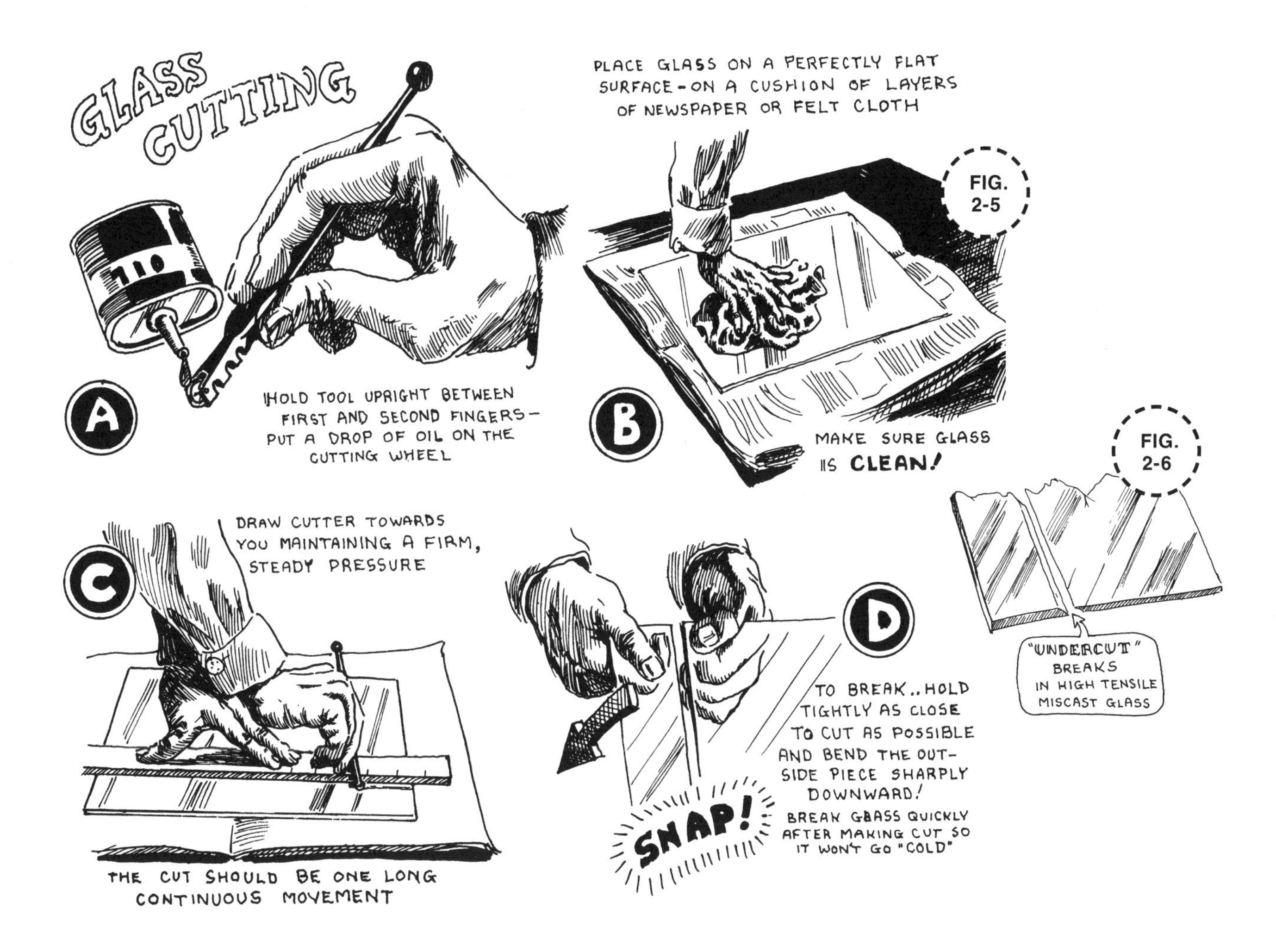

weather to come through. Where I made good cuts, I just stacked the glass vertically, edge to edge between the grooved studs, then nailed and puttied it in. Where the cuts weren't so straight, we put a ¼" to ⅜" lap on it just like a shingle. We put in one finishing nail on each side underneath the glass being lapped, one part way up to hold it in place and one just below the top of that piece to accommodate the above lapped piece and set in the putty. There was no leakage on the side walls and only a small amount of leakage from the overhead glass where the putty cracked.

The side wall two-by-four studs were rabbeted on the outside end to give them a ¼" tongue, just enough to keep the pieces of glass from touching each other on either side of the tongue to accommodate the glass. These laps were set in deep enough to allow a little space between the outer edge of the studs and the glass so nails could be driven in and putty could be spread (**FIG. 2-7 AND 2-8**). A space narrower than 21" was left near the end of each side wall to take the smaller pieces that we had left over. This prevented wasting our resources and spared us from having to compensate for the exact size of panes plus the distances between each panel, in figuring out the overall length of the side walls (**SEE FIG. 2-6**).

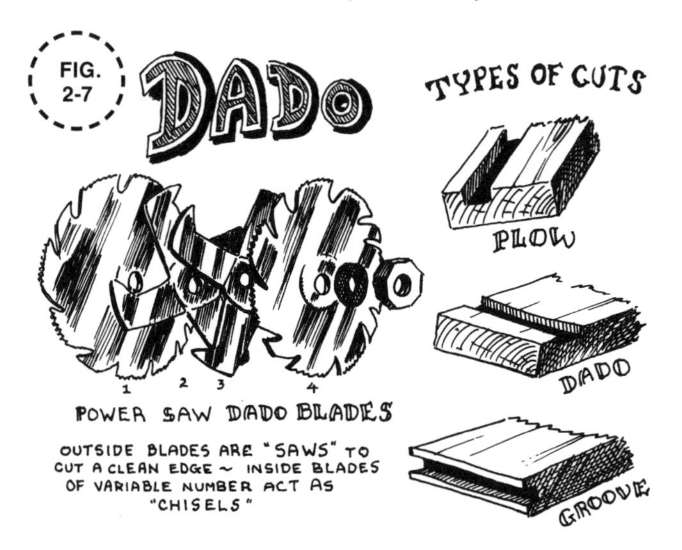

For a necessary artistic touch, I made a stained-glass window for the upper section of the Dutch door. The door provided the greenhouse with a softness that it wouldn't otherwise have had with its rigid glass and wood-framed composition. This was also an excellent way of using more of the scrap pieces of glass we had left over. It further tested my skill in cutting glass. But the came around the pieces was too small—a mistake. Consequently, I had to back it with unsightly welding rod for additional support. This door was hung on hinges to the double two-by-four stud that bordered it (**FIG. 2-9 AND 2-10**).

A Dutch door is an excellent door design in a structure such as a greenhouse where you want to allow the maximum ventilation on hot days, but without letting animals in. The Dutch door keeps the dogs out, but since there is no screen protection, the cats can still come in, walk across the flats, and make all the small plants collapse beneath their feet. They also like to use the flats as kitty boxes, scraping the soil, disrupting the seedlings, and doing their "business."

Screens are a necessity in our greenhouse because of a fly problem, caused by having our compost bin so near. We seem to be propagating as many flies as plants, but screens should eliminate this problem and also keep the cats out.

For the front windows, I constructed lightweight frames—21" square—out of 1½ × 1 inch material. I wanted to keep them narrow so they wouldn't throw off too large a shadow. In doing so, I sacrificed strength for transparency—which was a mistake. The frames should have been stronger to hold the heavy glass. They were grooved on the inside all around to snugly accommodate the glass pieces and were tenon-jointed at the ends. Handles were put on the lower sill pieces for ease in closing and eye screws were fastened for locking. I also installed locks at

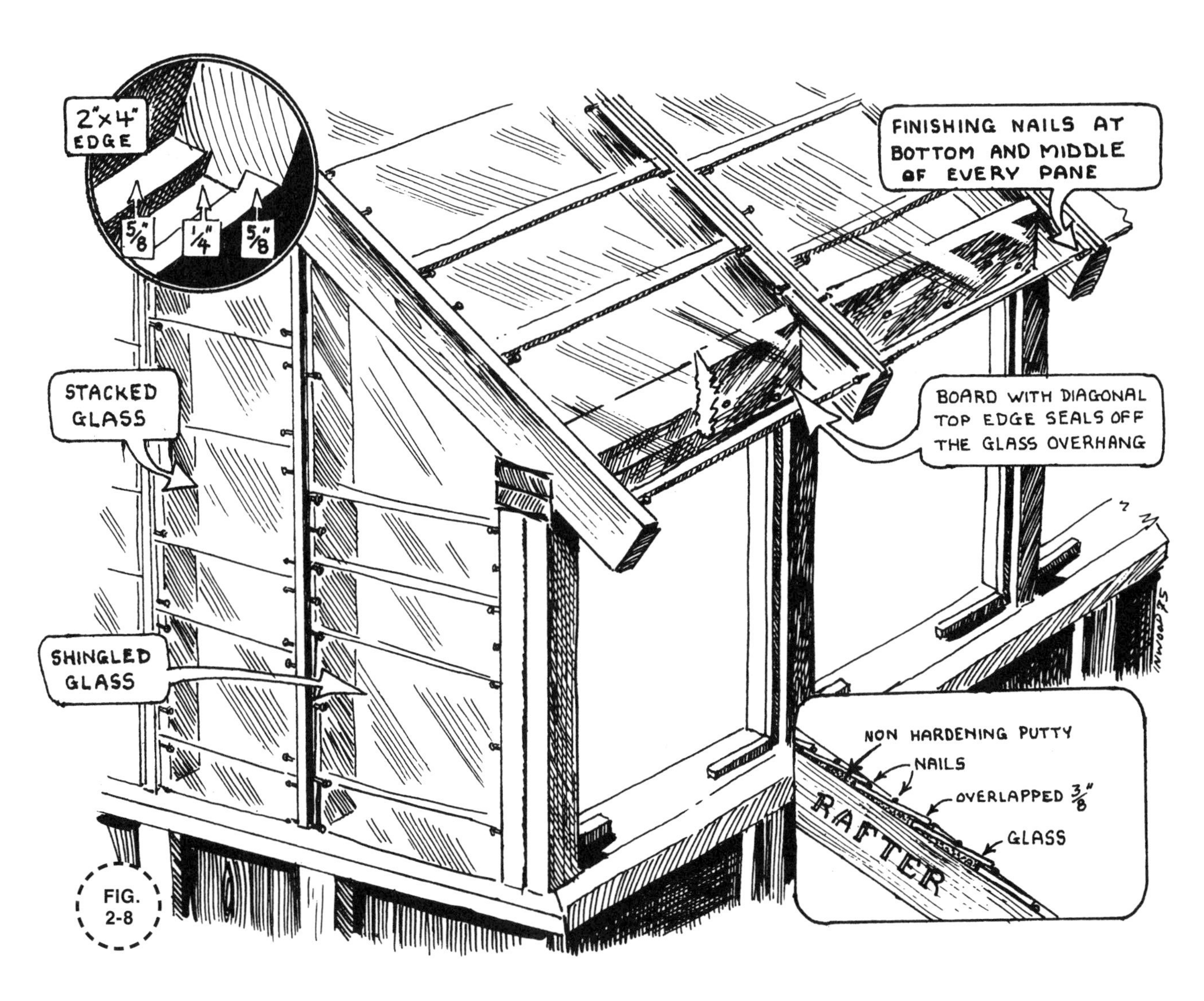

FIG. 2-8

FIG. 2-9

FIG. 2-10

the bottom to hold the windows open so they wouldn't bang shut during a gust of wind. Thin-culled cedar strips were then attached as stops, leaving the area between the ends of the sill open to conserve on materials (FIG. 2-11).

FIG. 2-11

There are eight windows along the front wall going from one end to the other (FIG. 2-12 AND 2-13). I used a lot of windows here for ventilation, because I didn't want to have to whitewash all the glass in the greenhouse to prevent the plants from burning on hot, dry days. As long as the soil in the flats is kept moist, the ventilation prevents the plants from burning. We watered first thing in the morning and an hour before the sun set each day, saturating each plant until a puddle formed. Then, as soon as the temperature dropped, we closed off the greenhouse so it would retain the heat.

FIG. 2-12

Directly above the window spaces, we put in a wall plate of double two-by-sixes laminated together (see FIG. 2-13). The lower rafter ends were cut flat at the bottom to meet this plate. These rafters extend from the front wall to the back wall, spanning 14' with a double-laminated beam supporting them in the center of their span. This beam is held in place above the center studs at the opposing side walls. Here, too, I feel the construction should have been more solid. This supporting roof beam should have been heavier and also be braced in the center by an upright of some type. But we were running out of material and wanted to finish the greenhouse for use. Perhaps an additional reinforcement will be put in later. Even the way it is now, the 33° pitched roof held a load of almost 3½' of snow at one point last winter without sagging.

FIG. 2-13

The tops of the rafters where the glass panes seat into them were rabbeted in the same manner as were the upright studs. All the roof panes were shingled above the rafters, starting at the bottom of each panel and working upwards with each, just like you'd shingle a roof. They were held in place at the bottom with finishing nails and putty, the same as the shingled wall panes were.

Square glazing nails would have been ideal for this job, but we could only work with what we had. In fact, the roof leaked because the putty I used soon got brittle and cracked. I couldn't find any non-setting putty myself, but I advise the use of this kind of putty for sealing in this type of roof. It remains soft and pliable, instead of hardening and cracking with age.

I would have used cedar for the rafters, because it is by far the most weather resistant of the local woods, but clear cedar in 14' lengths would have cost a fortune. I used fir, instead, and coated it with urethane to help preserve it. Hopefully, this coating will allow it to last about as long as the cedar would have. But the urethane has to be replenished at least every other year for full protection.

Next, we made the three top vents that allowed the hot, rising air to escape as the cool, low air entered to circulate. These vents were framed in the same manner as were the front windows, and were fastened to the top wall plate with hinges. The glass used was a heavy rippled, opaque shower-door glass. I used this heavy glass mainly to test my skill at cutting. Four long upright braces were then constructed above the high back wall and a two-by-four crosspiece was nailed in above to connect them. Three pulleys were suspended from this top piece for the cords that are connected to the bottom of the vent on one end to one of a series of hooks on the other. These hooks were spaced at different levels: the highest closed the vent; the one below it allowed the vent to be held halfway open; and the bottom one allowed the vent to be opened all the way. This lower hook was another mistake, because it held the vent open too far. Consequently, a strong wind broke one of the panes (**FIG. 2-14**).

The interior of the greenhouse was designed to fit our needs and was constructed out of the materials we had left, mostly culled cedar. The benches are 30" high along the front and side

walls. They were very simple and economical to build, utilizing a lot of scrap. The inner ledges were nailed to the wall at the 30" height. Several two-by-four upright braces were then set up 24" from the walls to accommodate a front crosspiece at the same 30" level. Then, 1" slats were ripped with a power saw and nailed at 1½" intervals across the two ledges. These racks held the starter flats, which were filled with prepared soil. The flats were also made from the scrap cedar. They were 14" × 21" and 4" in height. Their bottoms were slatted, with ⅛" to ¼" of space between each slat, to allow them to drain down onto the dirt floor (**FIG. 2-15**).

The plants are started in the flats on the front and side wall racks. Once they are developed enough to take care of themselves, they are shifted to the back shelves and more seeds are started, taking their place on the racks. The back shelves are also slatted for drainage and economy of materials. They are braced against the back wall with diagonal supports, which extend from that wall to the front crossing piece (**FIG. 2-16**). The higher shelves are for the more self-sustaining plants, like maturing cantaloupes and cucumbers. To water them, we just bring a ladder and spray them with a fine mist spray from a hose attachment. The cucumbers are set in deep flats along the top of the back wall. They just drape over the shelves and vine down. They were an excellent crop last year. The cantaloupes didn't make it, though, because there wasn't much direct sun to nourish them. They got to the size of baseballs and that was it.

An excellent attraction of our greenhouse is the hotbox that enables us to start plants as early as February or March. We keep the primary flats in here until the plants sprout their true leaves, then they can go out to the racks. This gives us a great jump on the season. This hotbox is 2' wide by 4' long and is 34" deep. It is powered by a series of four 60-watt bulbs located at the bottom. This is a pretty minimal source of heat, so the box was built as tightly as possible. It has two doors: a drop door on bottom hinges to feed the lower slatted shelf, and a lifting lid on top, with framed glass to allow sunlight into the top shelf (**FIG. 2-17 AND 2-18**).

FIG. 2-15

FIG. 2-16

FIG. 2-17

FIG. 2-18

Even when the temperature outside was just below freezing (26°) the hotbox maintained a good 70°. It got us over that early spring fluctuation of cold spells that snap in, threatening to wipe out all the flats, even once the plants have pretty well started. Now, in the warmer months, we just use it at night for the beginning flats. We put the flats in, then turn it on. It holds the temperature at about 75° to 80°. It's almost a better environment than the greenhouse itself. In the greenhouse, you have to keep watching the temperature, while the hotbox always maintains a pretty constant temperature.

The hotbox will take eight flats, but we get a better circulation of the heat when we use just seven—three in the bottom and four on top. The top ones are double tiered, two above the other two. As long as the plants are young, the top flats won't interfere with the growth of the lower tier.

To get rid of the snow on the greenhouse roof, we just turn on the hotbox and within twenty-four hours, the whole building gets warm enough for the snow to start melting. The only problem is that the greenhouse is so low in front that we have to do a lot of shoveling to get rid of all the accumulated snow.

Another necessity in a greenhouse is a large, central mixing box where you have plenty of room to set your flats and mix your soil. We have a large box in the center of the floor for this purpose, with a shelf built in below it for storage (**FIG. 2-19**). Here we prepare the soil. About 40 percent of our soil comes from the bottom of the compost. It gets set out on plastic for six months to air out and is used the following year. The remaining 60 percent of our soil is mostly peat moss with a handfull of finely crushed bonemeal in each 14 × 21 inch flat. This mixture works well. I haven't done a pH test on the soil, but from the way things are growing in it, I'd say it is more acidic. Most plants can take a soil that is more acid than alkaline.

FIG. 2-19

When we close the greenhouse off, it retains moisture so well that we get vapor from the humidity on every pane. I don't think I need to introduce another source of humidity, or it might begin producing mold. I sometimes get some surface scale, mainly because the soil is rich with compost and bonemeal. But a little scale can't really hurt.

For our needs, this greenhouse is proving to be perfect. The only flaw, which I want to eventually correct, is the cracking putty, but that's minor. In an area like the one we live in, where the spring weather is so unpredictable, and where several people must eat from the crops that we grow, a greenhouse such as this one is indispensable to ensure that those crops will be substantial. I wouldn't want to risk being at the mercy of the elements after witnessing the loss of crops by several neighbors last year due to late frosts.

I suppose we could have gotten by with several outdoor hot frames and cold frames, but it's not that pleasant in the early spring to be working outside, bedding your plants and whatnot. And in the middle of winter, the greenhouse is a pleasant, peaceful space to go into and just relax. It's always warm enough in there even when it's below freezing outside.

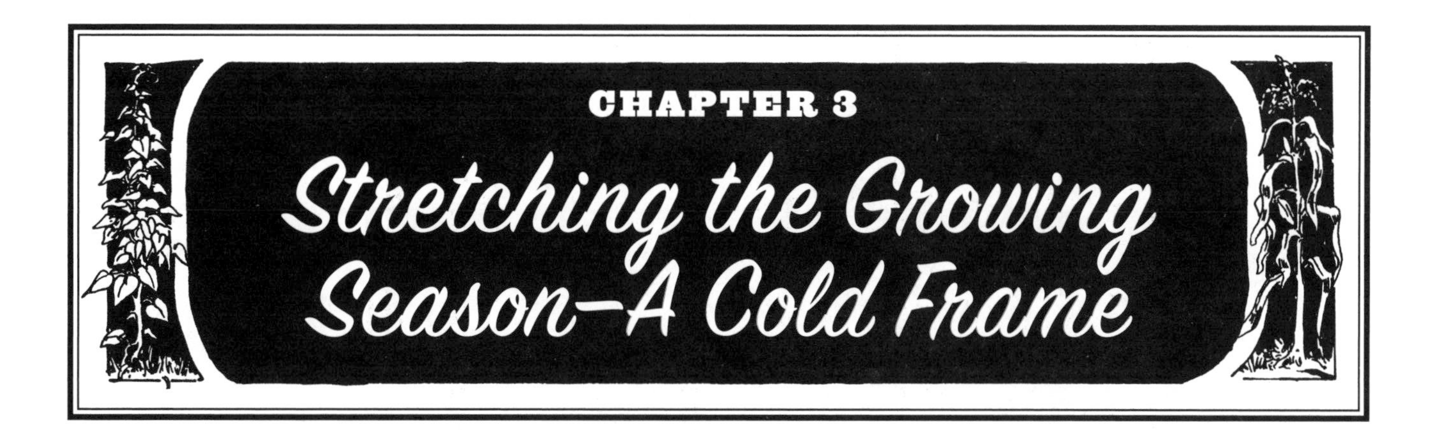

CHAPTER 3
Stretching the Growing Season—A Cold Frame

I built this cold frame to have more control at the beginning and end of the seasons, since the optimum growing period is so short in our northern valley. I wanted to be able to transplant in early May, getting a sufficient head start with the tomatoes, cucumbers, and peppers, and have it all extend longer into fall, so that most of the fruit could properly ripen on the vine. My original design called for a concrete enclosed frame which slopes slightly to the south. This frame was to be covered by a glass-paned roof. But I didn't have any old windows around that I wanted to use for the job, so I used 1 × 2 inches lath sticks as frames and covered them with plastic. This of course didn't work very well because the plastic ripped too easily and the thin wood strips didn't hold their shape. The whole mess was too flimsy but it made it through the first season, barely. The next season we tried the present design which has been working very well.

A concrete frame was used primarily to create a barrier against the rapidly growing couch grass which plagues all the gardens in the area. Also concrete is very solid and permanent. It allows moisture in as it sweats to produce humidity within the frame. The plants love this warm, wet environment. I constructed forms around the 14 × 5 foot perimeter of the frame using one-by-eights with two-by-four stakes to hold them in place and thin braces spaced out along the sides, inside and out. Holes were drilled into the tops of these braces for the wire spacers to go through. A nail was placed at the outside of the form to hold the doubled wire and another nail was twisted between the inside ends until the forms were held exactly 6" apart. They were then leveled with a carpenter's level. These forms were built high enough so they would be above the couch grass, 4" above the surface at the low end and 8" above at the high end to create the slope. The enclosed area was dug out 8" deep for added protection and the inside form was extended down to that level. The 6" wide space between the forms was then dug out for 18" under the outside surface so the concrete pour would go well below the roots of the surrounding couch grass to prevent them from working their way into the cold frame (**FIG. 3-1 AND 3-2**).

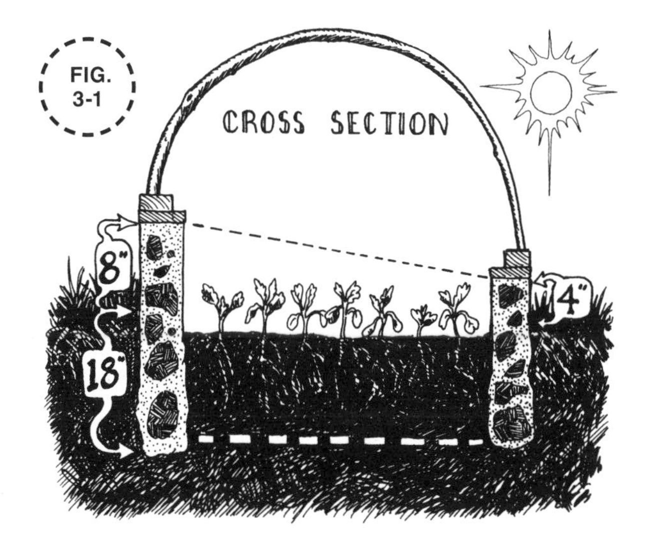

FIG. 3-2

FIG. 3-3

The concrete was mixed in a wheelbarrow and was poured into the forms. An additional horizontal brace was put in, extending between the inside forms of the two sides. The wet concrete was then tamped rigorously to make certain it thoroughly filled the space between the forms (**FIG. 3-3**). Within the forms, along the ends, a 1" board was set in on an angle to create the contour of the north-to-south slope and the remaining concrete was carefully shoveled in from the high side. The setting concrete was then troweled smooth and level at the top. Nails were embedded around the frame with their pointed ends sticking out long enough to accommodate the two-by-six sills and to bend down over them (**FIG. 3-4 AND 3-5**).

I didn't want to get into a lot of nailing and constructing of individual frame covers that would require some kind of hardware hinges for easy lifting. Hardware hinges seem to have a tendency to screw up in such outdoor type uses. And no other flat type covers seemed suitable. Finally, I started thinking about bent-pole "hoop" construction that I've seen around. The arched poles seem to provide a strong support with gravity and downward pressure working against the tension of the arch. Anyway I wanted a high, arched type roof over the top to allow

FIG. 3-4

FIG. 3-5

the plants plenty of room to grow and still be protected within the cold frame during the fall frost periods.

To get the poles, I went out into the woods where the second growth is really thick. In between the medium-size trees were a lot of little guys that would grow up to be no more than 10' tall, and not get any thicker than 1½" to 2" at the butt. I searched for trees with a 1½" butt that had very little taper—less than ½" from top to bottom. This is about as small as you'd want to work with. If they were any fatter they would be hard to bend over into an arch the size I wanted.

I was finding only fir trees for the job. Most other species tapered too much, even at that tiny size. They were cut in early spring when the trees were just getting the new sap (**FIG. 3-6**).

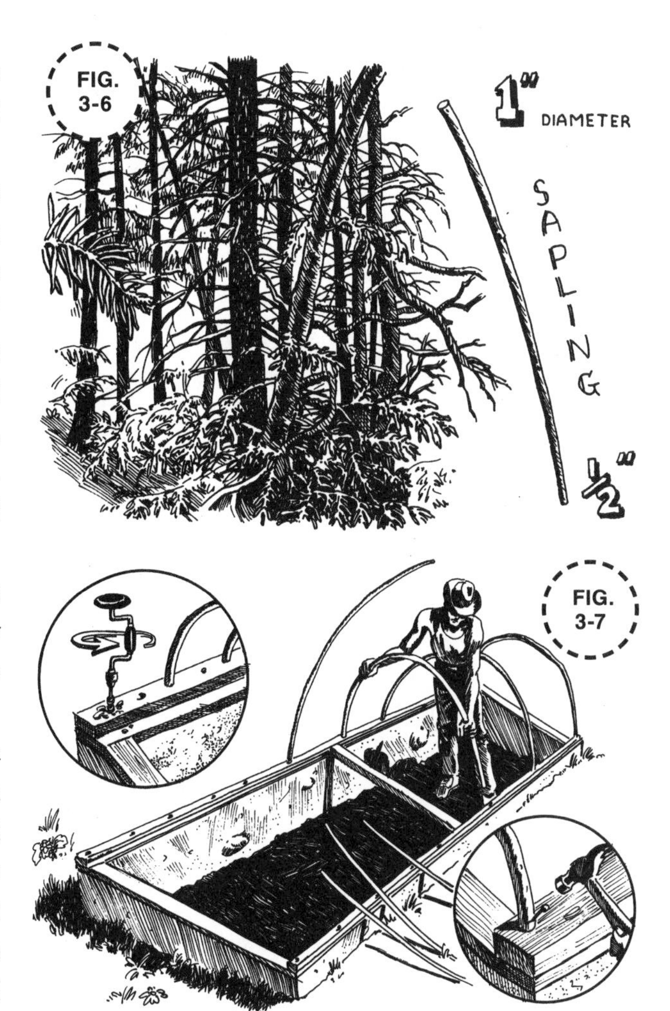

To figure out the length I needed I found a piece of black plastic pipe and arched it between the sides of the frame. I liked the height of that arch, approximately 3' above the concrete sides in the center, and decided to work with that length. In order for the poles not to break, there has to be a true arch. It can't be too shallow. I cut the saplings to the proper size and peeled strips of bark off with a knife. Then I drilled holes into the sill boards on the north and south sides, making a larger 1½" hole at the top side for the butt and smaller 1" holes on the bottom side. The poles were bent into position and their ends were diagonally nailed with thin finishing nails to keep them in place (**FIG. 3-7**).

This process worked great. Almost all the poles took shape immediately as I bent them. When finding poles for this job you should use green instead of dried-out or dead saplings. The greener they are, the more spring and durability they have. Actually, a couple of the poles cracked while being bent in place. I guess this could be attributed to the fact that it was so early in spring and the sap hadn't really had a chance to flow yet. But I could tell which ones were going to break through because they had knots in the wrong places and were less solid than the others. Once all the arched poles were in place, I lashed a long horizontal pole across the framework to provide it with greater lateral stability (**FIG. 3-8**). Diagonal braces would probably even keep it in place better, but actually the tension of the arches makes it pretty strong. It doesn't even sag or bend even when I lean on it.

I covered the framework with a long sheet of 4-mil plastic. It worked out that the arch was exactly the width of the plastic. I just wrapped the plastic around it instead of making any fancy

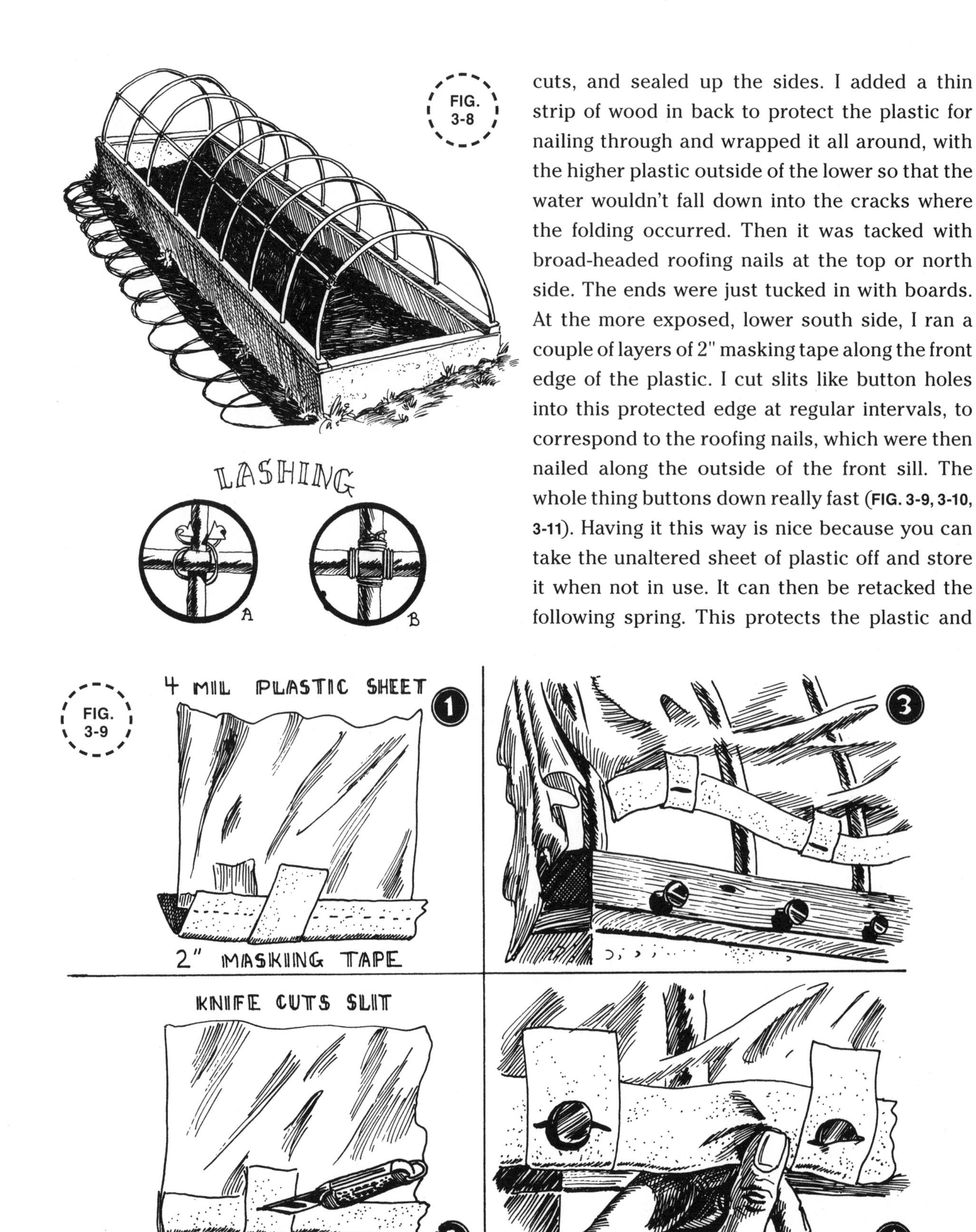

cuts, and sealed up the sides. I added a thin strip of wood in back to protect the plastic for nailing through and wrapped it all around, with the higher plastic outside of the lower so that the water wouldn't fall down into the cracks where the folding occurred. Then it was tacked with broad-headed roofing nails at the top or north side. The ends were just tucked in with boards. At the more exposed, lower south side, I ran a couple of layers of 2" masking tape along the front edge of the plastic. I cut slits like button holes into this protected edge at regular intervals, to correspond to the roofing nails, which were then nailed along the outside of the front sill. The whole thing buttons down really fast (**FIG. 3-9, 3-10, 3-11**). Having it this way is nice because you can take the unaltered sheet of plastic off and store it when not in use. It can then be retacked the following spring. This protects the plastic and

FIG. 3-10

FIG. 3-12

FIG. 3-11

FIG. 3-13

makes it last more than one growing season. This cold frame design works really great. All our tomatoes ripen on the vine. The peppers, cucumbers, and tomatoes still have fruit even this late in the year (late September), well after we harvested the garden (**FIG. 3-12 AND 3-13**).

There is only one problem with this setup, and I've noticed that it is a common and reoccurring problem with many people that I talked with who raise tomatoes and peppers in a greenhouse. Some of the ripe tomatoes and peppers have a tendency to fall off or develop large black rotten spots. I'm not sure what this is from, but I suspect it has something to do with the damp, humid condition inside the cold frame. But even with this slight problem, the tomatoes and peppers grown in the cold frame are always bigger, juicier, and ripen a lot earlier.

CHAPTER 4

Tender Loving Care—Biodynamic Gardening

A biodynamic garden is one that incorporates the principles of companion planting, planned yearly crop rotation, and replenishment with natural composting. A well-managed biodynamic garden should be well composted and replenished with large supplements of fresh manure every year. The gardener layers well-decomposed cow manure 4" thick throughout the whole garden and tills it in well. A few of the fundamentals of companion planting are:

a. Try to put plants that require plenty of light near those that prefer partial shade.
b. Alternate shallow-rooted plants with deeper-rooted plants, because they get their nutrients from different areas of the soil.
c. Protect plants that have few flowers and a largely exposed terminal bud (e.g., cabbage) with plants that have many blossoms (e.g., nasturtiums).
d. Do not grow heavy feeders together. These include cabbage varieties, leaf vegetables, celery, squash, cucumbers, and corn. Interplant them with light feeders, such as carrots, beets, and turnips (specific cases excepted).
e. Use plenty of aromatic herbs and flowers that repel insects and otherwise aid in the production of the garden without taking away from the health-giving qualities of the plants.

Planned yearly crop rotation should take into consideration the following principles:

a. Heavy feeders should be followed by nitrogen-rich legumes, such as beans, peas, clover, or alfalfa to replenish the soil.
b. Light feeders such as bulb and root vegetables (carrots, beets turnips, etc.) should follow on the enrichened soil provided by the legumes.
c. Plants that do not stand up well against competing weeds should follow those that were well mulched or relatively free of weeds.
d. Any plant that is prone to disease or malformation of the roots should not be put in the same place two years in succession.
e. Root plants should follow plants that loosen up the soil and aerate it.

Another important and helpful measure is keeping a record of the garden from year to year, to make sure the same vegetable is not replanted in the same bed and to properly plan the crop rotation. A good practice is to use the same garden layout and planting plan each year. The sections of this plan should be divided and numbered. Each section represents a bed. Each area of the plan representing sections of the garden should be filled in with the names of the plants that were planted in that section or bed (**FIG. 4-1**).

There is a particular biodynamic garden we know about. This garden is the best example we

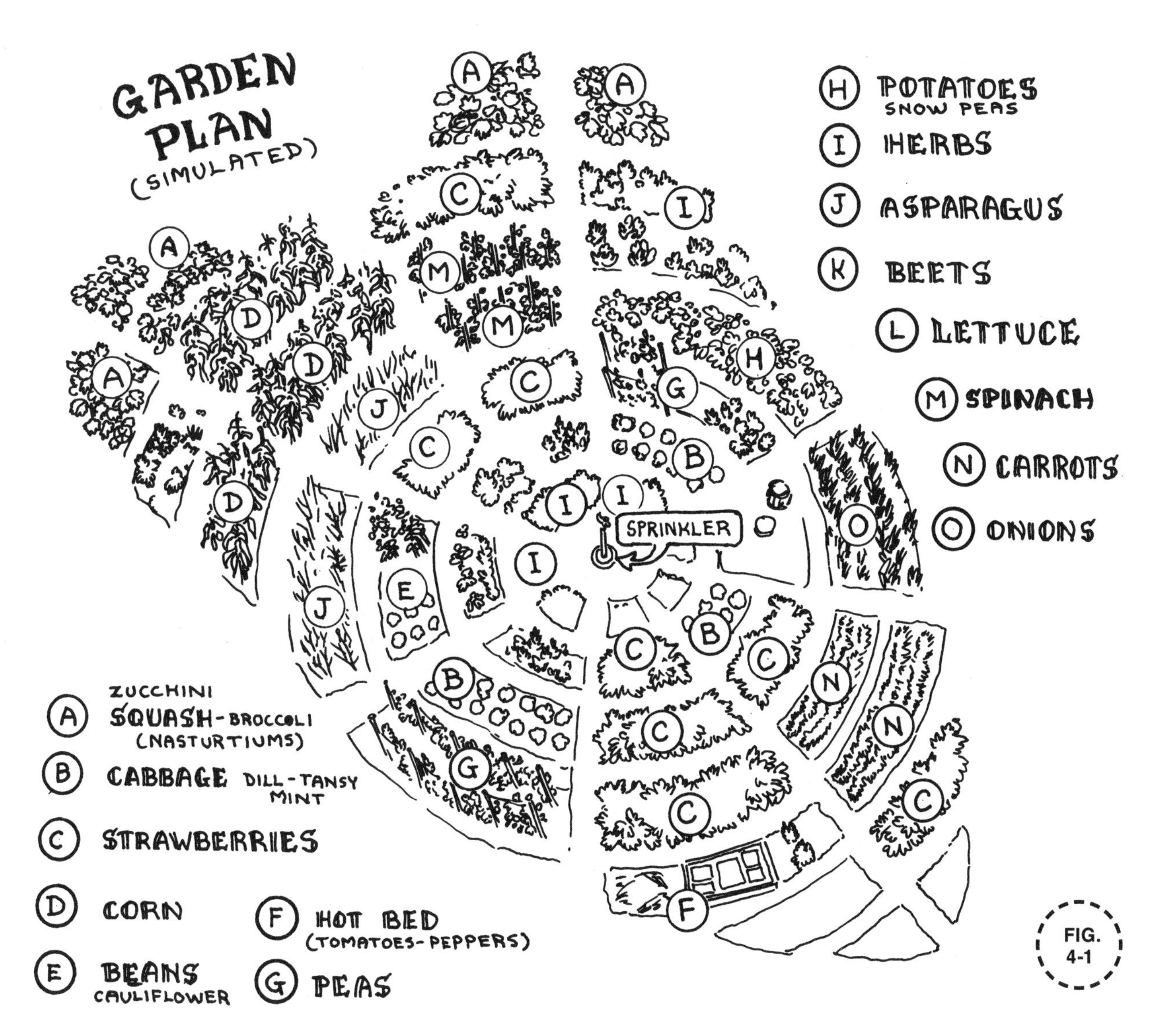

could find of a working biodynamic garden. It truly reflects the tender loving care of the lady who owns it (we'll call her "Rose"). Rose believes in the principles of crop rotation because she realizes that otherwise the plants would utilize the same nutrients year after year and consequently soon deplete the soil.

For example, if you planted potatoes in a particular bed, you don't want to replant potatoes the next season, because they carry diseases and the second time around, the potatoes are much more susceptible. If peas were grown in one area, something else that requires a lot of nitrogen will get nutrients from that pea bed in the next season. Plants that are bulbous, such as beets and carrots, can use a set of nutrients that are different from those of spinach or cabbage.

One exception—tomatoes can be planted in the same bed year after year. They thrive on a compost made with their own stalks and leaves.

The planting method that Rose uses divides up the garden plot into beds instead of rows. It is known as the "French Intensive Method." The beds are narrow enough to permit easy weeding and harvesting. Bed lengths vary with crop and location and do not have to be confined to any set measurement. Within the beds are combinations of one to four plants. She claims that all do well under the principles of companion planting. Planting thickly within the beds is an important aspect of this method.

Plant thickly, and then thin by harvesting as needed. Eventually the plants will displace the weeds. This biodynamic gardener had virtually no weeding to do and there was no problem as long as there was enough manure.

Generally, Rose has extremely good success with her garden, considering the variables of the climate in her location. (The garden is in a northern location with a very short growing season.) The last spring frost occurs as late as June 10th and the first heavy fall frosts come as early as mid-September. The garden is in a frost pocket and gets frost earlier than most surrounding areas. Since this particular climate is better for potatoes and cabbages, Rose grows an abundance of these and trades them with a friend down the valley who grows "terrific" corn, beans, and squash. Rose has a hard time growing these latter crops in the conditions she has to deal with.

Up until the time that Rose planted asparagus behind them, her potatoes were doing very poorly. They were really weak plants. She dug in a lot of manure and planted asparagus in the same bed. She obtained a prime crop of asparagus and had incredibly huge potatoes.

She had great success with cabbage, too. She planted two rows of cabbage, one row of dill in the center, and then two more rows of cabbage. The cabbages were planted close enough so that they were almost touching. That way, less time was spent weeding because the leaves shaded the ground, which prevented the weeds from growing.

In the first years, many of her cabbages were lost to root maggots and cutworms. She finally had to use a chemical to control the maggots and got rid of the cutworms with cardboard collars. These collars are 1½" to 2" high and 6" long. They are bent into a circle around the young newly transplanted cabbages' stems and stapled twice to hold them in place. The collars are pushed firmly into the ground to create a barrier around the plants. Cutworms have to come up to the surface for any great lateral movement. When they hit the cardboard, they can go no farther. She also used similar collars on tomatoes and peppers (**FIG. 4-2 AND 4-3**).

Cutworms live on dying weeds in fall, and if the garden is completely cleaned out, leaving no dying refuse, there is less likelihood of the cutworms living through fall.

Members of the cabbage family are prone to rapid decay because they have been developed into an enormous bud and their flowering process has become insufficient to protect them. They are helped by plants that have many blossoms and those that are strongly aromatic. These characteristics compensate for the cabbage's weaknesses. Rose added tansy and dill to the bed to maintain a balance (**FIG. 4-4**).

In her biodynamic garden, the carrots were almost completely surrounded by strawberries. Both crops did very well. The carrots were great.

The beans and cauliflower did exceptionally well, too. She planted purple beans. They do not rot on cold ground and mature quickly. By August 1st she had beans for dinner.

She planted nasturtiums and dill with the zucchini and broccoli. The zucchini didn't do too well because one zucchini plant was shaded by a huge broccoli plant. She maintains that nasturtiums should always be planted with zucchini. The purple beans and the white cauliflower are beautiful together—complementary in color—and nicely balanced.

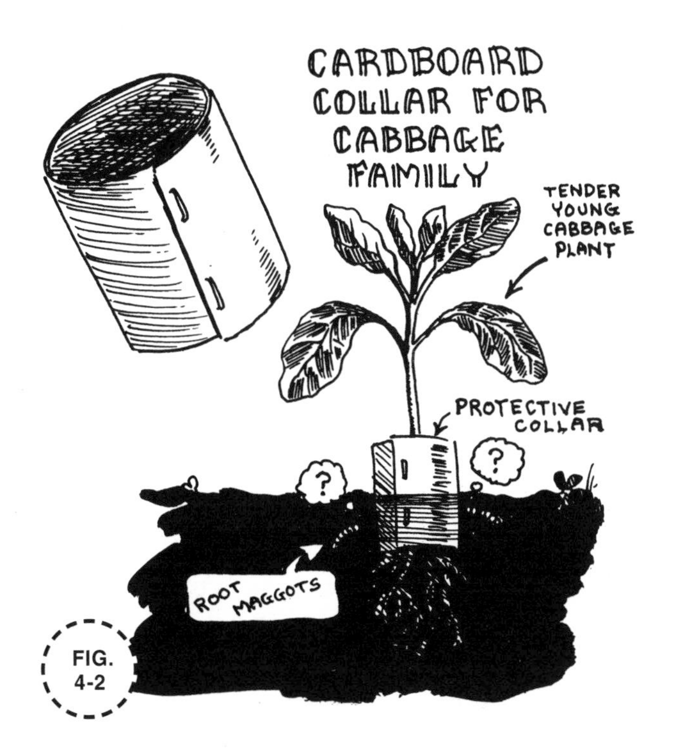

FIG. 4-2

FIG. 4-3

One is a long, purple pod and the other is the chunky white cauliflower. And it is aesthetically pleasing to see the huge heads of cabbage with long shoots of dill right in the middle. Likewise, in the middle of the enormous squash plants are the brilliant, bright red, orange, and yellow nasturtium flowers. Nasturtiums balance the squash flowers. Part of planting is seeing how you can make a bed look its prettiest. According to our biodynamic gardener friend, if you make it as beautiful as you can, you'll want to be there amongst it.

Unless they are tied, cauliflower will bolt—go to seed early. Tying protects them with shade cover and prevents the rain from entering and rotting them in their final stages of growth (**FIG. 4-5**).

Rose planted New Zealand spinach one year and was very pleased with it. She said it did not bolt at all like the other types. Its growing season is from June through frost. There was so much

FIG. 4-4

FIG. 4-5

spinach that there was enough for every meal throughout the growing season.

Where the potatoes were nearest to the snow-peas, they were the biggest, healthiest, and most plentiful. The surrounding brilliant purple, pink, and white pyrethrum daisies were planted for protection against the bugs. This is based on the theory that if the powder made from the plant is an insecticide, then the plant itself must be able to offer protection.

The garden is set up with herbs in the center and around the sides. These plants are beautiful and many are believed to discourage various detrimental insects. Mint was planted in the cabbage bed one year to discourage the cabbage moth. It seemed to do the job. Mint should be planted in small cans within the bed to prevent it from overtaking the bed with its tremendously fast-spreading growth.

Garlic in a bed discourages insects that do not like strong smells (nasturtiums and garlic repel aphid). Garlic is also easy to grow and is a very useful plant. The original settlers in this region are said to have taught that garlic should be planted in October, before winter, in really rich chicken manure. Leave the garlic plants until late summer when they lop over. When they lop over, they are ready.

Then harvest them, pulling them up with the stock still on them, and tuck them in underneath that stock. For best results in preserving garlic, just take a bunch of them and roll the stock back over them in the garden for a couple of days while it's sunny; then, take them out, rub them down (to clean them) and braid them. According to our biodynamic gardener, braiding them is the only way to keep them for any length of time. It keeps them in the air, and they don't gather moisture. She says that if garlic is kept in a dry place (but not in a root cellar), it will last about a year (**FIG. 4-6**).

Tomatoes and peppers need to be kept in a hotbed because the season is too short to allow them proper development time in the biodynamic garden. In the hotbed the tomatoes and peppers can be started as early as May. Ordinarily, they can't be put into the ground in this garden until mid-June.

The protective frame around the hotbed is approximately 3 × 8 feet. The size of the opening was determined by the dimensions of the three windows comprising the top of this enclosure. The hotbed itself is merely a pit, dug about 2' deep. This pit is filled to a height of 1' or 18" with hot compost, consisting of half cow manure and half chicken manure, mixed in layers with straw or other organic material to allow air into the mixture when stacked. The mixture is left beside the pit for two or three days before it is put into it. During this period, it is turned once. After this initial turning, the center of the pile is checked for temperature. Do not just stick your hand into the center to check as it can get too hot to touch. Put a thermometer into the center of the pile. If the reading is 90° or higher, put the pile into the pit. If it isn't, either turn it again and wait another three days, loosen the mix with more straw, or get some fresh manure (**FIG. 4-7 AND 4-8**).

FIG. 4-7

FIG. 4-8

HOTBED

Sprinkle a little lime over the mixture to help keep it cooking, then throw in an 8" layer of loose dirt. You want the dirt to be quite porous for aeration to allow the heat to rise to the plants. Check the temperature of the dirt after leaving it for a while under the glass cover. If the daily outdoor temperature remains a constant 80–85°, it is ideal. But if the temperature falls below freezing, which it sometimes does in this area (even in May), the hotbed should never go below 45° or else it won't function properly. To efficiently check the temperature of the hotbox during the low and high temperatures outside, a high–low thermometer should be used. This device has a magnetic registering system that holds the low and high throughout a given period with two black indicators. If the low indicator goes below 40º, the hotbed isn't working, consequently the young plants will not be adequately protected against the frost (**FIG. 4-9**).

The hotbed of the biodynamic garden has a 3'-high rail in front of it to drape a plastic cover over, to protect the mature plants from an early autumn frost (**FIG. 4-10 AND 4-11**).

The new blossoms of the tomato and pepper plants should be pruned in fall to keep them from forming any late fruit. The late fruit wouldn't have a chance to ripen anyway and the blossoms take juice away from the existing fruit.

Rose's cucumbers are also grown separately from the garden. They are grown in a cold frame with a protective wall behind it, which blocks off the chilling north wind. There is a network of string trellises that go up the wall for the five hills of cucumbers to climb. She lets her cucumber plants climb up like beans. They are easy to pick and there is considerably less chance of rot when they are up off the ground. They remain fresh (**FIG. 4-12**).

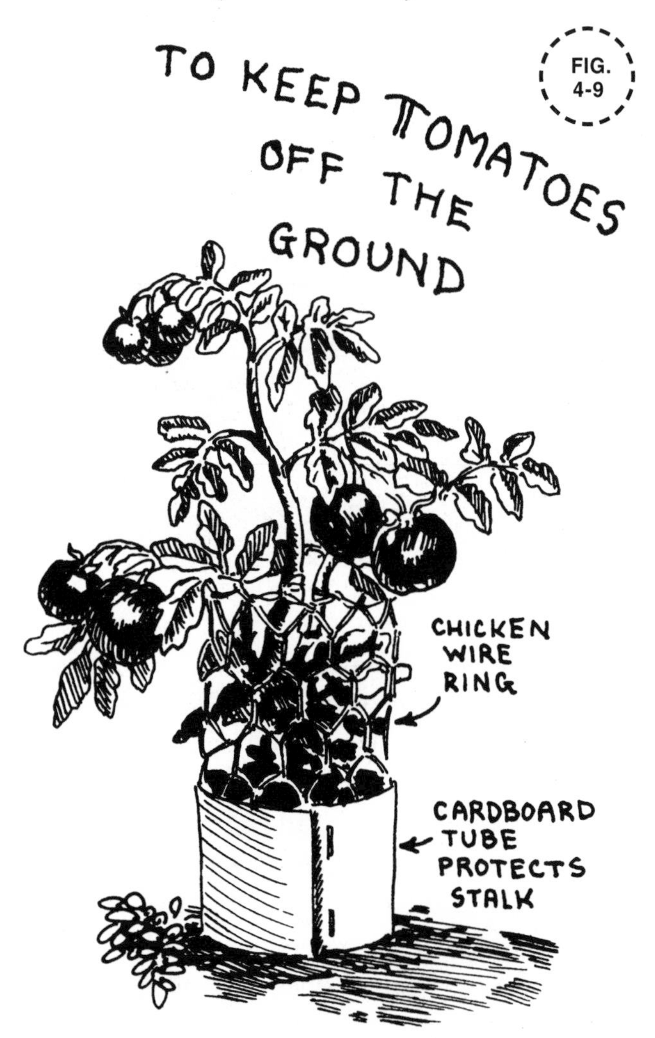

FIG. 4-9

FIG. 4-10

FIG. 4-11

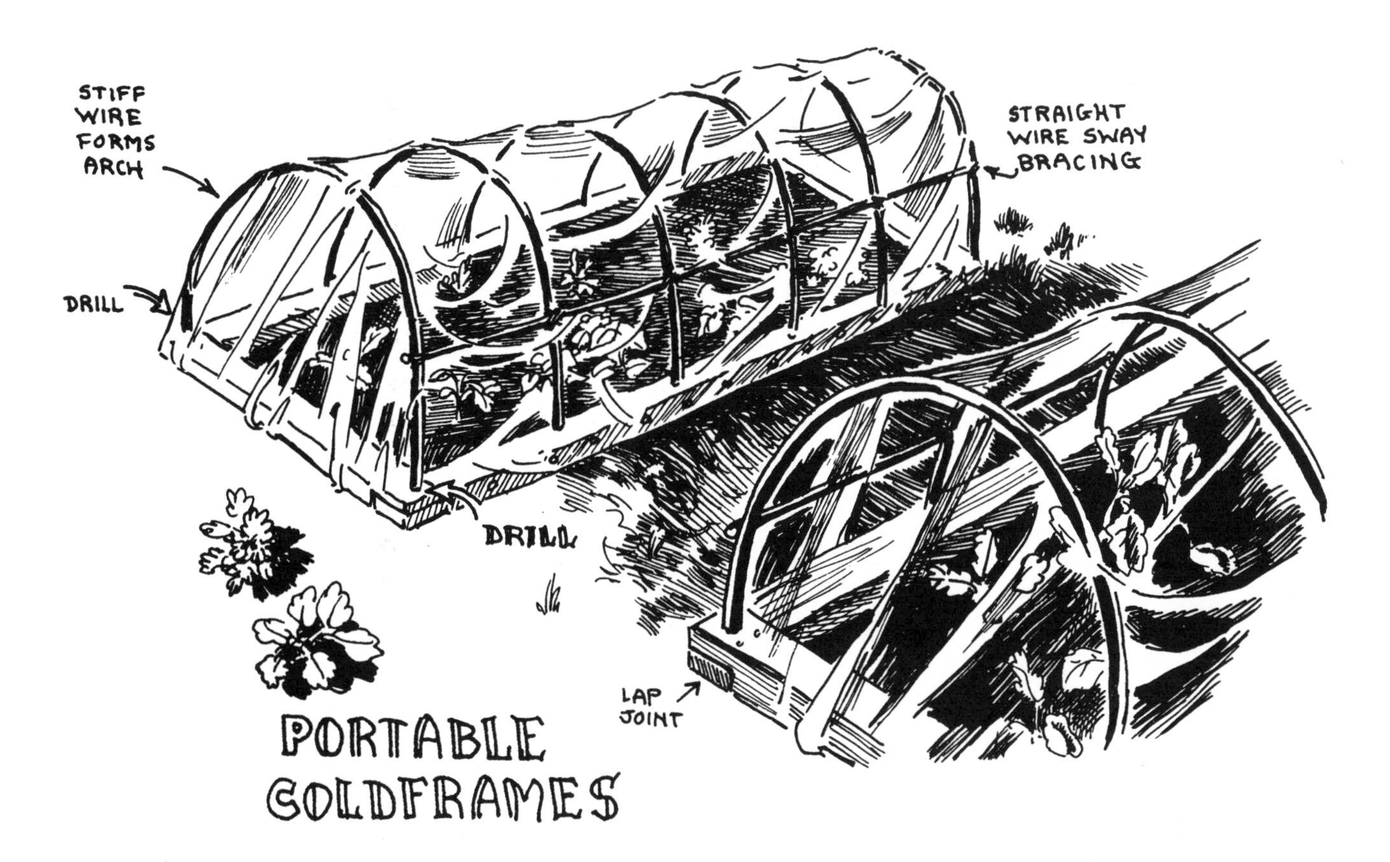

These examples were the only ones described as being successful combinations for biodynamic gardening. However, other suggestions have since been made and should be researched. It takes years of experimentation and study to develop workable combinations, and an equally long time to test the ones which are available. Several variables have to be taken into consideration and each garden plot has to be tested individually. Only through tried and true personal experience can any gardener really know for certain, beyond the obvious companion plant combinations and helpful hints, what will really work in their garden.

*Source consulted: Philbrick, Helen and Gregg, Richard B. *Companion Plants and How to Use Them.* The Devin-Adair Company. Old Greenwich, CT. 1966.

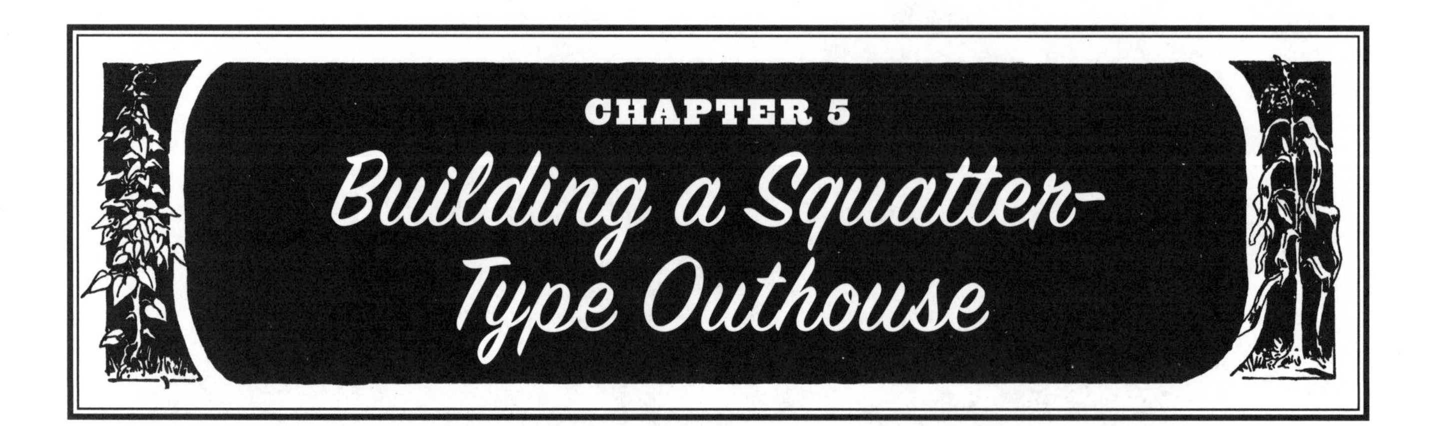

CHAPTER 5
Building a Squatter-Type Outhouse

After traveling around in the Middle East and North Africa for a year, I got used to squatting instead of sitting when eliminating, and began liking that position much better. It makes a lot of sense, too. Elimination is more automatic when your torso is in a straight line. When sitting, your body is sort of squeezed together in a rather unnatural position, making it more difficult to perform the natural function. When squatting, there is less pressure on the intestines and things usually happen very quickly. There is a tendency, when sitting, to take time to relax the constricted muscles for a while before eliminating, thus sometimes dragging out the performance longer than necessary and also making it needlessly laborious. A previous hemorrhoid sufferer once told me that his problem went away shortly after he began squatting.

A squatter-type outhouse is rather simple to build. It is post-and-beam construction with diagonal bracing and is set on skids to make it portable when the hole becomes filled (**FIG. 5-1**). To make our outhouse, the hole was dug first. Its size was relative to the average approximate cubic footage eliminated per year per person. Supposedly, the average person eliminates a cubic foot per year of compressed waste (3 cubic feet uncompressed). The hole was approximately 7 cubic feet, so the squatter would not have to be moved very often.

The cedar half-log skids were placed about 4' apart, with the hole centered between them. Two cedar sill logs were nailed across them at front and back, and a two-by-four flooring was laid across the sills, leaving a long narrow opening in the center. Then, one-by-four boards were nailed across the outer ends of the two-by-fours, narrowing down the hole to the approximate size

FIG. 5-1

I saw in the Middle Eastern and North African countries—about 6" wide by 12" long, slightly wider in front than back. This size seems to be adequate for both men and women, but it may be a little short for women if they don't squat right back on the foot boards when urinating, since they tend to hit the front boards.

The foot rests were set up to my measurements. They were adjusted back and forth until I found the right position that would be the most comfortable and would allow me to miss the wood. This design is a hassle for little kids, but is quite comfortable for most people. The way it is now, kids get on it and are afraid they might fall in. I'm not sure how I'd adjust it to accommodate small children—maybe a removable seat would be the answer.

I wanted the floor space to be rather large and up off the ground. There is enough room on it for some storage and for a container of ashes that get thrown into the hole every couple of days. The ashes help cover the odor of the excreta and enrich the decomposing waste with lime and other minerals. This large platform also provides the base to the loft storage area above. I believe in having as much as possible under one roof. The 4" floor sills allow for an air circulation space under the platform to prevent odors from building up in a stagnant space and seeping up through the platform hole. Also, I ask my visitors to avoid using the squatter for urinating whenever possible, because urine mixed with excreta promotes foul odors. Another preventative measure against unpleasant odors is a hole cover, which I shaped to fit snugly into the platform hole.

I believe that the more open you have an outhouse, the better. A closed building brings on insect and odor problems. It attracts flies and other insects because it lacks proper air circulation and also traps these pests inside. An open outhouse allows a person to feel as one with the surrounding beauty instead of being closed off from it. To keep the squatter open yet protect it from chilling winds, I set up posts at each of the four corners and supported them with 45-degree diagonal braces. A 3' railing was put up around three sides and the sections below the railings were sheathed with vertical slabs. The posts were notched out a bit where the diagonal braces met them, to provide a small shelf under the brace as a ledge for it to rest on, so it isn't entirely relying on the nails to keep it in place (**FIG. 5-3 AND 5-4**).

The squatter was then roofed with a gable roof to allow for storage of wood and other materials. The end rafters were extended out at slight angles at the peak to give the building more overhang on the ends and the eaves were gradually brought out with an increasing thickness of nailers to give the roof a slight sweeping affect for aesthetics. The roof was then finished with leftover number-four-grade shakes, which would be plenty good enough for this small building (**FIG. 5-5, 5-6, 5-7**).

I'm very pleased with this squatter and consider it a very healthful alternative to the conventional seated outhouse. Surprisingly enough, even my folks, who were in their mid-fifties at the time, enjoyed using it as a refreshing change from what they were accustomed to. And any type of temporary outdoor setup is far more efficient than the indoor toilet system that empties into a closed-off cesspool or sewer, because the compost that is being produced can be used for many purposes after it has been given the proper time for anaerobic decomposition to ensure the destruction of pathogens and insect eggs. This takes at least eight months of undisturbed decomposition. To speed up this process while the excreta is accumulating, I run a hose into the hole about once a year and throw in a bunch of lime to help it spread out and break down. Instead of collecting this as compost to be taken elsewhere, I cover the almost-filled hole with good soil, wait the proper period for decomposition, and plant a tree over the hole. Where I've done this, the outcome has been amazing. I also have mint and rhubarb growing over a few of the previous holes. These plants just take right off every year.

FIG. 5-3

FIG. 5-4

FIG.
5-5

FIG.
5-6

FIG.
5-7

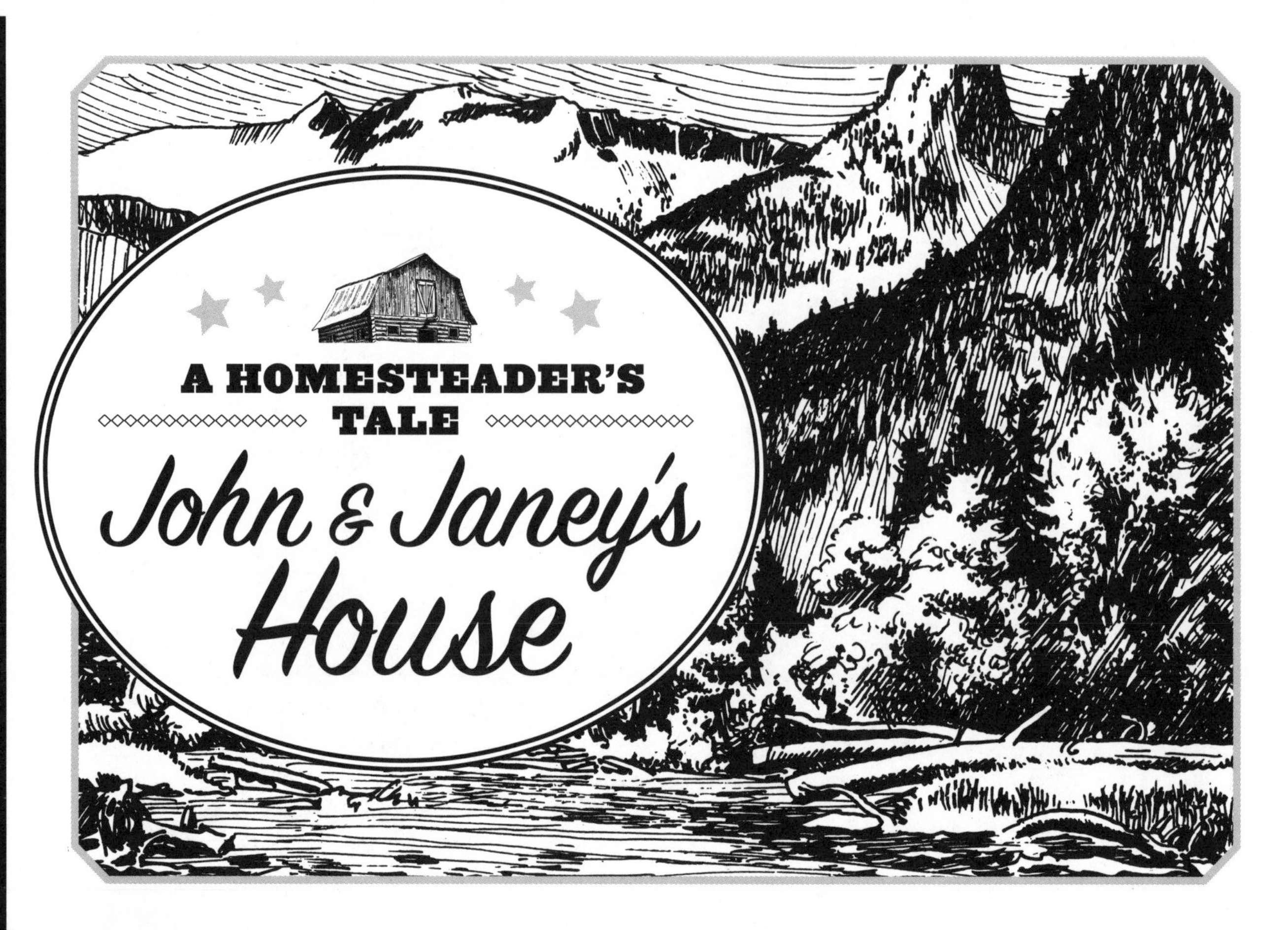

There we were on top of it all, sitting at our picture window, sipping cocktails in our plush high-rise apartment in the heart of a huge eastern city. The fifteen thousand people residing on our block shared its conveniences, but very few of us cared even to pass the time of day with each other. We were like bees in a filing cabinet, anonymously struggling for the betterment of the hive. We were servants to the protective security it offered us.

To maintain our personal sanity, we surrounded ourselves with hundreds of material conveniences. There always had to be an immediate compensation for what we were sacrificing in terms of our health and emotions, of pleasure received from the manufactured world which surrounded us.

Our greatest enjoyment came when my wife and I were lucky enough to share the same day off together. Then we'd jump in our sports car and join the multitudes who also found it necessary to periodically escape from the noise and pollution of urban life. After several hours of driving, we would find peace in secluded, forested spots, beside small clear-water lakes. There we'd picnic and skinny-dip. We'd forget our cares and appreciate the natural beauty of the place. We'd lie in the tall grass and smell the pine-scented, refreshing air. We felt more comfortable and secure there than in the locked confines of our expensive apartment. But always too soon, the beehive summoned us to return.

We began craving those carefree excursions into the country more and more. Thoughts of actually changing our lifestyle began to dominate our spare moments. We spent two years reading available information about living in the woods and preparing a homestead that would provide us with food and shelter. In that period, we realized how important it was to find a place of our own, a place where the sounds we heard would be the flowing of creeks, the chirping of birds, and the whistling of the wind through the trees and not the roaring of subways, the honking of horns, and the hollering of poorly matched married couples blaming one another for their torment.

We searched for a long time, and we finally found a piece of land to our liking. Though the searching itself was very enjoyable, and put us in touch with many fine people who were living in the style we wanted to live, it left us short of cash.

We had just enough money to buy the land and rent a small cabin until we could build our own house. I found a job at the local mill, in order to raise the money we needed for our supplies. For months, I sweated on the "green chain," lifting wet boards and transferring them from one place to another. For some strange reason, though, I rather enjoyed that job. It readied me for the hard work I had ahead of me on our own homestead. It also put me in touch with several of the local people who were later instrumental in the construction of our house. In fact, a few of those fellow mill workers have become my closest friends and working companions.

We bought our place in the winter, but we didn't really get started with the building of the house until late spring. Those first few months were filled with planning the layout of the homestead. Before we made any decisions of placement, we had to prune some of the wild growth so we'd know what the area looked like.

I took on the chore of limbing each tree in our thick, six-acre forest. I cut off every branch as high up as I could reach with an axe. This may sound like a lot of work, but it was very satisfying. Limbing the trees opened up the forest and turned it into a park with many paths and walkways. I became sensitive to the land and looked forward to every task it required with a Zen-like enthusiasm. I knew I was where I wanted to be.

As far back as February, we observed the arc of the sun. We wanted the house positioned so that in the cold winter, the kitchen and dining areas would receive the warm morning sun through the windows. It was also important that the house be in such a position that the summer sun's light would be blocked by the roof eaves.

After much experimentation, we finally located a site for the house that would provide proper sunlight. We kept this site completely surrounded by trees to keep it cool in the summer and make it warmer in the winter. Trees are a natural wind breaker that shield an area from the icy winter winds. Before we staked off the exact site for the house, we made sure to cut down any snags or standing dead trees that might possibly fall on the structure we would build. Every cottonwood tree was also removed because this species has a weak root structure and cottonwoods have been known to fall for seemingly no apparent reason.

The area we chose was also well out of sight of the public road, but close enough to it to be practical in winter. We have no problem carrying our groceries and supplies down the short path to the house, yet we are not assaulted by car headlights and constant road noises.

Long before we actually started construction, we put stakes around the proposed perimeter of the house. Every morning we came to the site from our nearby cabin and inspected that area. Some mornings the area would seem very tiny, so we would enlarge the perimeter. Other mornings the area would seem huge. Those mornings, we'd make it smaller. We soon began "playing house" in the area. We figured out where the various sections would be and how much space would be necessary for each. In the "kitchen," we measured areas for our refrigerator, stove, sink, cabinets, and cupboards.

Then one of us would pretend to make a meal and the other would "go into the living room, stoke the fireplace, and sit down on the couch." We'd pretend to entertain friends. We'd pass each other in the "hallway" and sit down in the "dining room" to eat. Though we didn't stick to plan for the finished product, we had a good idea of the space we needed. It turned out that 20' x 24' with an adequate loft area would be sufficient space for us—as we had no immediate plans for enlarging our family.

We wanted a house where the rooms were defined by layout of furniture, not by walls and partitions. In conventional houses, the kitchen is separated from the living room and other areas. If a conversation is happening in the living room, the cook in the kitchen cannot participate in it because of the space barriers. We are gregarious people, and we like being with others—so we eliminated partitions on the ground floor. The only separate space was the mudroom, sealed off to keep out drafts.

After we set up the batter boards to make sure the 20 × 24 foot area was square, I dug twenty holes, four rows of five, for the cement piers which would hold up the sills. There wasn't any span greater than 6' in any direction between them. I felt it was necessary to have

Path of Sun

that many piers to prevent the floor from sagging and to make sure the foundation was good and solid. It's a real drag to have to reinforce a poorly planned foundation as an afterthought.

Each of the holes for the piers was about 2½' deep and was wide enough for the forms that framed the 6 × 6 inch piers. The frostline was only 18" below the surface, but the soil was clay, which is not as stable as rock or other porous soils. Because of this factor, we had to dig an additional 1' to support the piers on a firmer, sand-type soil. The piers stuck out above the ground about 10" to prevent the sill logs from coming in contact with the clay soil.

We used a mixture of 5 parts clean sand and 1 part Portland cement for the piers. Many larger rocks and pieces of scrap metal were tapped into the concrete to save on cement and to add to the strength of the mixture.

Concrete piers

The area was relatively level, but there were slight differences in height among the finished piers. We made up for these differences by setting various lengths of log rounds above the piers. Then we placed small squares of black building paper between the log rounds and the piers. The squares of paper acted as vapor barriers to protect the logs from the dampness of the concrete.

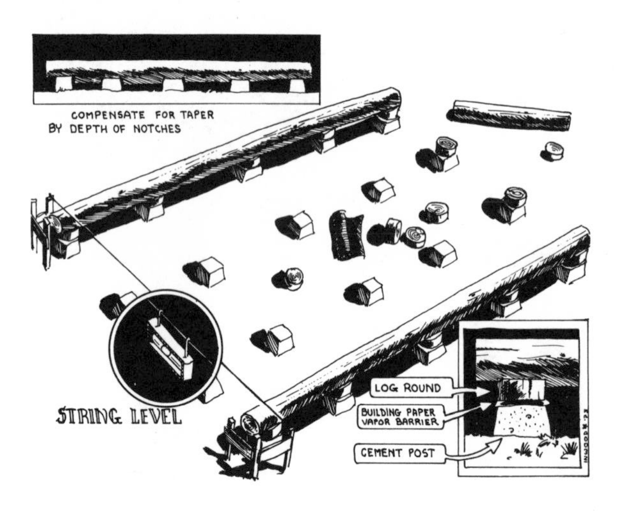

Notching the sill logs into place

The four sill logs were notched in place. We dealt with the two outer sills first. These were raised to the height of two string levels, and both ends were notched onto the pier rounds. The depths of the notches were determined by the log's relationship to the other sills. The notches made them level with each other. If the log being worked was higher than the other sills already in place, its notch was made as deep as necessary until the sill was lowered to the height of the others. The space under the sills could be used for storage.

Before we went any further on the construction of the house, we dug out the drainage system. Because the soil is clay, which does not allow for proper drainage, we had to dig a deep dry well 8' wide, 10' long, and 12' deep. This hole was located about 5' east of the kitchen and shower areas. It was far enough from the house to prevent flooding the foundation piers, but close enough to save on expensive drainage pipe.

After the drainage system was dug out, we refilled 9' of it with many pickup-truck loads of large rocks. The bigger the rocks, the better the drainage. One end of the 2" plastic drainpipe was placed above the rocks, then the hole was covered with large cedar logs. A layer of 2" boards and a black building-paper vapor barrier were placed above the logs, and a final covering of dirt was shoved over the other materials to protect this system. For additional foundation support, and to seal off the area under the sills from cold air, we built a rock retaining wall around the perimeter of the house. For this wall, I dug an 18"-wide and 2½'-deep hole between the outer piers. The bottom of the hole was filled with 4" of mortar, 3 parts sand and 1 part masonry cement. Masonry cement, when dry, produces

a white finish which is more appealing to me than the darker finish of regular cement. It also takes less time to dry. Large rocks were then fitted into the wet mixture, and other rocks were mortared over them until the wall sealed off the area between the ground and the sill logs. The rocks touching the sills were then wedged in for a tight fit. We left a 5' space at the west wall for a storage entrance.

The two-by-eight floor joists were then placed over the sills at 16" centers. Over the joists we nailed a subfloor of low grade two-by-eight tongue-and-groove cedar. Fiberglass insulation, in 4"-thick strips, was stapled under the floor joists and the subfloor. You should take care when installing the insulation. If the fiberglass is not properly sealed, the 4" space between it and the subfloor will act as a wind tunnel instead of a dead-air space.

FLOOR JOISTS

To seal off this space properly, overlap the ends of the paper backing whenever possible and make sure the open joist ends are covered. Though there is zero heat loss through the floor, this area must be well protected to prevent cold air from entering.

One of the subtle room divisions that we designed was an 11"-high raised platform for the dining area. It is roughly a 9 x 12 foot area wedged against the south wall. The height of the raised area was determined by what we considered a comfortable setup.

Insulating the floor

In the early spring, long before we began the foundation, I went out searching for logs for the stockade-type walls we had in mind. A farmer up the road wanted some acreage cleared for a pasture. He told me to spread the word that

Raised dining area

anyone who wanted firewood or logs for building was welcome to cut on his property; the wood was free for the taking. I told everyone I knew about it, and before long, all the usable timber was cleared away. My first plan was to cut down live trees and buck them to 7' lengths, then to peel off the barks and let them dry a season. But when I arrived at the site, I came upon a pile of aged, fallen timbers. They were mostly pine, tamarack, and fir. By the end of the day, we transported home enough logs to do all the walls. I was really pleased with my find.

Since the logs were already aged, I could use them immediately. They were light to handle and I wouldn't have to worry about any drastic shrinkage. One thing I didn't figure on, though, was the problem of peeling aged logs. After struggling with a drawknife for many hours and only completing a couple of logs, I decided there must be a better way to get them peeled. I rounded up several spuds and drawknives and invited all the able-bodied teenagers in the area to come to my "peeling party." They accepted my invitation, ate my food, drank my wine, and listened to my music, but by the end of the day, only a few of my logs had been peeled. In the days that followed, a couple of the more industrious youths peeled most of the remaining logs. I paid them 50 cents a log, and believe me, it was well worth it to have that chore out of the way.

A lot of the dark pitch from the bark did not come off the logs. At first I thought the discoloration was ugly, but once the walls were up, I began liking the effect of the dry, peeled texture and the color patterns of the logs.

One day after the subfloor was finished, we sat around drinking wine. We were celebrating our achievements to date and were so inspired by what we had already completed that we decided to begin on the walls soon.

We nailed the two-by-six bottom wall plates in place, then set up boards outside each of the four perimeters of the foundation. These were actually long batter boards that were nailed into the sills. Each of these eight two-by-fours extended upright over 7' so a string could be attached near its top. The boards were positioned so the guide strings would be half a log's thickness, or 3", inside the perimeter of the building. This would make the strings intersect at half a log's diameter at every corner. These points of intersection would designate the exact center of the corner log.

A black line was then drawn along the center of the two-by-six bottom wall plate. A log was centered over this plate. To make sure the upright wall log was exactly plumb, I attached a weighted string to one of the intersecting strings which crossed above the log. A fishing sinker was used to hold this vertical string plumb. The wall log was adjusted according to the plumbness of the string. It was then braced with a diagonal board, and the weighted string was transferred to the intersecting corner to check the other side of the wall log. The log was again braced, and was considered to be plumb.

Using a level is an inaccurate way to check how plumb an upright log is. No log is perfectly straight from end to end unless it has been hewn or milled that way. All logs have protrusions and low spots. What may be level in one 3' section does not necessarily give a true reading for the entire length of a log. I had doubts about my method of plumb-reading until I saw the electric company setting up our power pole. They put it upright with all their expensive equipment but had to do a final check with a string and a bolt, which they hung from a stick beside the pole.

By the end of the day, we had one log up. It protruded from the flat subfloor like the monolith from the movie *2001*. There was something

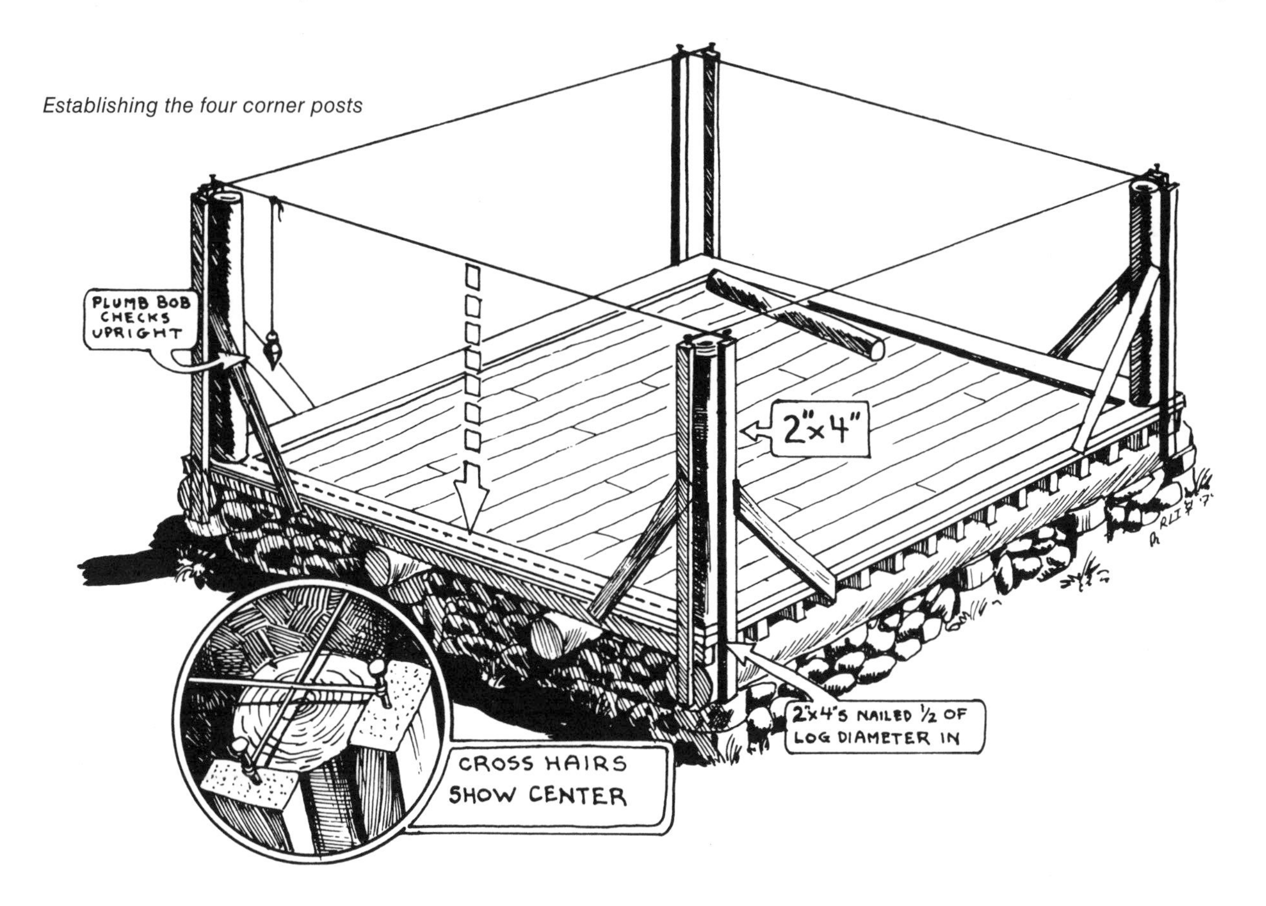

Establishing the four corner posts

spiritual about that log and what it represented to us. We didn't put up another wall log for days after. We did other chores instead, and just marveled at our "monolith" standing against the forest background.

The following wall logs went up quite fast. Each of the corner logs was put in place and checked for plumbness. Then each was nailed to the bottom wall plates with 7" ardox nails, using one nail on the inside and one on the outside. Ardox nails were chosen for their incredible holding ability. The only frustrating thing about them is having to pull them out if you make a mistake—forget it! We then braced each corner log with diagonal two-by-fours.

I laminated two two-by-six top plates together and nailed them over the corner posts. These double-laminated plates were nailed together every 16", and the nails were staggered to add strength to the plates. In putting up these plates, I overlapped an end of each so that the plates of one wall interlocked with the double-laminated plates of the intersecting wall.

After the top plates were in place, we filled in the full-length wall logs of the west wall, leaving adequate space for the windows. A 3" strip of fiberglass insulation was stapled to the bottom and top plates and along each log. The log to be placed was pushed in as close as possible at the bottom plate by one person, while a second person toenailed it in. We made sure the log was plumb before nailing through the top plate.

All the full-length wall logs were put in place and the window spaces were roughly outlined. To secure the corners and strengthen the walls, two-by-six diagonal bracing was used at each

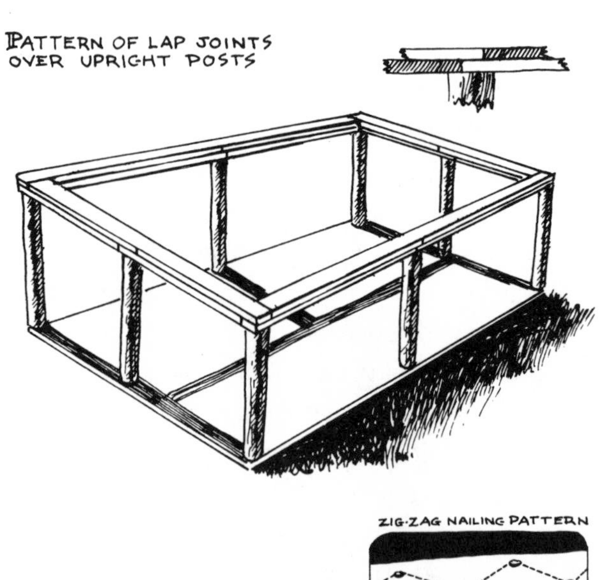

corner except for the dining area, or the southwest corner, where it would remain exposed if used. The bracing is hidden in the other corners by a brick wall, a shower stall, and a kitchen stove. These two-by-sixes start at the bottom plates of each wall, a few feet out from the corners, and meet at the corners 5' above the floor. Each is notched into the wall logs. The other full-length logs went in by the same manner.

Most of the log structures I've seen in this area have small windows. Their interiors are dull and dingy because not enough light enters to brighten them up. I asked many of the builders of these houses why they wanted such small windows. They all gave me the same answer: they wanted the small windows to prevent heat loss and to conserve on wood and other heating materials.

We designed our windows big because we wanted light and we desired a view of our beautiful homestead. I would rather cut an extra cord of wood during the winter, when I'm in the house most of the time, than have small windows and be enclosed by four dingy walls. Anyway, large windows at the south wall bring in more of the sun's heat even during the coldest part of winter.

At the west wall, we framed a 6 × 4 foot window space high enough to be able to look out when we are eating. To frame a window when using vertical logs, you simply cut your top window header two logs longer than the desired opening, so the header log can rest on the vertical sleepers on either side of the window opening. Since there is no wall stress on the bottom windowsill, it is cut to the size of the window opening and toenailed to the sleepers on either side. This two-by-eight sill is tilted dsownward toward the outside so water slides off and does not collect and seep into the house. After the sill is nailed in place, cut short uprights and fit them into the space between the header and the bottom plate. Then staple insulation between these short upright studs.

West wall windows

Franklin Fireplace on cold north wall

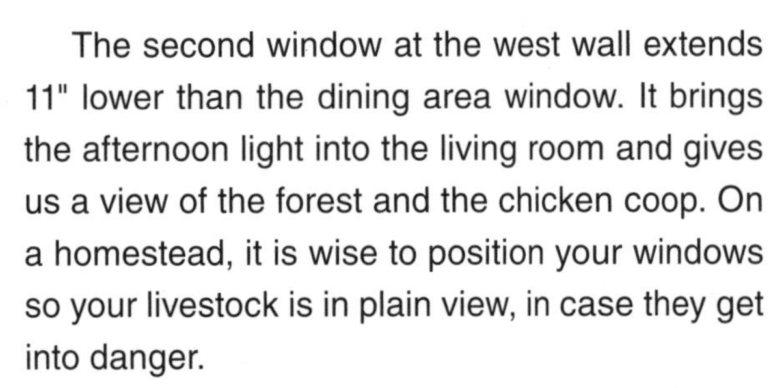

The second window at the west wall extends 11" lower than the dining area window. It brings the afternoon light into the living room and gives us a view of the forest and the chicken coop. On a homestead, it is wise to position your windows so your livestock is in plain view, in case they get into danger.

The north wall has a large window in the center, above the woodbox. We close the window off with a heavy curtain in the winter because no sunlight comes in from that direction. Since no sunlight enters from the north, it is the coldest wall and should be the most protected. Our Franklin Fireplace is located here. It has a brick wall in back of it to shield the log wall from its heat.

Built into the north wall, beside the Franklin Fireplace, is a fire box. It has a large outside entrance at which we load the box full of cord-wood. This entrance is covered by a latched door to keep the cold from entering. The box has a lid enabling us to take the cordwood out of the box on the inside after we load it from the outside. Having the fire box helps prevent the heat loss which would occur if we were opening the house door to bring firewood in. It also keeps the floor clean from the wood chips and pieces of bark which plague many a wilderness floor.

There is a temporary door at this wall which we used until the main door and front porch had been constructed. This door is very light in weight, studded with two-by-fours, and sheathed with cedar one-by-twelves. The main door is at the east wall. This wall is double log between the shower room and the kitchen to frame in the mudroom and storage area. We used logs for this purpose because they were a cheap material and went well with the interior design.

The mudroom, which is 5' wide and 11' long, is a necessity for homestead living. It is a separate area where people can enter without disturbing the other occupants of the house with a cold draft and lots of noise. It is a place to hang overcoats and take off muddy boots. It is a reorientation corridor to prepare a chilled and road-weary traveler for the warm and cozy indoors, or to prepare a sedate, wine-filled guest for the journey which awaits him or her.

To save on space, both the inner door to the living room and the outer door of the mudroom open toward the outside. The outer door opens to a large enclosed porch which is 6' wide and 20' long. This porch is framed with logs and has log rafters which extend 2' past the porch as eaves.

To make the stairs of the raised porch platform, I hewed flat the tops of two log pieces. One was imbedded in the ground, and the other was perched 1' higher up on log uprights.

The front porch, kitchen window, and the windows in and above the dining area.

The south wall has many large windows for heat and light. The huge 4 x 6 foot window in the kitchen brightens up that area, so whoever is cooking does not feel trapped in a dingy little space. This window almost makes it seem like we are cooking in an outdoor kitchen. It makes us feel that we are part of the forest which surrounds us instead of closed off from it.

For the same light and heat purposes, we have two larger 4 × 4 foot windows in the dining area behind the table and two elevated windows in the dormer area above that space. These elevated windows bring in the higher-arched summer sun to both the living room and dining areas.

Also bringing light into these areas are the windows of the west wall. They are very important because they allow the afternoon sunlight, which filters through the tall evergreens, to warm the house and prepare it for the cooler evening.

Though all the window spaces were framed and ready, we did not put in the glass the first

Elevated windows

year mainly for two reasons: We wanted to give the walls time to settle and we were running out of money.

The last task, before the walls were completed, was covering the insulation strips and sealing in between the wall logs. At first I was thinking of cutting out quarter-rounds from larger poles for this purpose, but a nearby friend had a grove of young cedars. He let me thin out the grove and use as many of the narrow poles as I needed for the house. I brought home a truckload and proceeded to peel them. Even though they were green, they were as hard to peel as the dry logs because it was late in the season and the sap was low. But after a few days of constant scraping, they were ready. I beveled the ends of each pole and nailed them into the wall logs tightly over the insulation.

With the walls finished, it was time to get started on the loft and roof. I hashed over various roof designs with a few friends who had volunteered their labor and ideas. We decided to cover the kitchen and entryway areas with a bedroom loft, and leave the area above the living

Saplings covering insulation

room and dining room open as a high, cathedral-style ceiling. A 45° pitch for the roof seemed to be the best because it made for a spacious ceiling, allowing a distance of 10' between the top wall plates and the ridge peak. The 45° angle, as we found out while working with it, was the easiest angle to use. Rafters could be simply measured and cut on that angle, and even insulation pieces and sheathing boards for the gable ends could be cut out with the minimal amount of waste and planning.

It was time to begin the loft construction, and we did not even have a center pole or a cross-beam for the joists. That problem was quickly taken care of, though. A hemlock tree was cut down and two logs were sawed from it. The logs were quickly peeled and brought to the house site. The 7' log was positioned upright near the center of the floor and the 14' one was notched in above it. It spanned between that upright and the inner east wall, where the other end was notched in place. As we pounded the spikes in, the logs spit back at us.

The longer log extended well past the halfway point of the house, which was designated as the end of the loft area. Since the pole was longer than necessary, we decided to modify our loft design and add a 36" extension to the north side of it. This space was used for a bookshelf and a reading area. The north end of that additional platform was later closed off for a small linen closet. The loft joists were spaced 16" on center and spanned between the north and south walls. Every other two-by-eight joist was doubled, to add support to the loft floor. These doubled joists also acted as braces to prevent the walls from spreading out under the weight of the loft and roof.

A subfloor of fir two-by-sixes was nailed across the joists and a finish floor was laid diagonally above it. This diagonal finish floor is another precaution against the walls collapsing. It prevents the stockade walls from twisting by bracing them against leaning in any direction. Some people accused me of overbuilding, but I don't think I did. I just wanted to make sure our house stayed around for a long time.

Next came the rafters. The rafter sets were put together assembly-line style, using a simple matrix (jig) framework consisting of blocks nailed to the floor. First we laid out one set of rafters and cut their tops at 45° angles. The A-brace was

Joist placement beneath left extension. To enter left bedroom, you climb the ladder and step through curtained entranceway.

then cut and positioned and the rafter bottoms were notched to fit the top plates. Blocks were nailed around the rafter set in all the necessary places: two at the feet, a few along the sides, and two near the top of each. This method assured us that all the rafters were uniform and it gave us a guide for each of the necessary cuts.

After a few sets of rafters were finished, we laminated two of them together as a double set for the end rafters. One helper stood on the floor with a long board extended, raising the set by the A-brace, while another helper and I spiked the rafter bottoms to the top wall plates.

At this point, Janey and I stood on the loft floor and tried to figure out how much space we would have for our bedroom under that simple gable roof. Neither of us was satisfied. We wanted a larger room. We recalled how some friends of ours had turned a small attic space into two

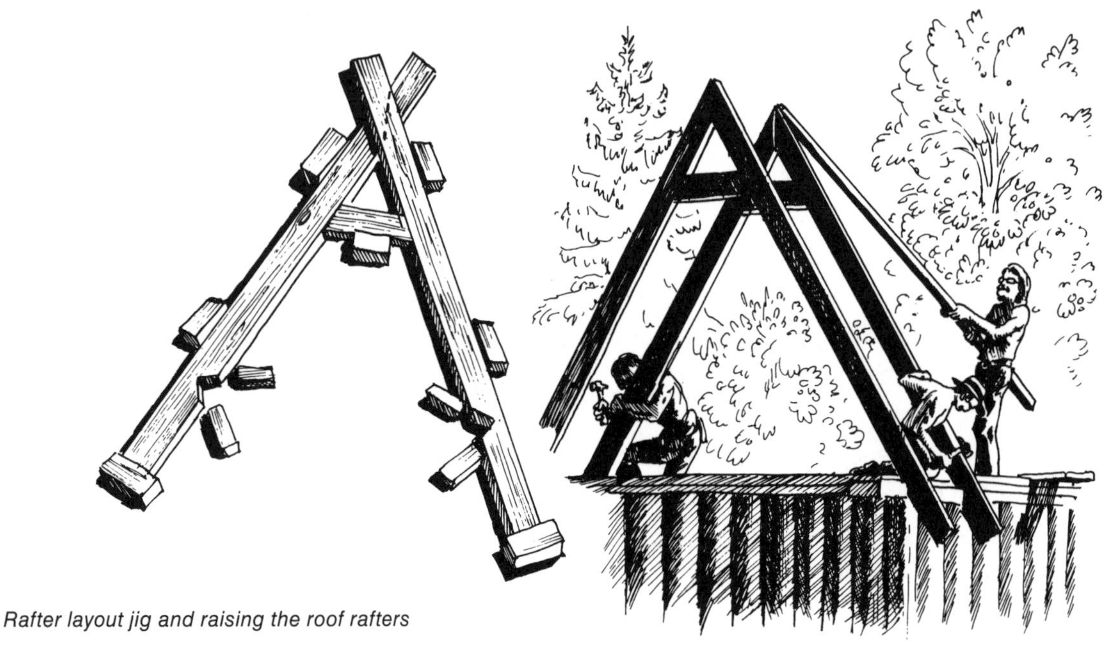

Rafter layout jig and raising the roof rafters

good-sized bedrooms by incorporating a large shed dormer into the roof of that structure. We liked the idea and designed our roof to include three shed dormers, one on either side of the bedroom area and one at the south wall just above the dining area, to bring the morning sunlight into the lower spaces.

We wanted the shed dormers in the bedroom to be 8' wide. Since we had to be able to stand in them, 5'9" was a good height. This came to about a 22° pitch. I am glad we stopped long enough to decide on that extra roof space, because at that point the dormers were easy to install. At a later time it would have involved tearing out rafters or maybe even replacing an already completed roof. We found over and over again that it pays to stop and live in the house at every stage of construction.

Roof and dormer framing plan

Exterior view of north and west walls showing completed dormer and asphalt shingles

All the rafter sets were then spaced at 16" centers and nailed into place. Wherever the rafters braced the ends of the three dormers, they were doubled like the first two sets. The others were single sets. To frame the bedroom dormers, four jack rafters were cut at 22° angles, and each was nailed to a peak of one of the four double rafters. These extending jack rafters were then supported by several two-by-four upright studs spaced 16" apart. At the peak, the jack rafters were connected by an 8 double two-by-six which spanned between them. Another 8 double two-by-six was then placed between the bottom end of each of the jack rafters, and these headers were braced in the center, between the spaces for windows, by two two-by-six uprights.

The living room shed dormer was framed in the same manner. Being a smaller dormer, its jack rafters were cut at 35 degrees. This allowed for a space 8' long by 2½' wide in which two windows were put.

After the roof structure and gable ends were framed, they were sheathed on the outside with low-grade one-by-six hemlock. To protect the roof sheathing, we covered it with a finish roof of asphalt shingles. Asphalt shingles were chosen because they are safer and last longer than cedar shakes and they are more aesthetically pleasing than are aluminum. Cedar shakes are more natural and pleasing to the eye, but they are dangerous and are known to ignite with the least little wild spark. In fact, a friend of ours one winter lost his house because a spark set fire to his cedar-shake roof.

With a protective roof over our heads shielding us from the autumn rain, we turned our energies to the loft interior. Again we stopped and figured out the spaces we needed. For comfortable, uncluttered living, plenty of closets and cupboards are essential. With our shed-dormer design, there was adequate cupboard area under the dormer windows.

Bedroom window

Shed-dormer design

For a closet, we set up a divider wall between the bedroom and what became the reading platform. This wall closes off the bedroom from the rest of the house, making it a separate area. The closet space is two rafters (32") wide and extends from the north wall to the center crossbeam. It was also partitioned off at the loft doorway entrance.

Because the house is specially designed for togetherness and has no totally separate spaces except for the mudroom, there is no door to the bedroom, just a heavy velvet drape hung from over the entrance to control heat. When the drape is open it allows the rising heat from downstairs to enter and warm the loft area. When it is closed, it prevents the heat from entering that space.

The inside ceiling of the bedroom and the open cathedral-style ceiling were sheathed with one-by-eight boards, using the lap-siding method. This method was used to prevent having spaces between the boards and to seal off the insulation more efficiently.

Because more heat escapes through roofs than through any other section of a structure, I took care to make sure ours was well insulated with 3½"-thick fiberglass. To find out if our roof had enough insulation, I checked the snow load on it. If the snow melted and iced at the eaves, I knew that heat was escaping. If snow stayed on the shingles, the room was properly insulated because no heat was escaping to melt it. It is also important to block off the sections between the rafters where they meet the walls. I stuffed loose insulation in those ends and nailed on 15"-board sections to protect the fiberglass.

To figure out a design for a set of stairs to the loft, I positioned the temporary ladder at various angles. In this way I found out which degree of incline would be the most comfortable. An angle of about 45° seemed to be about the best. I put up two 3"-diameter rails at that angle and made notches in them at about every 16" on center for the 2½"-diameter pole steps. The rail on the right was longer and extended 3' above the loft floor to enable the climber to make a graceful entry onto the loft platform in whatever condition he or she might be in.

Since we have completed the basic construction of the house and have become familiar with it, we have been finding ourselves preoccupied with putting various finishing touches on it to make it a visible representation of our personalities. We are making much of the furniture out of log pieces to complement the house's natural setting. The dining-room table was built out of extra 2" cedar boards. Wherever we found bargains on usable materials, we would contemplate how we could incorporate such finds into our home. For instance, as I was driving into a nearby town this winter, I spotted a stack of broken chimney tiles in front of a builder's supply store. I stopped and asked the store manager what he was going to do with those tile pieces. He told me they fell off a delivery truck and asked me what I would want to do with them. Instead of telling him that I wanted to use them to tile my kitchen floor, I told him I needed them for something less beautiful, so he would give me a better deal. He told me I could have them all if I would just clear them out from in front of his store.

I took the many pieces of the formerly 12 x 12 inch chimney tiles and carefully tapped the curved ends with a hatchet blade to flatten them out. There turned out to be enough pieces to cover our kitchen floor. I then bought some ready-mix and mortared them in. I found out that for the best results, the tiles should be kept free from moisture when they are being laid because they adhere better when they are dry. So, for the price of two 60-pound bags of ready-mix, we now have a beautifully tiled kitchen floor.

I also want to mention our wood stove for those who are handy—or who have friends who are handy—with a welding torch. This oil-drum stove was made by a friend. It is very efficient, and even holds its fire overnight. Its flat top makes an excellent cooking surface, and it uses a lot less wood for the amount of heat it puts out than the Franklin stove.

Now, after just a few months of working on our homestead, we are a lifetime away from our plush city apartment and the beehive which surrounded it. We've made a home for ourselves, both beautiful and spacious. It is surrounded by nature's wonders and a community of people who believe in working and learning together. Our house is just one of the many examples in this immediate area of what can be achieved through cooperation. It was built and designed by a few people who each had a little specialized knowledge and were willing to share that collection of information. Without the assistance of those fine people, I doubt that the winter would have been as comfortable as it was. And I am sure that had we built limited to just the experience and technical know-how we had when we started by ourselves, we would not be living in a house that we are so proud of. We not only built a house, we built confidence in our own resourcefulness. We also built many lasting friendships with people like ourselves who value the importance of making their needs known and who are always willing to help their fellow human beings.

Franklin stove

We were living in a small flat in a large West Coast city when my wife, Kathrina, announced that she was pregnant. The thought of parenthood and all its responsibilities sparked a new light in me. I suddenly began to recognize a previously dormant need for security. We both wanted to settle somewhere and have a home we could call our own.

We decided to buy an old house just outside the city limits and fix it up as an investment. The house, with payments, tax, and insurance cost us about 30 percent of my monthly income. As we started fixing up the house, we realized the sad shape it was in. The plumbing was rusted out and leaking, the wiring was dangerous, the whole thing needed much repair. What really got us, though, was good old compound interest. Out of the monthly mortgage, only about 20 percent of that was going to the principal. At that rate, we would be paying the house off for the next thirty-three years. Even with this hassle, we still thought it a good investment and we knew we wouldn't have any trouble getting a good price for it when we decided to sell. But even from the very beginning, for some reason, this house never really seemed like home.

The property was surrounded by a beautiful wooded area that provided us with nearby peace and serenity—two luxuries rarely found so close to a city. While we considered this an asset, so did someone else. Soon a logging company came in and carelessly wiped out our little neighboring forest. What used to be a wooded playground was transformed into a graveyard of slash and stumps. We luckily unloaded the property for a little above what we had paid for it and decided to look elsewhere for a home.

Some friends of ours were thinking about settling on some undeveloped rural land. They had heard about a place near a large lake, surrounded by picturesque mountains. Since we were planning a vacation in that direction anyway, we offered to go see the place for them. We went to the area but it was still early spring, and because of the muddy runoffs, we could not get up the road leading to that parcel of land. But the valley we were in was so incredibly beautiful that we decided to ask around to find out what else was available.

We soon found just the place we wanted. It was bordered on one side by a large river, and on another side by a rapidly flowing creek. It was deep within a thick forest of second growth. This land and the immediate area that surrounded it had been logged out many years ago and was left forgotten. We camped near the river and stayed until the mosquitoes finally drove us out. But they didn't stop us from returning.

We came back with our friends and found that they shared our enthusiasm. Within a few weeks we made a deal for the land and bought it. Throughout the next few months, I spent as much time as I could in the library learning about architectural design. I had visions of the home I wanted to build. It was an arch-shaped house with a courtyard in the front. I was preoccupied with this vision because I had never really had a home before, not even as a child.

On returning to our land when we felt prepared enough to do so, we set up a tent and waited for the rains to stop. It rained for several weeks. We put up a quick A-frame to shelter us from the miserable weather. As we finished the roof, the sun came out and shone for seven weeks straight. Kathrina and I spent many days walking around our section of the property, trying to decide where we were going to build our home. We wanted to be near the river and we wanted to have a southwest exposure so that we would get as much sunlight as possible.

Kathrina put markers up in a line, designating where the sun rose over the mountains on a certain date (in this case, it was May 1). She put another set showing where the sun set on that date. From these markers we calculated how much sunlight we would have at a given spot in the different seasons of the year. In May, the sun's arc was still pretty low, so if a site received plenty of sunlight at that time of year, it would receive even more when the sun's arc was at its peak in the summer. After several trials, we finally located the ideal building site.

While I was working on the plans for the house, I also figured out the quantities of lumber we would need. I checked the local mills and priced the boards. Prices are always on the rise and materials are not always available. I knew that in order to get the best deal, I had to act immediately.

At that time, I did not know anyone in the valley except for our partners in owning the land. I went out looking for a Caterpillar to make a road into our property. As I was searching, I met many fine people, especially the man with the "Cat." I asked him to do the work for me and I told him I would pay what he asked, but he said he had hurt his back and could not do it. I knew how to operate his machine, so I asked if I could borrow it. (It never hurts to ask; and a person doesn't get what he needs unless he makes his needs known.) To my surprise, he let me use his machine. I made the road and returned the Cat to him. I told him that I had used it for fifteen hours. He saw the road I had made, and said that it couldn't have taken that long. He asked me for $20 for the use of the Cat. I couldn't believe his generosity. Here was a man who was unspoiled by greed, a man who genuinely liked people and wasn't out to get all he could from them.

As I searched for a truck to haul the lumber, I met a family of people who were just as helpful as that man. Instead of making an impersonal bargain of renting their truck for cash, Kathrina and I traded our labor for the use of the vehicle. We worked in their nursery, and we learned much useful information from these experienced homesteaders who unselfishly traded their knowledge and good "vibes" for our respect and friendship. That was one of the best deals I've ever made. We are still reaping the benefits.

Whenever I had time to spare, I went around and searched out more of the local inhabitants. Never have I met so many compassionate people who were willing to share themselves so freely. (Sure, there were a few grumps and such, but there's no need to focus on them.) Most of the folks had two things in common: the willingness to communicate their needs and knowledge, and the awareness that such sharing is a vital element of their lives.

In these local travels, I also met a fellow who was building his house out of scrap lumber. I asked him where he was getting the material. He showed me an old lumber mill that was being torn down. The owners were going to set fire to it soon, but they were allowing people to salvage whatever they needed. By the time I got there, all the usable 2" material had been stripped off. I walked around, feeling disappointed—I had missed the boat. Then I realized that there beside me was a huge 23' long, 8 × 12 inch truss beam. Near it were several other large timber pieces that had been left to be burned. There were eight-by-eights, ten-by-twelves, four-by-fours, and six-by-sixes. Altogether, we got six pickup-truck loads from that old mill.

Shortly after that, we found an old planer mill that had caved in and been discarded. The owners let us dismantle it and take what we needed. Then, a few days later, the railroad was rebuilding an old trestle and was going to burn the old timbers. We loaded the timbers up and took them home.

Before long, we had a mountain of such timbers. We measured everything and made a list of what we had. I then tried to figure out how we could best utilize the material in our carefully thought-out plan. The timbers could not be efficiently figured into the plan. Many of the pieces were too short, and the nice long ones would have to be cut. So the original plan was thrown out, and a new one was drawn up on the basis of the lengths we had. The dimensions of the house were changed in order to prevent wasting any of the long beams. The west wall, originally planned as 20' long, was changed to 23', the length of the longest timber I found. The north wall stayed 20', the east wall became 16', and the south wall became 21'. So the house took on a trapezoid shape.

Because of the massiveness and structural soundness of the timbers, I discarded the frame-construction techniques that I had studied for so long and decided to use the post-and-beam method. The post-and-beam method would give the walls more structural strength and would be a better solution than having to clutter the house with all the interior upright supports our original plan called for.

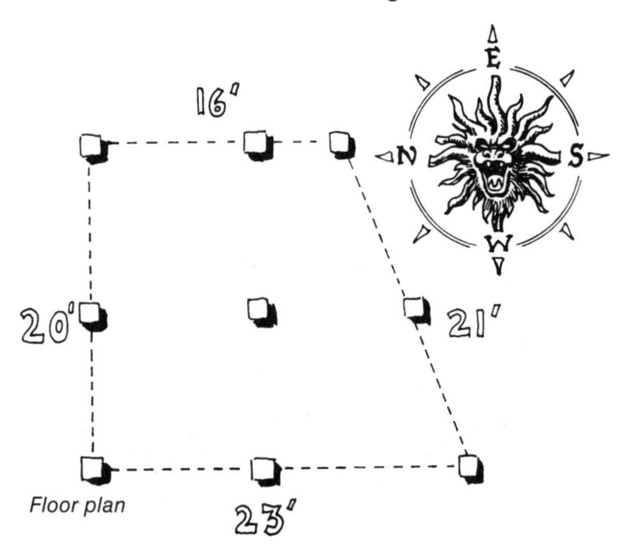

Floor plan

It was already late in June before we started our foundation. Since the property was all level riverfront land, we didn't have to deal with a slope. We were free to pick the best site on the basis of sun exposure and view of the river. Instead of having a foundation wall around the perimeter of the building, I chose to use a network of sill logs which would be supported on nine cement piers. The cedar logs were mill runaways given to me by a neighbor who had snagged them out of the river as they flowed downstream from the local mill.

I laid out the position of the foundation with traditional batter boards and checked the trueness of the two square corners. If those corners were true, the others would be also. At this point, I checked the levelness of the ground with a surveyor's level that I had borrowed. Then I dug the holes for the foundation piers. A hole 2' wide and 2' deep was dug at each corner, in the center of each side, and in the center of the building. These holes were well below the 18" frostline.

The bottom pad of each pier is 24" square and 6" high. These pads were poured first, without forms. A 48"-long rebar rod was then stuck into the pads. When the pads were nearly dry, one-by-four cedar boxes that were 12" square and 33" in length were centered over them. Concrete was then shoveled into the boxes until each was filled. When the concrete set, the boxes were removed. Each pier stuck up 15" above the ground, to lift the sill logs well above the carpenter ants. It is said that these little devils won't climb over 18"

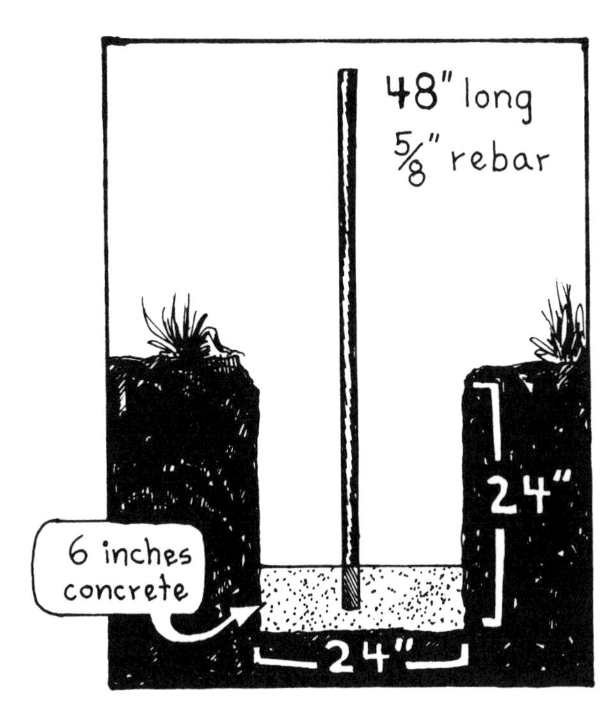

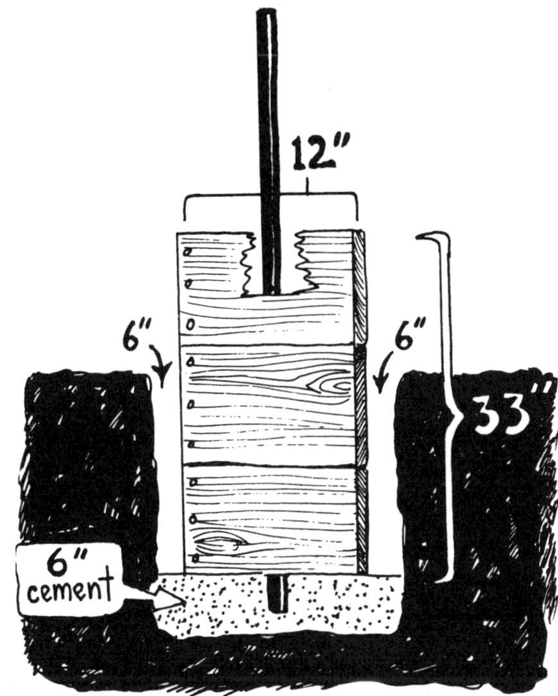

Foundation hole with steel bar set in place

of concrete to make their nests. The rebar pin extended far enough above the pier to go through the cedar sill beams and about 4" into the upright posts which rested on them.

The sill logs ranged between 9" and 14" in diameter. They were each notched so that there was 8" of beam above the piers to make them level. I placed the larger diameter butt ends directly on the piers and then notched them on both top and bottom to prevent too much from being taken out for one notch. You can cut up to a quarter of a given log's thickness for each notch without weakening that beam and losing the structural value of the material. If more than a quarter of its diameter is cut out, there is a tendency for the log to split lengthwise under pressure. I tenon-notched the thickest beams over the piers and lapped the perpendicular sills so they would fit over them. A hole was then drilled

Severe checking in a ceiling beam

in each end to accommodate the rebar pins, which would secure them together.

Since I was working alone most of the time, I had an incredible amount of difficulty moving the first few beams. Such physical punishment just didn't seem necessary. I thought about various simple machines which could help me in my work. I remembered a device I had come upon while traveling. I had seen it being used beside a mountain stream that was being mined by hand, gold-rush style. Pyramids of rocks were piled on the banks; some of those rocks were huge boulders. They had been taken out of the streambed and piled on the banks in order to get to the stream's gravel bed. I asked an old prospector how they lifted those stones up and stacked them in such high piles. He told me about the gin-pole device they used. It consists of a stationary support pole and a swinging pole that is notched into its butt. The swinging gin pole's tip is suspended out in a 45° angle by a guy wire, which is secured near

the top of the stationary pole. A come-along, or a block and tackle, is attached to the tip of the pole to lift the huge rocks. The swinging pole is then guided to the bank by ropes. For the support pole, a substantial tree is used.

If there is no strong tree around, you can use a scissor (a two-member tripod). It is made out of two logs latched together at the top. Hold the scissor almost upright, leaning it a bit toward the stream, by a guy wire which is attached to a solid stump or tree in the background. Then suspend a block and tackle down from the peak of the scissor into the water. Guide the block and tackle line to the bank by hand and lift the boulder, swing it around, and lower it onto the pile. This second method is less efficient than the first, but it prevents many an aching back.

For my stationary pole, I used a 50' cottonwood, which was beside the building site. The tree was going to be felled anyway, so I cut a deep wedge-shaped notch into its butt for the seat of the swinging pole. The butt of the 35' tamarack swinging pole was trimmed and put into the notch. It was suspended at a 45° angle toward the house site with a guy wire secured to the stationary pole about 10' above the notch. This tree for the gin pole was later used for the roof's ridgepole.

A chain block was hung from the swinging pole's tip, and a pair of log tongs was attached to the end of the tackle to grip the log or beam. I soon found the tongs to be a dangerous and inefficient method of gripping the logs, because they only secured them at one center-balance point. I ended up using a chain that circled the log at two points and was lifted in the middle by the traveling block's hook. This pole could swing in a 150° radius and could reach any part of the building site. When this device was in action, I could pick up the biggest log or beam, crank it up, swing it into position with guide ropes, and lower it down to where I wanted it. This was a slow process because of the time it took to raise and lower the beams with the chain block, but there was nothing difficult about it and I didn't break my back with the lifting that had to be done.

After the sill logs were in place, the two-by-eight floor joists were put in across them, running east and west. Though they were about 20' in length, they never spanned more than 8' between supporting points. A good rule to follow when putting in floor joists is to make sure the joists are at least 1" thick per foot of span. If the joists span 8', you need to use two-by-eights.

To allow for a split-level main floor which would be a subtle divider between the living area and the dining and kitchen areas, I hewed the inside of the west wall sill and the side of the center sill that faced it. These surfaces were made flat to accommodate the double two-by-four lips which were spiked to the bottom of these sides. The lower joists of the dining and kitchen areas were notched over these lips to split the level of the floor. We then nailed in joists for the upper level over the crossing sills, making that floor 8" higher than the lower one.

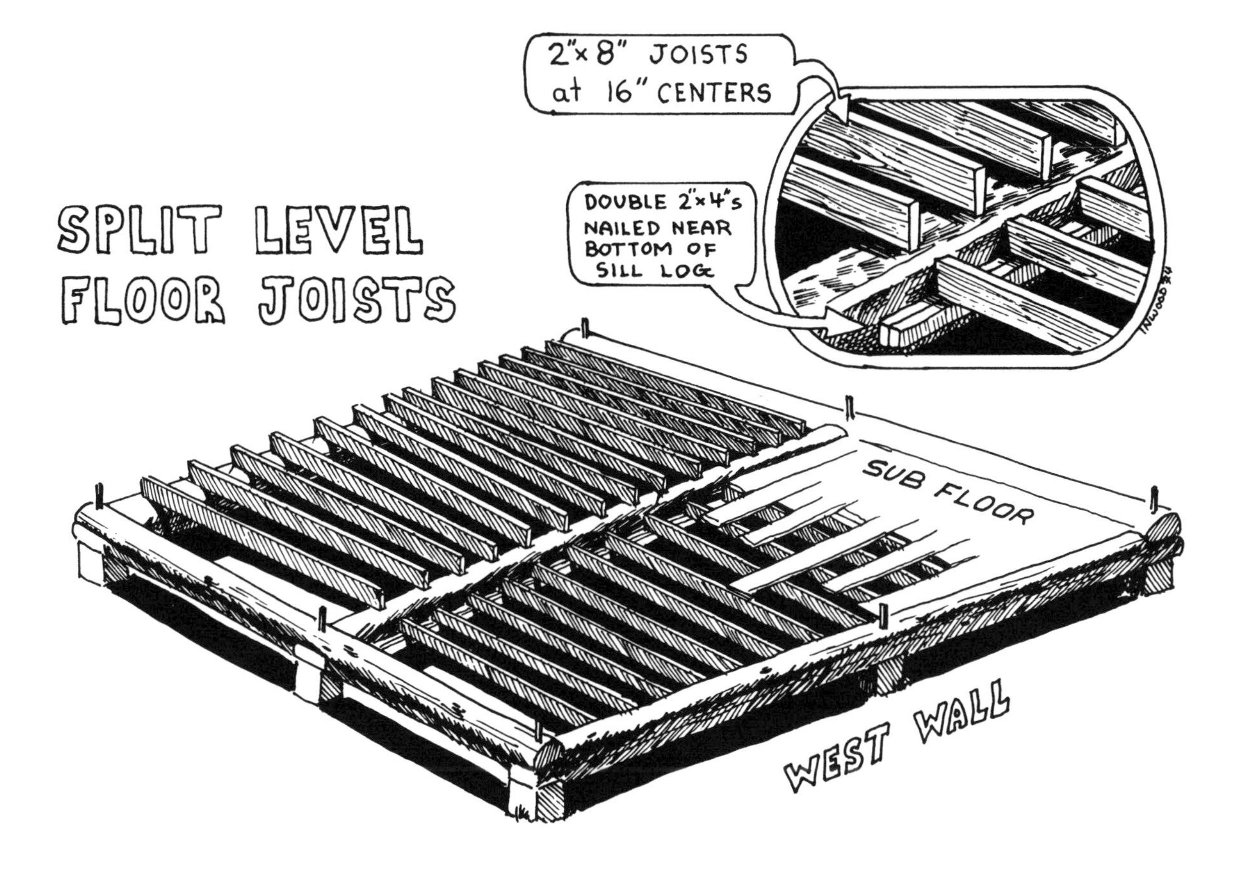

The one-by-twelve cedar subfloor was then put down so we would have a nice surface to work on. We nailed on all the flooring with ardox, spiral-type nails. This type of nail prevents floor squeaking because it holds better and does not have a tendency to pull away from the joists as the floor is walked on.

The eight upright support posts were put in above the sills, over the outlining piers. A hole was first drilled in each of their bottom ends for the 4"-long rebar that stuck out above the sills. The four-corner posts were ten-by-tens that were salvaged from the planer mill and the four inside posts were eight-by-eights that were runners for the carriage in the old sawmill. They were sawed on two sides. The sawed sides were used as flat surfaces for the wall logs or lumber frames to rest against. The round sides could be seen from the inside and outside of the building. These fir posts carry all the structural weight. The wall logs, or studs, do not carry any of it. Each of these posts could support as much as 100 pounds per square inch. So an eight-by-eight has the structural strength to support 6400 pounds above it.

The posts were checked for straightness with a level and were braced with diagonal two-by-fours. The perimeter beams were then raised with the gin pole and lowered into position above the posts. The gin pole worked well, but it was a slow operation. We guided the suspended beams with poles and ropes and stood clear until they were in position.

We notched the ends of the beams and tops of the posts and augered a deep hole into each of the ends. Steel pins were then driven into each of these joining corners to secure them. The steel pins were lengths of ½" and ⅝" bar stock that I found at a local junkyard. Most of this material I made from tie bolts that I got out of an old boiler and which I hacksawed to length. These super-hard steel rods were only 5 cents per pound and were a lot stronger, and much cheaper, than the softer rebar stock.

Lowering beams onto the braced posts with the gin pole

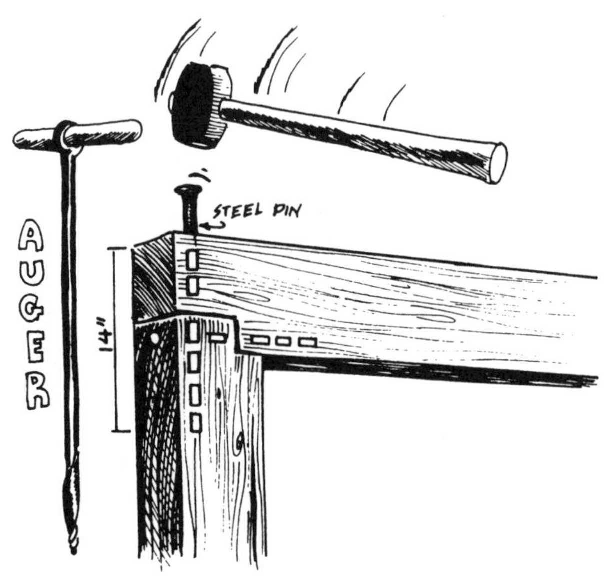

Next came the cedar-log center post to brace the ceiling beams. This was my first attempt at hewing a log on four sides. Eventually, I wanted to carve the pole and make it an ornamental finishing touch to the house. To start the hewing process, I centered a 10" square on one end of the log and an 8" square on the opposite end, letting all the crossing lines go out to the edge of the ends. A chalk line was snapped down the length of the log from one line to its corresponding line on the opposite end. With a chain saw, I then made a series of scoring cuts into the side of the log to the depth of the lines.

After one side was scored, the log was dogged in place and the sections between the cuts were knocked out. This technique was repeated on each side, and the whole log was trimmed with a shipbuilder's adze.

Instead of raising the log so the 10" square end was down, I put that larger end up to give the center post a tapering effect, making the building seem lighter. This effect would be even more pronounced if the center post were longer.

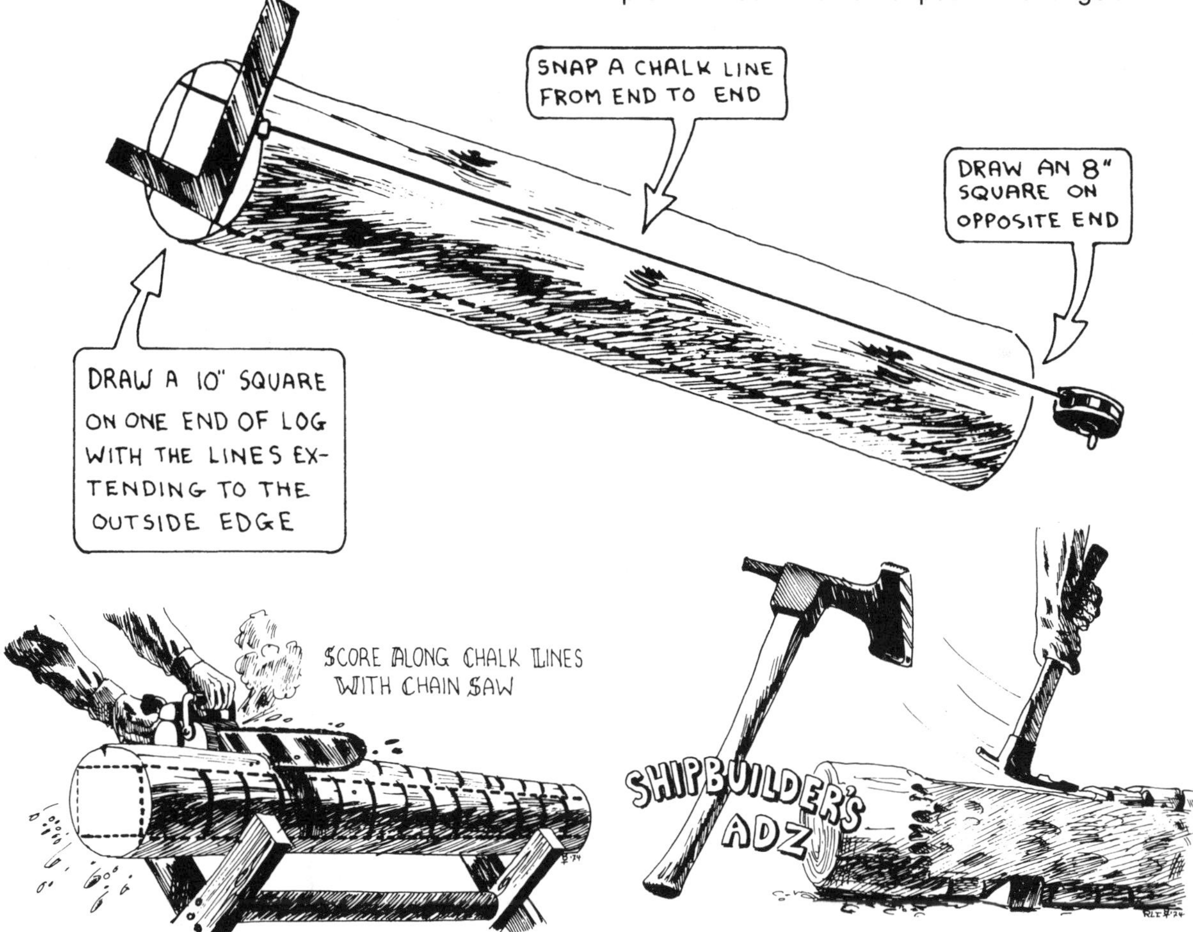

Forming the hewn-tapered center post

Centerpost

The four center eight-by-eight ceiling beams were put in and joined over the center post. Each beam spanned only from the center of the wall to the top of the center post. I found out later that this was a structural error. Because the separate pieces joined in the center instead of making a continuous tie across the house, the design lacked sheer strength. (Sheer strength is the tying together of the separate parts of a whole section to reinforce it.) For example, in order to gain the greatest amount of sheer strength when tying a wall together, the sheathing boards should be put on diagonally. The diagonals create several triangles; this is the strongest form of sheer bond. If the ceiling beams were two continuous pieces notched over the center post, instead of four separate sections, their sheer strength would be considerably greater because they would, in fact, tie the walls together. After putting in the center beams, other ceiling beams were put in as braces for the two-by-four joists which span the three loft sections of the roof area.

The two-by-four loft-floor joists were nailed in at 16" centers above the ceiling beams and a floor of one-by-six tongue and groove was fitted together over the joists. I used tongue-and-groove stock because a subfloor is not necessary with this interlocking material. No dust comes through between the boards, falling to the lower sections of the house. And with the joists at 16" centers, this one-by-four flooring is sturdy enough not to bounce or sag with weight.

We did not want a full second story, but we did want a loft space built into the sloping roof. The only way to have a full, open space between the ceiling and roof, without cluttering that space

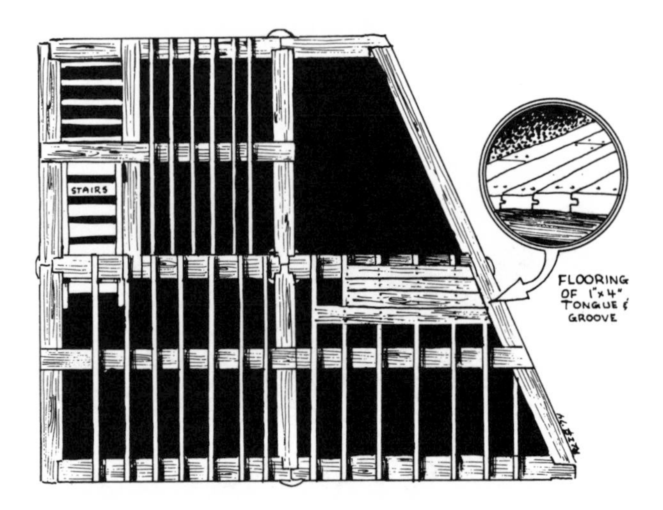

Loft floor plan

with a network of roof trusses, is to support the rafters at the peak with a ridgepole. The ridgepole ties the rafter tops together and braces them. This prevents the rafters from sagging under a heavy snow load and spreading the walls with an outward force of pressure.

Because of the structural importance the ridgepole has in this type of design, it has to be a good-size log at least 8" to 12" in diameter. Our ridgepole is the 35' tamarack which had served us well as the gin pole. We just lowered it onto the loft floor and disconnected it from the stationary cottonwood tree. Five feet were then sawn off its tip to make it 30' long. This provided 5' overhangs on either side to keep the walls dry and to shelter the porch area below. This ridgepole runs east and west and is held 9' above the loft floor by posts.

Putting such a beam in place is always a struggle. Four of us broke our backs trying to lift that huge pole above the 9' uprights, but even with all that force, it refused to cooperate. After several clumsy attempts, we gave up this method of placing it. As I cursed and tried to catch my breath, a welcome solution came to mind. I put two two-by-six rafters in place, overlapped their top ends, and joined them together as a scissor brace. I secured them with diagonal supports and attached a block and tackle to their crossing peak. Then, with the aid of my helpers, one end of the beam was hoisted up. I was able to hold what they gained by lifting while they set up another pair of rafters and raised the other end of the ridgepole in place at the peak. Both ends were then spiked to the crossing rafters, and the three supporting uprights were notched into the underside of the ridgepole. The overhanging sections that protruded past the walls were braced by diagonal poles which were notched into the ridgepole ends and the uprights.

After that mighty beam was put in place, I framed in a 5'-high pony wall on the north side

Northeast corner

Hoisting the ridgepole into place

of the roof directly over that wall. On half of the south side, above the floored loft area, another 5' pony wall was framed in. The other half of that side was left open and the rafters for that section were notched in place over the south wall beam. They spanned from 2" beyond that top beam to over the ridgepole and were spaced at 16" on center.

The rafters were spaced at 16" on center to efficiently carry the heavy 4' snow load that sometimes collects on roofs throughout a long winter. This design enables these two-by-six rafters to support between 150 and 175 pounds per square foot, which I hope is more than they will ever have to carry.

The rafter bottoms were notched over the pony walls and the top wall beams in the conventional manner. A wedge-shaped, double-mitered piece was sawn out of each rafter bottom, and the rafter was then spiked onto the top plate. To figure out the various angle cuts, since I was doing all the cutting with a hand saw (we did not have electricity), I used a sliding "T"-bevel. This is a square that can be adjusted to any angle. I figured out the angles on the first set of rafters by trial and error, and put those rafters in place. I loosened the thumbscrew which holds the adjustable bevel blade and pushed the tool into the intersecting corner of the top plate and rafter, with the body on the top plate and the blade snug

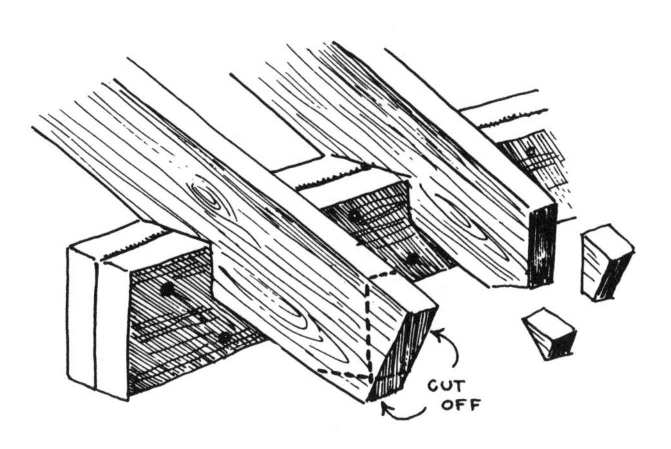

against the angling rafter. I then locked the blade in place by tightening the thumbscrew, and transferred this angle pattern to the next rafter.

I cut the bottom end of each rafter so it was vertical. This was partially a decorative effect, but also a way of prolonging the life of the rafters. This cut gives the rafter end more exposed surface, so it remains drier and the rotting process is slower. To lengthen the life of the rafter ends even more, I also cut off the bottom point, which would collect moisture.

By the time it came to sheathing the roof, I was exhausted from the past months of steady sunup to sundown physical labor. Some old friends from the city dropped by, saw my condition, and volunteered to relieve me of the chore. Within a few hours, all the one-by-twelve boards were up and the house was ready for the finish roofing of 90-weight asphalt building paper. For the next few days, I expressed my gratitude the best I could, by showing my friends around the beautiful valley.

The rolled asphalt went on very fast. As soon as the roof was finished and the house was protected from the fall rains, we brought the tent and cookstove inside and moved into our new home.

Next came the walls. We wanted to design each one a different way. I cut down some standing dead tamaracks which were bone-dry from age and filled in the west wall with log pieces while Kathrina framed in the north and east walls. Since whatever materials we used to fill the walls did not have to support any structural weight, I used pieces rather than full logs. That way, I could position and frame the windows as I built the walls. I took time to carefully plan the window placement: I set each window up on sawhorses and stood back to see the effects. Then, when I was finally satisfied, I framed it in and went on

Window placement in northwest corner

Dining area windows on the northeast corner walls

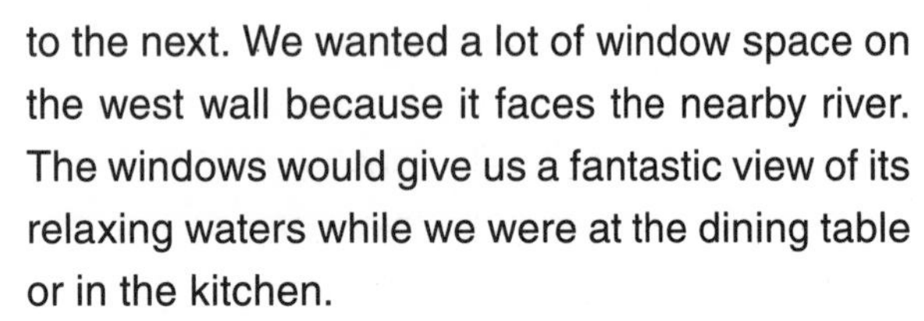

to the next. We wanted a lot of window space on the west wall because it faces the nearby river. The windows would give us a fantastic view of its relaxing waters while we were at the dining table or in the kitchen.

Since it was canning season already, Kathrina worked several hours a day both canning and framing. She framed in the north wall with two-by-fours at 16" centers. She built the frames on the floor, then put them in place between the posts and spiked them in. The outside one-by-twelve cedar sheathing was nailed on with galvanized nails to prevent board rot around the nail holes. Galvanized nails do not oxidize as regular nails do. We put 3½" insulation in between the studs, and covered it with a layer of black paper as a vapor barrier. The inside cedar sheathing was nailed on later. This wall does not have any doors or windows because it is the coldest wall and does not get any winter sun. The fireplace and staircase are located along it on the inside and a woodshed protects it on the outside.

Kathrina built the east wall in the same manner. Here she framed a large 5 × 5 foot window which gives us a view of the small, open area that we landscaped, and of the forest surrounding it. It brings in the morning sunlight to brighten up the long living room area.

The southern wall has two parts. The section west of the middle post is a log wall made of aged white pine and tamarack. The logs were extremely hard to peel, even with a drawknife. These short logs span between the middle post and the long 18 × 72 inch window to the far west. This long window helps to distribute heat. The cold air comes off the bottom and rolls the heat from the nearby wood heater to circulate it through the room. This oblique corner, with its two long windows, is a nice space in which to relax and read. It feels separate from the rest of the room and puts you more in touch with the outdoors.

And now we come to the experiment that failed—the east section of the south wall. Here we have a cordwood wall made of bone-dry cedar rounds which sat in the house for six weeks in 100° weather. Even though they were aged and dried, they still managed to check and crack within

their cement frame. I took extreme care every step of the way in making this wall. I even dry-mixed the 1 part lime, 2 part Portland cement, 3 part sand mixture before adding water, to keep it from shrinking. But the cement still shrunk. Even with every precaution exercised, before long, the cedar rounds were loose in the cement.

This wall took forever to make. I could only go 18" at a time because so much mortar was needed between each round. Then I had to wait for that portion to dry before going on to the next. The wall is pleasing to look at, but it doesn't have any insulative value because concrete does not hold in heat. When I would later look at the wall, I would say to myself, there is a month's worth of firewood that we could have used.

This cordwood wall has a window and a door. The window's frame was set into the mortar and seems to be solid in it. The door frame was built as a box and was also set into the mortar. Its sides and bottom are full-dimension, two-by-ten bridge timbers, and its top is a curved jack-pine nurse log which extends beyond either side to grip the concrete. The door is double two-by-six tongue-and-groove cedar. It has no strap hinges and is held together with headless 4" nails, which were driven in diagonally along the seams between the boards. It has proven to be a good, solid door.

After the walls were filled in, we traded the temporary ladder to the loft for a stairway which began near the middle of the north wall and went up over the fireplace. Due to lack of space, the stairway was laid out to be as steep as possible while still not being awkward. Its total run is about 5' for the 8' rise. I wanted each riser to be 9" and the stair treads to be 8" by 30". I calculated the cuts for the stringers so they could fit accordingly.

The roof section is closed in by the loft floor except at its smallest quarter. There, a high ceiling comes up from the living room to join the upper and lower spaces and make them one continuous space which is not closed off by any full partitions. It gives half the living room a cathedral-ceiling effect and brings light down to

Interior and exterior views of cedar cordwood wall

Cedar door

Cathedral ceiling over living room

Living room area

it from the high window of the east gable. This gable, along with the west one, is framed with two-by-fours and sheathed with cedar one-by-twelves.

The loft area has three sections that are defined by rectangular areas. The first section, coming up from the stairway, was designed as our daughter's sleep and play area. It had high railings and was safe for her to amuse herself in. The area directly behind that is a small storage space. There is a partition between that space and the next. Beyond the partition and the curtained doorway is our 10 x 12 foot bedroom. It has two low windows at the south wall which bring in light throughout the day and one low window at the west wall which overlooks the river. These windows are at a height which allowed us to look out while lying on our bed.

We found that a good way to bring extra heat into a small loft area is to have the stovepipe come up through it. You'd be surprised how much heat is radiated throughout our little bedroom from just that stack.

As an afterthought, I dug a 6 x 10 foot hole, 6' deep, under the back porch for a root cellar. The sides come up another 1' under the porch, making the cellar 7' deep. These short sides that come above ground are insulated with the fiberglass we had left over from the roof. This little cellar has bins on one side, shelves on the

other, and an aisle between. The three bins are each 2½' deep. They separately store the carrots, beets, and potatoes so their flavors won't mix. The carrots and beets are buried in sand so they will keep better. The shelves on the other wall provide enough space to hold two hundred jars, but we never have more than one hundred jars to put on them, so we also use them as a wine rack. Though we have not yet installed the proper vents, this root cellar maintains about a 46° temperature and stays dry. It is a very satisfactory storage space, but if I had it to do over again, I would have built it directly under the kitchen floor and added an inside trap door so we wouldn't have to go outside in the cold winter to get our supplies.

Though much of my designing of the house was oriented toward saving money, I was definitely not cutting corners on any of the structural materials. In fact, much of the house was overbuilt. It was designed to be functional, though

Kathrina was more interested in how things looked. The planning process was actually a push and pull between form and function. It was sometimes difficult to find a balance, but I think on the whole, we did very well except for a few mistakes—the cordwood wall and the separate crossing beams.

As I said before, I never really had a home, not even as a child. Because of this, I was sparing no energy in building my home the way I wanted it. It was something I really needed to do. Building the house myself took away a need or a pressure that had always been hanging on me. This pressure was not relieved when I bought that other house. I never felt I belonged in that place because no matter what I did to it, it would never be an extension of me. It would always be someone else's "trip," someone else's home. The experience of building my own home also built something in me. It gave me a sort of confidence in myself and other people.

Section 2

WATER

MAGIC
INWOOD '75

CHAPTER 6
Beat the Heat—Outdoor Rock and Iron Stove

Like most people, we find ourselves outdoors a great deal of the time in the hot summer. Since this season is so short and good weather is such a treat in our area, we take every chance we can to enjoy the sunshine and make excuses to be out in it. We are constantly building on to our homestead and working the garden and fields, so we are outdoors much of the time. But there are many indoor chores that we dread during this season. You can guess them—cooking, baking, and canning. All involve starting up the old wood cookstove (since we don't have any other fuel source). With this old stove at full blast, we had to remain indoors in the stifling heat. We needed to remedy this situation.

Upon visiting friends, we found some alternatives to tending the indoor cookstove. One group had an old rusted stove that they had set outside. They did their cooking on it. What a great idea. We thought about it and decided that we definitely needed an outdoor cooking setup, but we wanted something more permanent—more protected. We built a large porch area outside our cabin and roofed it to protect whatever heating device we would use. This porch would also shelter us from the weather when we cooked outdoors during the not-so-desirable seasons and as a change of pace from being cooped up in the cabin.

We decided to be creative and also add another aesthetic touch to our home. A design for a rock stove was roughed out and materials were soon gathered up. The gathering process became quite interesting. Our design changed to fit the materials we found. We found old discarded wood cookstoves and ripped them apart, taking the oven doors, dampers, and grates. One stove had a top that was in excellent shape, so we removed that for the top to our stove (**FIG. 6-1**).

Before long, we accumulated a quantity of junk that we could possibly use as vital organs for our creation, and trucked it home.

We brought in several loads of angular, granite rocks, which are a plentiful commodity on our homestead. Angular rocks are preferable to round ones because they are easier to stack and adhere to one another. Granite was favored over a softer type, since granite rock doesn't chip or crack as easily. These were to be the facing rocks for the stove.

Up to this point all the materials were free. All we really needed to buy was about five bags of Portland cement (type 1), two bags of lime, and a few fire bricks. The lime was added to the cement to give it a sticky consistency. But we had to go sparingly with it, because lime tends to weaken cement. I mix my own Portland cement rather than buy pre-mixed masonry cement. Portland cement is far stronger for the purpose, and if by chance you end up with any extra lime, it can go into your compost or be used in the outhouse. Nothing should ever be wasted.

I will explain the procedure of building our stove, but I will give only the measurements that are structurally necessary to make an efficient unit. All other dimensions should suit the individual taste of the builder and should not be copied from mine.

We began construction by first digging an 18" deep U-shaped trench for the footing. The open part of the U was where the front of the stove would be. This footing can be deeper or shallower, depending on the frost level in the area, but it should always be at least 6" to 8" in thickness below ground to properly carry the heavy rock load. It should also be at least 12" wide to support the mortared rocks.

The use of a form depends upon the ground. If the soil is claylike, no form is needed. If the ground is rocky or otherwise porous and loose, build a form out of two-by-four material and brace the outside with stakes. Have this form extend up past the surface another 18". If you do not use a form below the surface, start the 18" form at ground level, just outside the poured footing (**FIG. 6-2**).

At this point, the inner form boards and stakes are removed and a short front form piece is added. The height of that piece is dependent on how thick the ash pit is to be. At the open

end, hinges were sunk in the soft cement to accommodate an old oven door which I used as the ash pit door.

On top of the U footing, I laid a rectangular fire grate into a bed of fresh mortar. This grate rests 2" into the side walls of the footing for sufficient support. Our grate is a solid piece of ¼" plate steel, 2 × 4 feet wide, with a space left at the closed part of the U for ashes to drop down to the ash pit. The ashes remain there until there are enough of them to put in the garden or compost pit. The grate has no holes to supply the fire, so all the air comes from the front opening above the grate (**FIG. 6-3**).

A more airtight stove can be made if the grate has holes in it allowing air to come up from a controlled ash pit door, and if the firebox were supplied with a tight door. Our firebox does not as yet have a door. We are thinking that one would be a good idea because it would make this a more efficient stove. But since we've been burning pieces of wood that are longer than the firebox—and which stick out from it—we wanted to leave that space open for the time being.

After the grate was set in, we put in the firebox. We began with a collapsible wooden mold. The size of this mold form can be up to 4" less than the width of the grate, and just as long as the grate, depending on the size of the stovetop to be used. The height is determined by the size of the wood to be burned. This form will temporarily take up the area of the firebox. Next, we built a course of form boards around the top of the footing. This form does not have to be strong, because it is just a guide for the rock face. All the rock work can be done without a form, but I prefer to work with one to help me develop the proper taper for the stove. I wanted it broader at the base to carry the weight, and narrow at the top where it didn't need as much support. These forms consisted

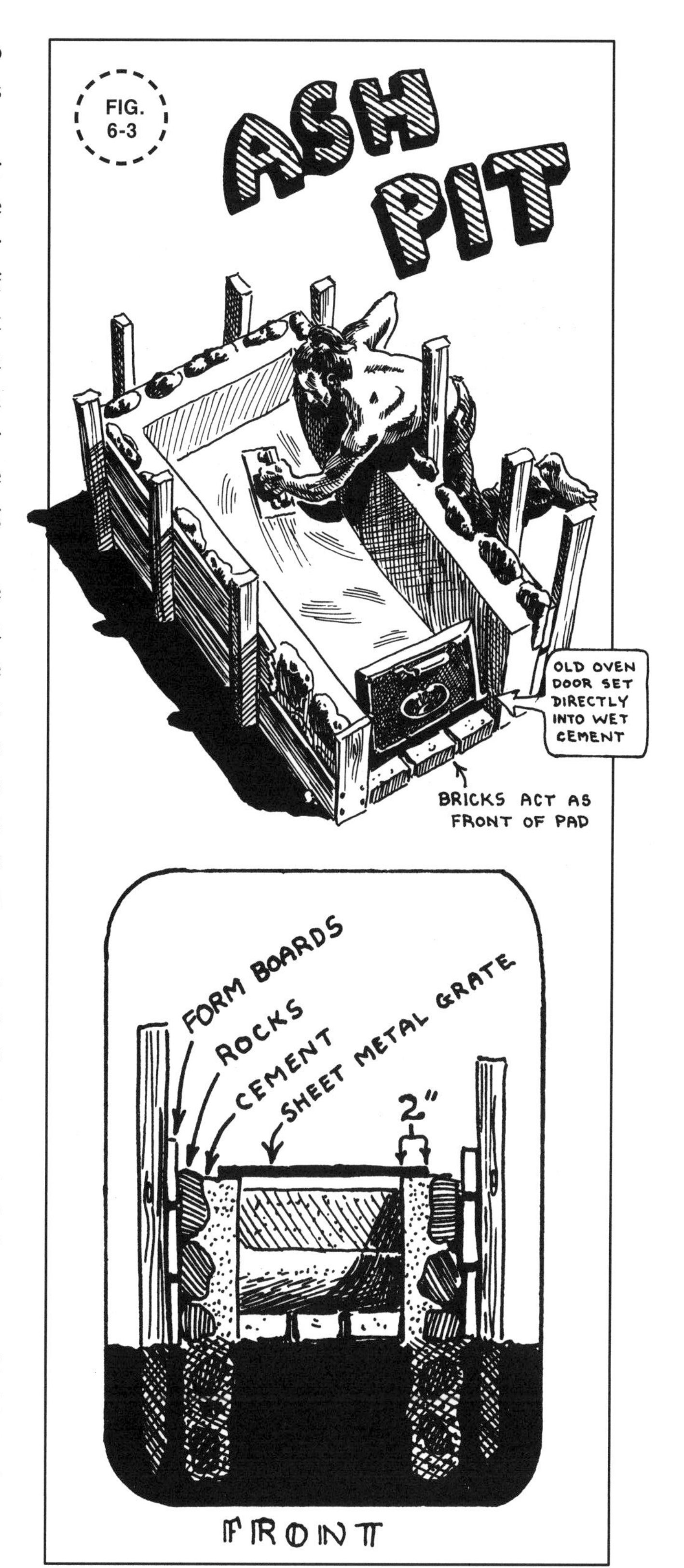

FIG. 6-4

merely of long two-by-four stakes set at the corners of the stove, which accommodated the horizontal form boards as they were nailed on. The boards were raised as I laid the courses of rock. This prevented having to deal with a high clumsy form when setting in the rocks.

The collapsible wooden form acted as the inside mold for the firebox area. A layer of firebricks stacked to the height of the mold was placed on either side of it. Then mortar and rock were set in to fill the gap between the outside forms and the firebrick. I used 1" firebricks, and can see now, after several years of stove use, that a few are already cracking. It would have been better to have used 2" bricks. They last much longer. Also, I did not use any type of mortar or fire clay between the firebricks. The firebricks should have been glued together with fire clay for optimum heat radiation. This would also have kept them secure.

Firebrick is the best of all possible materials and can be removed easily for replacement, because it is not interlocked into the structural work of the fireplace the way rocks are (**FIG. 6-4**).

In our chimney base, we installed an oven drum just behind the firebox. It has a door on the left side of the stack base, and is made from a cut-off section of a water heating tank bottom. It is 14" in diameter and about 30" long, made of heavy-gauged steel, and would last almost indefinitely if not exposed to hot flames as our oven drum is.

If we were to do this drum over again, we would put it higher into the stack—about chest height. This would make it last much longer because it would not be directly exposed to hot flames. This would also allow the heat to circulate around it more evenly. Now we have to watch it very carefully to get good results when baking with it.

To create the space for the oven, just make a temporary inside form box that is the size of the drum diameter, plus 8" wide, and the same length as the drum. This form should be set on short legs, suspending it between the grate and the wall of the closed end of the U footing. It should extend through to the outside left side to

FIG. 6-5

allow an opening for the drum door. Then build the rock work up to the top of that box, using an outside form as suggested before.

When the mortar has set, take the inner form box out and replace it with the oven drum. You can improvise a homemade door to this oven, which would be hinged to the rock work opening. The hinges can be sunk into the mortar while it is still setting.

For easy replacement, make the door opening large enough so the drum can slip out if it wears out or needs to be cleaned. It is also a good idea to sink a bolt support somewhere at the middle of the drum, above the firebox. That way, a bolt from the drum can be secured to the stove, keeping the drum in place. The drum must have a backing of 2" fiberglass insulation or other material wherever it directly comes in contact with the mortar and rock. This prevents the heat expansion of the drum from cracking the mortar and rock, and also makes for easy removal of the drum (**FIG. 6-5 AND 6-6**).

FIG. 6-6

FIG. 6-7

FIG. 6-8

As you build the stack, be sure to leave at least 4" of space between the top of the drum and the mortar and rocks so the smoke can easily draw up through the chimney. This space can be easily allowed for by improvising two small removable forms that fit into the areas to the front and rear of the drum, (**FIG. 6-7 AND 6-8**).

Above the firebox, about halfway up the drum, I placed a damper, which regulates the smoke around the drum. When it's in closed position, the smoke and heat travel only around behind it. This damper could be omitted if the oven was farther up the chimney. I strongly recommend the higher drum placement, as opposed to my present design.

Before building the smokestack and chimney, we prepared the area above the firebox for the stovetop and set that piece in. A fresh bed of mortar was poured over the rock sides around the firebox to sink the long edges of the old wood stovetop into. This cast iron stovetop is 32" long by 24" wide and has four removable plates. The plates are covers for the two 9" holes and two 8" holes. This stovetop can be taken completely apart. Its hole covers and all the crosspieces that hold them can be removed, leaving only the outlining rim as a stationary piece. It is the ideal stovetop, allowing us to enjoy an open fire for evening get-togethers. We sit around the fire pit with guitars, drums, and other instruments and play to its energy-giving inspiration. The flames and smoke from an open fire such as this are an excellent deterrent against pesty mosquitoes.

At either side of this stovetop, directly above the rock mansonry, is a 4" shoulder extension to give the top more surface area. To prevent these shoulders from cracking as the stovetop expands from the heat, V-grooves should be made between it and the metal top. This extra area is excellent

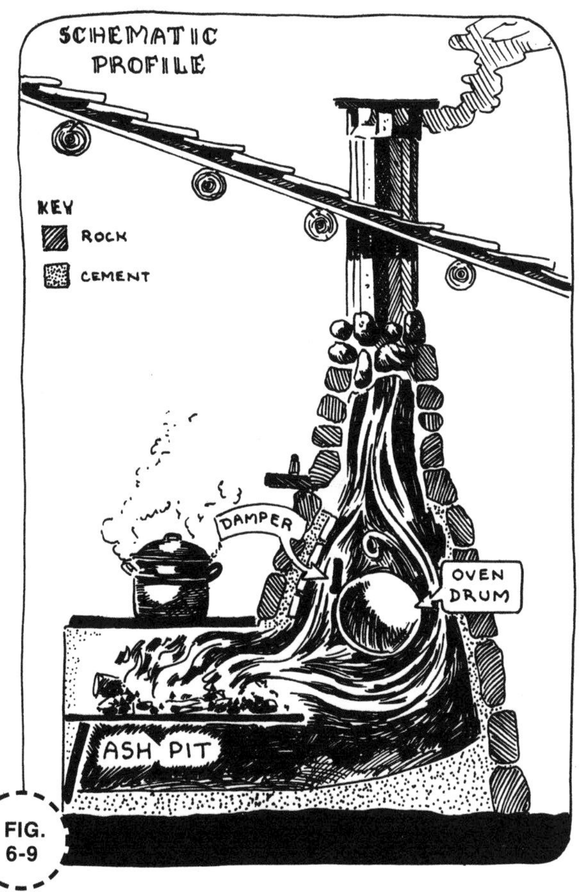

FIG. 6-9

for accommodating large canning pots, steeping kettles, and laundry tubs, which would not be able to fit on the smaller stovetop (**FIG. 6-9**).

Behind the stovetop, two supports were put in to carry the weight of the rock work above the firebox. They straddle the two sides and were set into fresh mortar. These supports were made from an old chain-saw bar and an old leaf spring. The leaf spring is still holding up fine after several years, but the chain-saw bar got soft from the heat and consequently drooped down into the firebox area—not doing any harm, in fact, preventing large pieces of firewood from being shoved down into the ash pit. So you could very well use either expensive 3" to 4" angle iron or the leaf spring as alternatives for this support (**FIG. 6-10**).

Next, the mortar and rocks were set up to approximately the 6' level, leaving a flat platform at the top. The inside should be a fairly square flue hole, 10" to 11" wide. The rock work has to

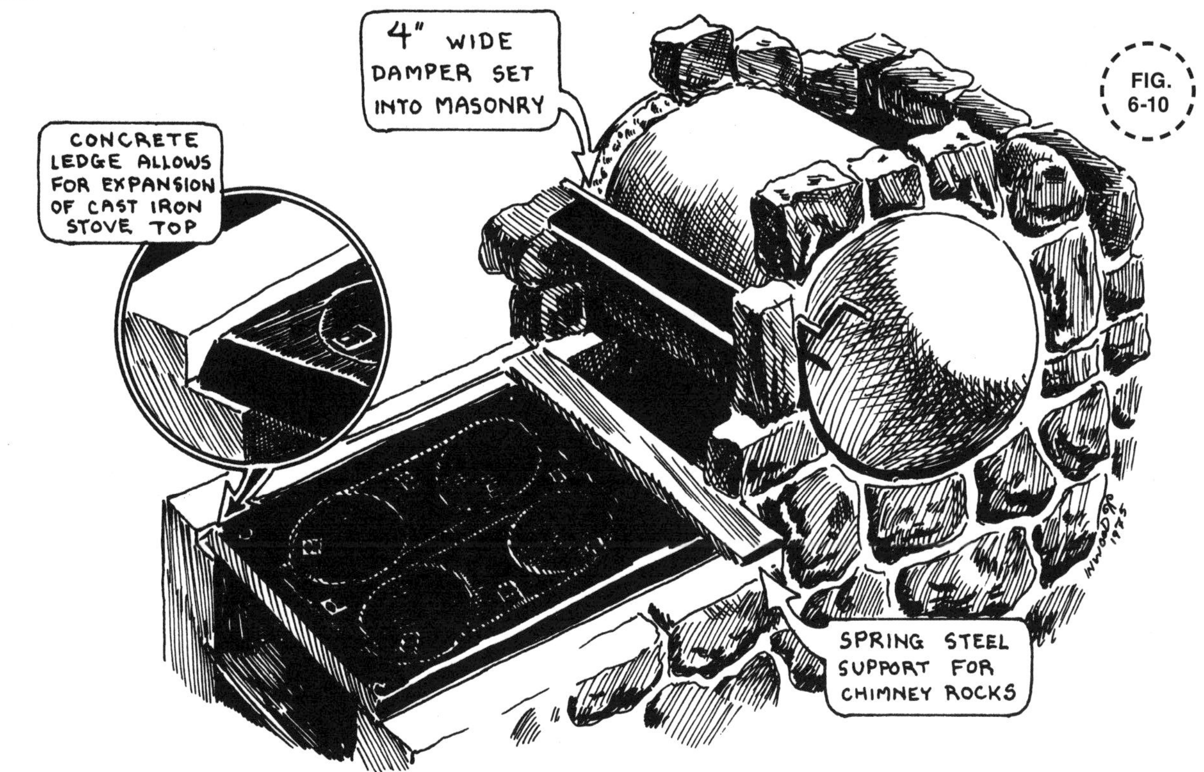

FIG. 6-10

FIG. 6-11

FIG. 6-12

FIG. 6-13

be at least 6" to 8" thick all around this area to provide proper support for the smokestack and chimney. Before this point, all inner forms should be removed so as not to obstruct the draw (**FIG. 6-11**).

Above this platform, I stacked a few 5-gallon oil cans, which I use as improvised flue linings and as inside forms for the rock and mortar chimney. Being tapered at the tops, these cans easily fit into one another and are held in place by a protruding lip about 4" down from the top. They are of a heavier gauge than commercial stovepipe and have a diameter of 11". This wider diameter is instrumental in providing a proper draft for the firebox, since the chimney is so short.

When using a metal flue lining, I suppose one should take the precaution of allowing an airspace the width of 2" fiberglass insulation (or another material) to absorb the heat expansion, but I did not take such a precaution. Yet, in several years of use, the chimney has only developed a few hairline cracks.

At this point, I stopped using awkward forms and laid the rocks around the cans as they fit. I inserted strips of doubled-up chicken wire between the oil can flue and the mortared rocks to reinforce against cracking or loosening. Since we didn't have a form over this area, we had much more freedom to be creative with the rock work. We made little shelf ledges for matches, thermometers, and other equipment that we would be using regularly. We also were able to clean the rocks as we laid them, providing a much more aesthetic finish to our labors (**FIG. 6-12 AND 6-13**).

This chimney extends up through the roof of the porch about 2', giving it an overall length of 11'. It shouldn't be much shorter than this or it might cause draft problems. It could have been capped with a 4" concrete top for protection against the weather. We omitted that precaution. But one precaution that should always be taken is to add a spark screen. We set in a piece of ½" hardware cloth above the chimney to protect the roof, the cabin, and nearby woods from any hot sparks that might otherwise jump out and cause a fire. This cloth should be shaken out often since it collects a lot of soot and ash (**FIG. 6-14 AND 6-15**).

As a finishing touch, fill in all the cracks between the rock work. This does not provide

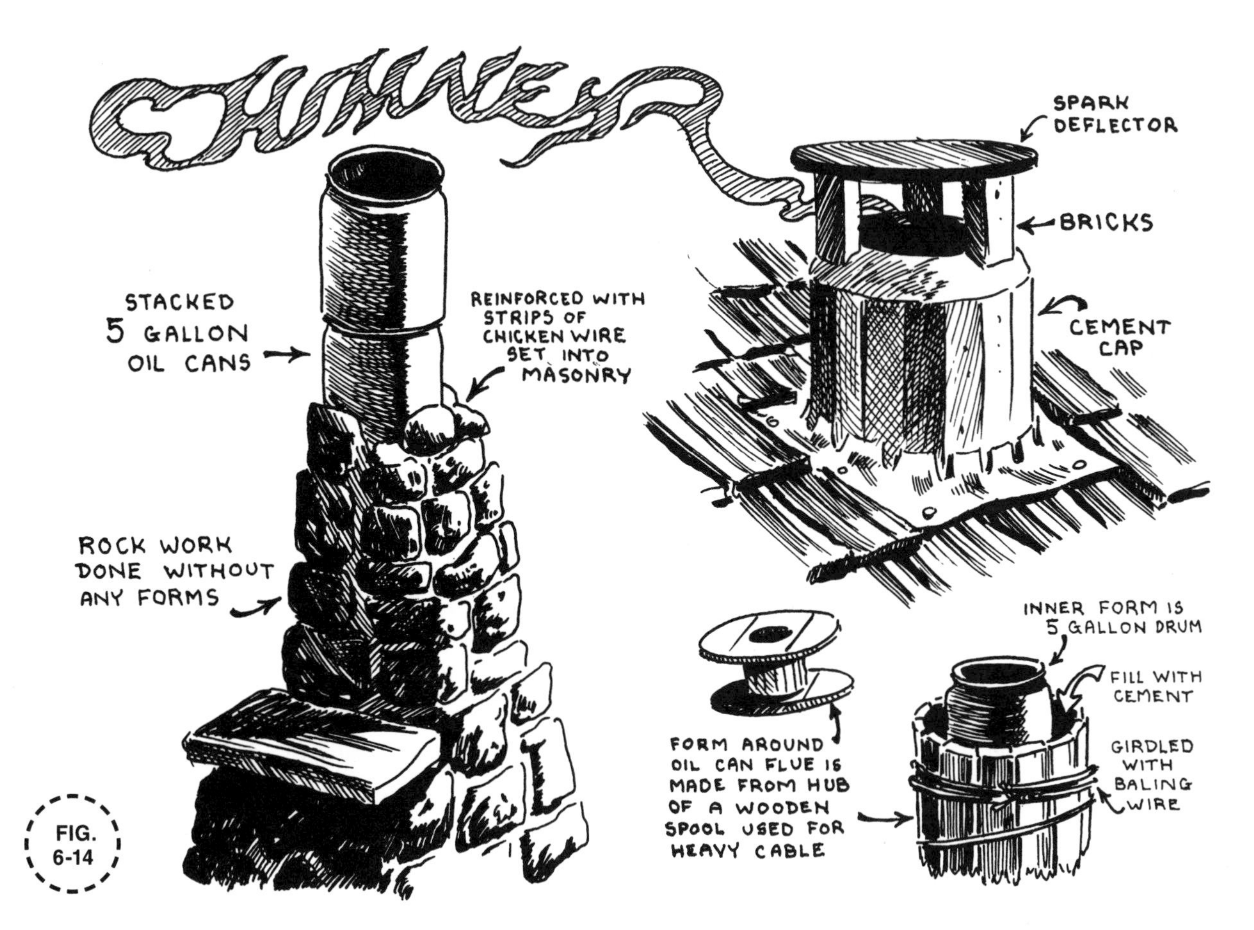

FIG. 6-14

any more structural strength, but improves the looks of the stove. Actually, you can do this pointing as you build if you devise a removable form system that would enable you to clean off the setting mortar and work around the freshly laid rocks.

Be sure to let the rock work season at least five to six weeks before starting your first fire. The mortar should be dampened at least once a day during this period so that it will cure properly. When starting your first fires, make them rather low instead of roaring, until you are certain that the mortar has seasoned long enough. And clean the chimney often, paying particular attention to the area surrounding the oven drum, because creosote will have a tendency to build up more in that area than the straighter areas.

So, we have developed an outdoor stove that enables us to heat and bake and cook all in one unit. It is a beautiful addition to our homestead and is invaluable during the hot summer months. Plus—we don't have to worry about spillage and keeping the floor clean. We can be as messy as we want, letting wood chips pile up and dirty feet go wherever they want.

FIG. 6-15

CHAPTER 7

Water is Life—Gravity-Flow Water System

The very first consideration on any homestead should be setting up a water system. We unfortunately bought an existing homestead that had been neglected for several years. The orchard had been neglected for over fifteen years, with sucker branches as big as main limbs. The huge logs that comprised the barn were so rotten that when I walked on a side wall, the bottom logs caved in. The house roof leaked and the floor was on enough of a slant from age that a ball placed in the middle would quickly roll toward the south wall. And when the winter came, the snows caved in the utility shed roof and the old water system froze, forcing us to haul water for ourselves and two cows.

So we were compelled to leave our warm, sheltering retreat and battle the elements until we successfully thawed out the water system. This sometimes took days of sitting by the water box in subzero weather, chipping ice, and burning through rusted galvanized pipe with a propane torch until the ice melted enough for the water to begin to flow again. I remember one time sitting up there, cursing the cold as my frozen fingers gripped the wrench, trying to disconnect an old galvanized fitting, only to have it freeze to the fitting before I could loosen it. Those are days I'd rather forget.

In the following spring, as soon as thaw set in, we dug up much of the old water system to find out why we were having so many problems with it. First, it hadn't been put deep enough into the ground. A water pipe should always be put in well below the frost level to prevent it from freezing. The frost level in our area is 18", but most of the pipe was only buried from 3" to 6". No wonder it froze.

As we were digging out the old makeshift system, we came upon a few 6' lengths of fir logs. They were connected to each other by a tapered male end fitting into a female end which was bored out larger than the hole that ran the length of the logs. The hole was 4" wide in diameter, continuing through the pipes. A wire was wrapped around the outside of the female end to keep it from expanding as the male end swelled from the water running through it. This sealed the lengths together. Most of this pipe was badly rotted, but I could tell that this must have been the original water system of the homestead (**FIG. 7-1**). I searched around for more traces of it and came across a few old crisscross trellises, which once held sections of flume (**FIG. 7-2**).

Curiosity soon got the best of me. I searched the neighboring farms for some of the homesteaders who had been around for a long time. My search led me to one extremely friendly pioneer who settled here long ago. He knew the people that originally farmed our land and explained the water system to us: It consisted of two systems that began at the same source. One system was of the buried fir wood pipeline for year-round domestic use and the other was a

FIG. 7-1

FIG. 7-2

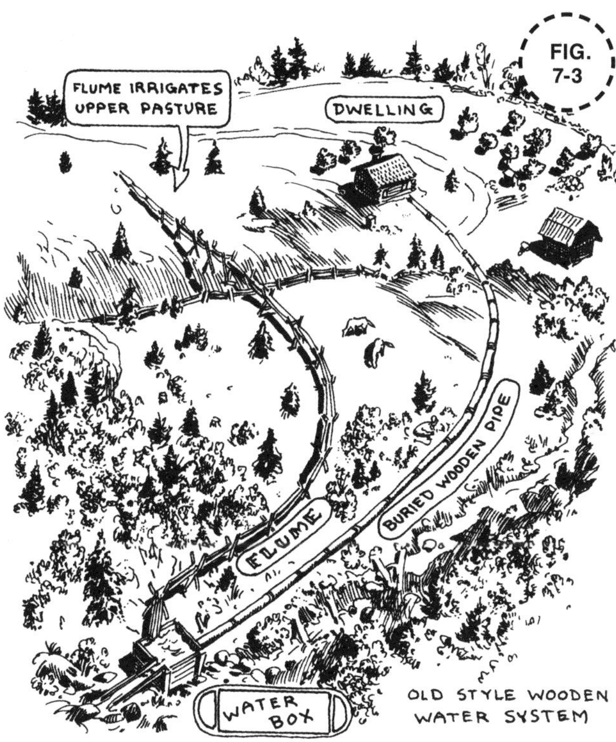

FIG. 7-3

series of trellises that carried a long flume across several hundred feet to the irrigation ditches of the farm's pasture. The flume came directly out of the creek on a slight decline, all the way to the pasture. The underground system began at a nearby collection box (**FIG. 7-3**).

The homesteader I talked with told me of the days, some forty years before, when he and his partner used to make these pipes. They rigged up a drilling device made from a 1927 Chevy overhead valved engine and transmission with a sliding carriage. Auger heads of various diameters were made from large, thick-walled, pipelike truck driveshafts and whatever other material was available.

The end cutters were ground and sharpened hard steel bits and the blades were simply ½" bar stock, spiraled down the entire length of the auger body. They were periodically screwed into the body to hold them securely. The smaller augers were made from the driveshaft that originally fit into the plate of the transmission, to avoid excessive play while boring the holes (**FIG. 7-4 AND 7-5**). The sliding carriage was made from an old sawmill carriage mounted on angle iron bed rails, which guided it straight into the stationary engine and direct drive auger. The

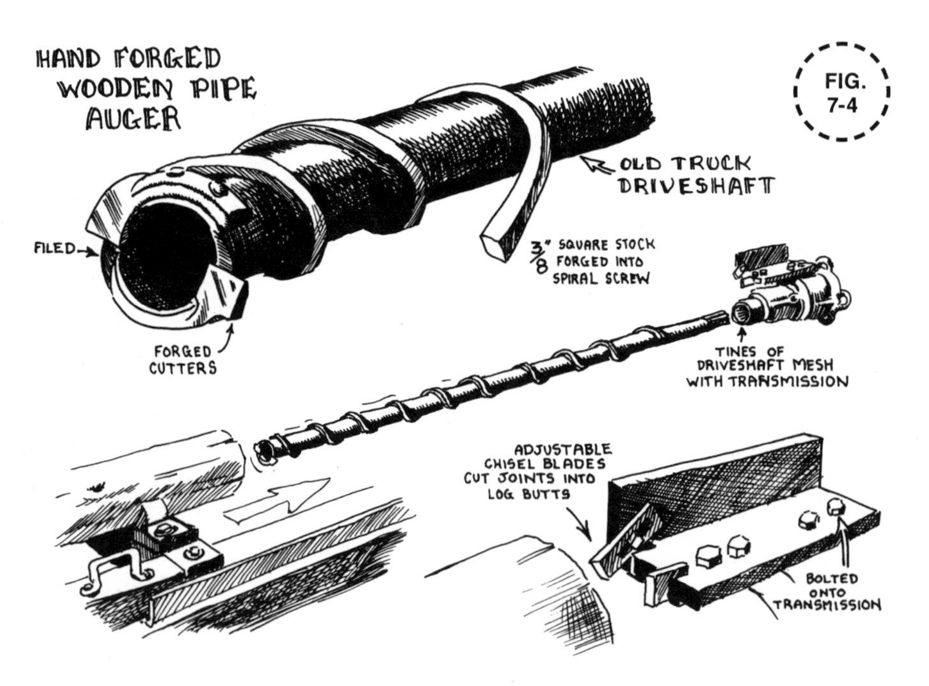

carriage was at a nice height so the workers didn't have to bend over. The 6' length of log was placed on the carriage and held with two wood levers. These levers consisted of a wooden crossbar with a rounded notch on the underside that had two short spikes coming from it to hold the log in place. A spring was attached to the opposite end of the handle to allow the crossbar to remain horizontal over the log, thus providing equal pressure on the log. There were two of these levers, one for each of the workers (**FIG. 7-6**).

The hole was bored through in several stages, going about 1' at a time. The auger was brought out each time and the shavings were cleaned off. This operation was repeated until the shaft was in 3' or more. The log was turned around and was brought through again until the hole was completed. The female end was then augured out a bit larger to accommodate the tapered male end, which was chiseled out in a conical shape on a revolving lathe, also attached to the motor. The male end was checked with calipers

FIG. 7-5

FIG. 7-6

to make sure it would be a good, tight fit and the wire was put around the female end. All the leftover shavings were used for kindling; nothing was wasted (FIG. 7-7).

This pioneer told us that he made very little money from selling the pipe, but he didn't need much money in those days because there was nothing to spend it on.

For our water system, in realizing that our present system was inadequate for our purposes, we bypassed the old system and started a new one at a higher elevation. We needed approximately twenty-six outlets for the fields, the house, the orchard, the garden, and the barn. This meant that we needed more volume and more drop. The drop would be provided by constructing the intake collection box much higher up the hill than it had been and the volume would be supplied by larger diameter water pipe, over a longer distance.

Our water source is a substantial year-round creek that comes gradually down the mountainside at approximately a 7½ percent grade. Even in the driest years when many of the surrounding water supplies had dried up, this creek kept flowing. A proper drop needed for our situation would be at least 50', working with a constant volume supplied by a 2" pipe. In order to obtain that 50' drop, the water line had to be run well over 1000' down the slope before the first outlet. In other words, with an average gradient of 7½ percent, we would get about a 50' drop in

1000', supplying the pressure we needed. Each 2.31' of head or vertical drop produces about 1 pound per square inch (psi) of pressure.

When using most brands of low friction plastic pipe (check with manufacturer's specifications for exact friction loss), there will be approximately 40 percent pressure loss to friction if the gradient is an average of 5 percent. At 25 percent gradient, the loss will be only 8 percent. The steeper the gradient, the less the overall loss by friction. This means that if you are running a length of pipe 1000' at a 5 percent gradient, instead of getting a total of more than 50' of drop, you will only be getting 30' of head or a loss of 40 percent by friction.

You can figure your own gradient with a clinometer by taking several readings of equal distance no longer than 50' each. Add up the readings and divide them by the number of readings to get the average slope over a given area. Then figure out the average percent. That is your drop (FIG. 7-8). The amount of drop required depends on your particular situation. Check with your local water rights branch for that information.

We required a large drop but were careful in our calculation not to have too much pressure in our system or we would be constantly repairing broken pipe. With all the outlets turned on, we have enough pressure to keep the 18-to 20- pound pressure sprinklers going and still have enough flow for the house and barn. Yet the total psi pressure is not greater than 60 psi at any point. The reason for this is that we compensate for pressure by supplying volume. For the first 1000' or so, the line is a 2" plastic pipe, which provides sufficient volume for the entire system. From this one 2" line, we gradually tee off to four 1" lines without losing any of the

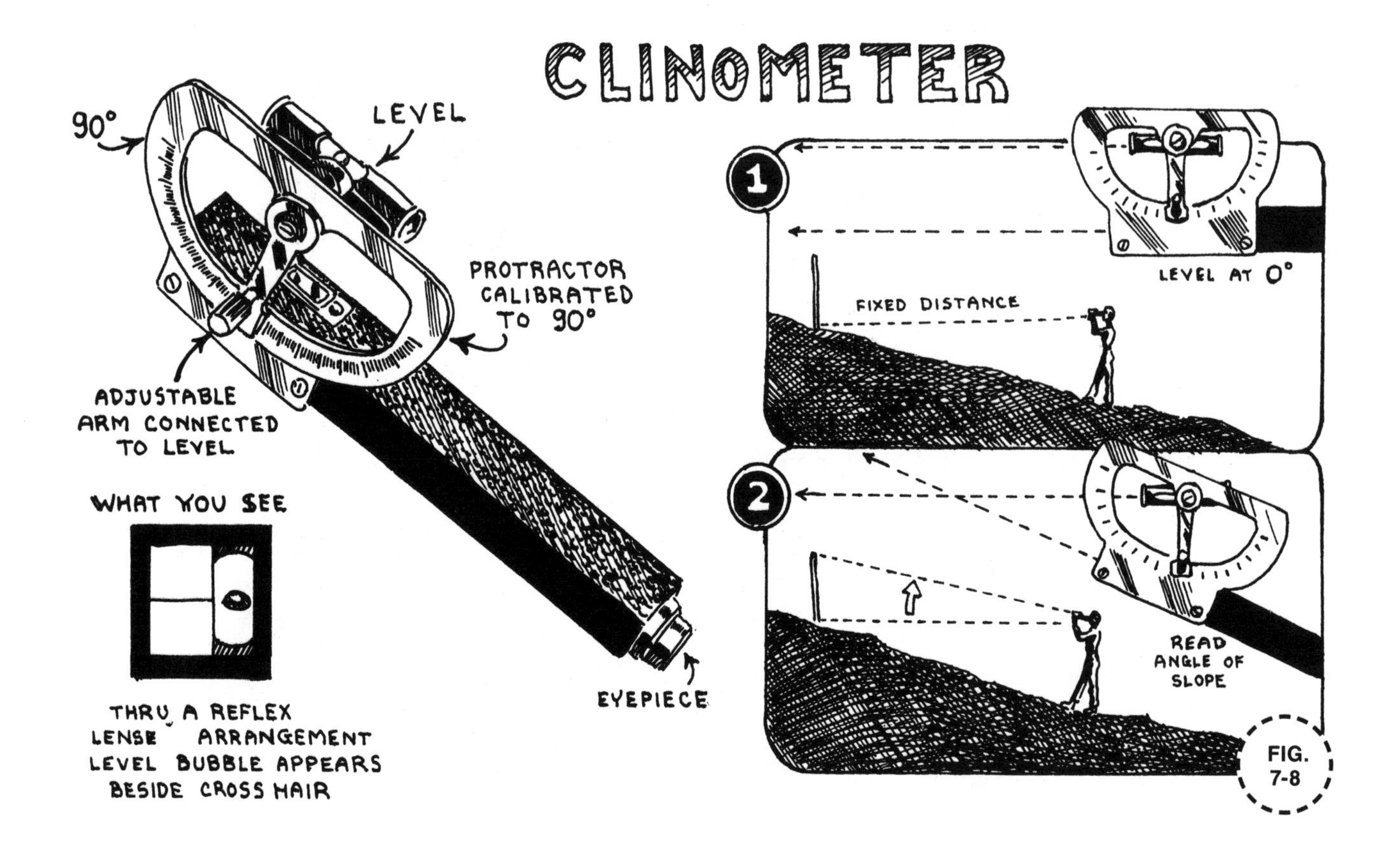

FIG. 7-8

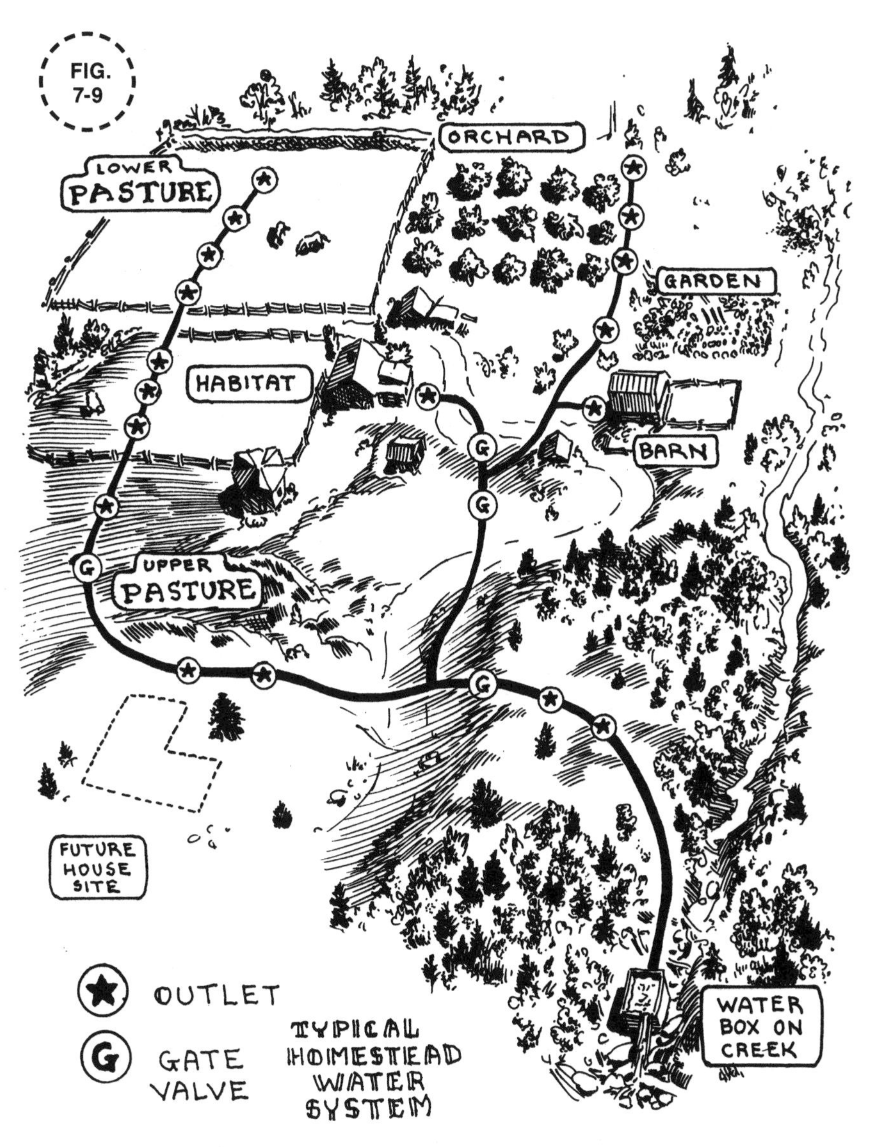

For the pastures we began with a 1½" pipe and gradually reduced it down to 1" at the bottom, thus providing it with half of the entire volume. When that volume is forced into a smaller diameter pipe, you get much more pressure. For the house, barn, garden, and orchard, we didn't need quite as much pressure. So these usages could share the other 1½" line, ¾" pipe going to each (**FIG. 7-9**).

Type 2 plastic pipe, varying between 50 psi for the 2" line and 80 psi for the 1" and ¾" was used. Since the 2" pipe is closest to the source and higher up the slope, it doesn't require the pressure rating required by the smaller diameter pipe. Because of the expense of plastic or any other pipe, you don't want to purchase any thicker gauge than necessary. We chose to use type 2 plastic pipe because it is durable, long lasting, and flexible. It requires no maintenance and is easily taken apart when it has to be. It has plastic fittings with stainless steel clamps to keep them in place. The fittings are very versatile and inexpensive compared to any other type fitting. For instance, you can make a tee from a 2" line to a ¾" line with only one fitting. With galvanized steel you need several fittings to gradually reduce down to the ¾" size. Also, this material rusts in time, making it nearly impossible to separate the lengths, short of using a blowtorch. It is very rigid, requiring fittings at every bend. Special tools are also needed for cutting and threading.

volume, because the volume of one 2" line is equal to the volume of four 1" lines. Figure it out. In other words, we can tee off to two 1½" lines, one going to the pastures and the other going to the house, barn, garden, and orchard, without losing any pressure, because two 1½" lines have the same volume as the 2" line. If we wanted to, we could then reduce the 1½" lines to two 1" lines and still have the same pressure. But we decided to reduce the pipes differently because we wanted the greatest amount of pressure in the pastures.

I guess another alternative could be the wood pipe, but I doubt that you'll be able to find anyone who is making it. If you want it badly enough, maybe you can try making it yourself. It could actually be done with a hand auger set up without the use of a motor-driven machine (**FIG. 7-10 AND 7-11**).

Then again, there are other easy ways to bypass the tremendous expense of a water system such as the one we put in. The pipe and fittings cost only a total of about $80. We simply applied for a retail license and bought in quantity for several families, charging all of them just slightly more than the wholesale cost of the pipe. This saved us all a considerable amount of money.

So, with our water system, the first thing we did was locate the intake and collection box site. Next, we acquired a backhoe operator to dig the trench for the system. (Be sure to take a lot of time to clearly think out the route because this job is expensive and should never have to be redone.) Then we laid our pipe into the ditch, carefully thinking out all the outlets, tees, and gate valves.

Our outlets for the pastures were simple tees above the main line, each ¾". Above that tee, we added a short length of ¾" pipe that brought the outlet above ground level for the hose valve. This hose valve then takes a regular ¾" wide thread, standard female coupling for the irrigation pipe to the sprinklers (**SEE FIG 7-1**).

The barn, garden, and orchard outlets are done in the same manner. The pipe for the house is simply coupled to the male end of a galvanized steel nipple made to receive plastic pipe.

Gate valves should be installed throughout the system wherever there are any major tees. We have one before the first 1½" tee, one after the second 1½" tee, and one after the tee to the house and the tee to the garden and orchard. The first gate valve can shut off the entire system in case there is trouble anywhere along the line or in case the line has to be drained for some reason. The second gate valve shuts off the water to the pasture for the winter. You don't want to chance water in the lines throughout the winter when the system is not going to be in use. Water could go up the short, vertical pipes and freeze, bursting them. This gate valve is located just after the tee because a 1½" gate valve is half the price of a 2" gate valve. The third valve shuts off the water to the house and the fourth shuts off the water to the garden, barn, and orchard for the winter.

Note: Be sure to leave a valve open near the shut-off gate valve to prevent vacuum from collapsing the empty pipe.

These gate valves also adjust the volume of water. In summer, when the system is in full use, all the valves are left open, but in winter only the house system is being used, so the main (first) gate should be shut down to at least half or the pressure might be too great for the house's plumbing (**FIG. 7-12**).

Next, we considered a collection box at the intake. The old collection box was inadequate in size and was poorly constructed, made from ½" plywood, which bulged with the weight of the water it held. It was definitely not worth transporting to the new site. I went to the local mill and purchased some low-grade two-by-six cedar tongue and groove. A lower grade could be used because only short lengths were needed for the box. The box is 6' in length by 42" in width, and 38" in height, including the bottom. The volume of this box provides a constant pressure to force water down into the pipe. It keeps the water flowing steadily without danger of drawing air into the line. If there wasn't such a collection box, there would be the problem of dealing with the varying water level of the creek.

The 6' sides were cut and dadoed exactly 2" in from the ends to accommodate the ends of the sides. These dadoes were carefully cut ½" deep by 1½" wide so that each of the interlocking tiers would stack properly. The dadoing was done with a radial arm saw, using a regular blade. Then each of the boards were marked and were clamped in place along the box. The horizontal hold of the saw carriage was loosened and the blade was brought over the wood a few times, back and forth until the cut was the proper width.

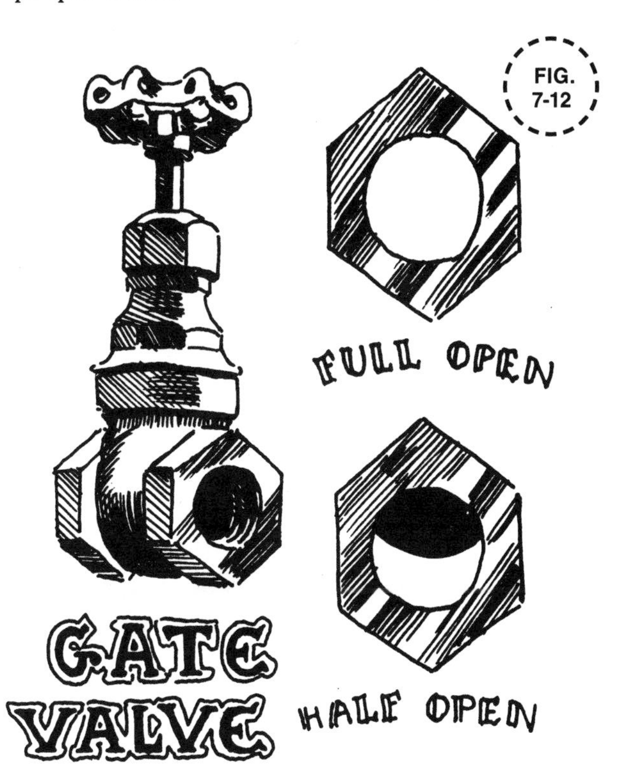

Each of the box tiers were then nailed together through the dadoes and were stacked above each other so that the surrounding grooves fitted snugly over the protruding tongues. No nails were driven in to hold the tiers to each other. A two-by-six tongue-and-groove bottom was then nailed to the bottom tier.

Six 48" long two-by-fours were cut and placed along the length of the box, one at top and bottom at either end and at the center. Holes were drilled into these boards just outside the 42" width of the box to accommodate the 48" long threaded ready rods, which hold the tiers together. The rods were pushed through the holes, then nuts and washers were screwed onto both ends of the vertical rods. These nuts must be tightened very securely to press all the tongues deeply into the grooves. At least 2" of ready rod was left sticking up from the box to accommodate the overlapping 48 × 76 inch plywood top. This top prevents leaves and other organic matter from dropping in, and consequently clogging and contaminating the water box and outlet (**FIG. 7-13 AND 7-14**). It is also a shelf in winter allowing snow to stack on it. Snow is an excellent insulator, usually building up to a substantial height before the subzero cold spells of midwinter really hit.

With this construction, the cedar sides swell enough when the water is put in so that no water leaks out from the box. Some water does leak out from the bottom, however, carrying out accumulating silt that comes in through the intake

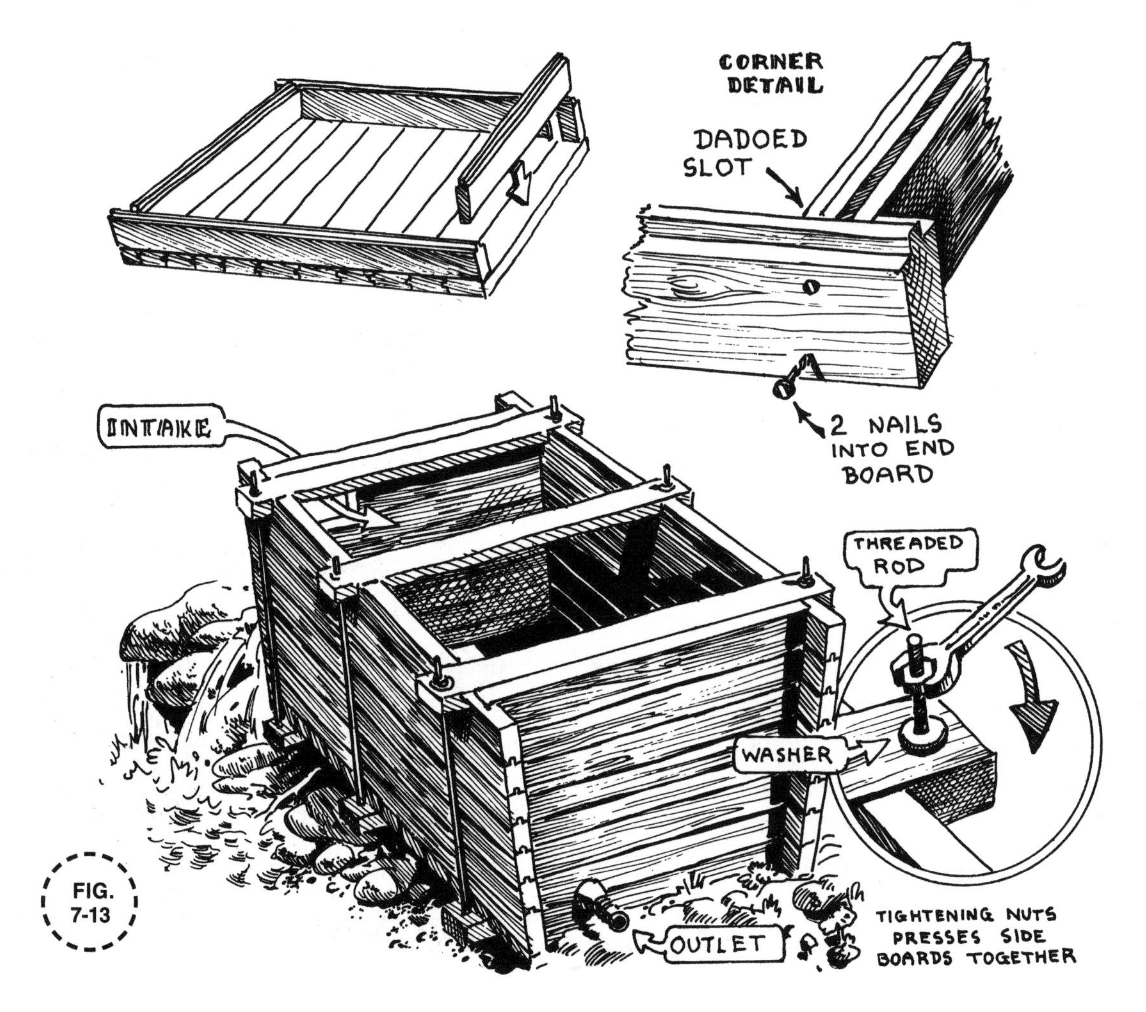

FIG. 7-13

FIG. 7-14

FIG. 7-15

FIG. 7-16

screen. In the last three years, only 6" of silt has accumulated at the bottom and I've had to shovel it out for the first time only this season. The only maintenance problem we do have with this design is cleaning off the leaves and rocks that collect around the intake. I don't know of anyone who has been able to successfully alleviate this problem (**FIG. 7-15 AND 7-16**).

Our intake consists of a double-screened 16 × 16 inch framed opening and a second double-screened opening, 24" long by 4" wide. The 16 × 16 inch opening is the actual intake from which water enters the collection box. It is located at the upper corner, closest to the creek. It is screened on the inside with a heavy-gauge, wide-mesh screen, which protects a lighter gauge, narrow-mesh outer screen from collapsing under the pressure of the water. These screens prevent large matter from eventually clogging the outlet. They are framed with 1½ × 2 inch strips of cedar, as is the other opening (**FIG. 7-17 AND 7-18**).

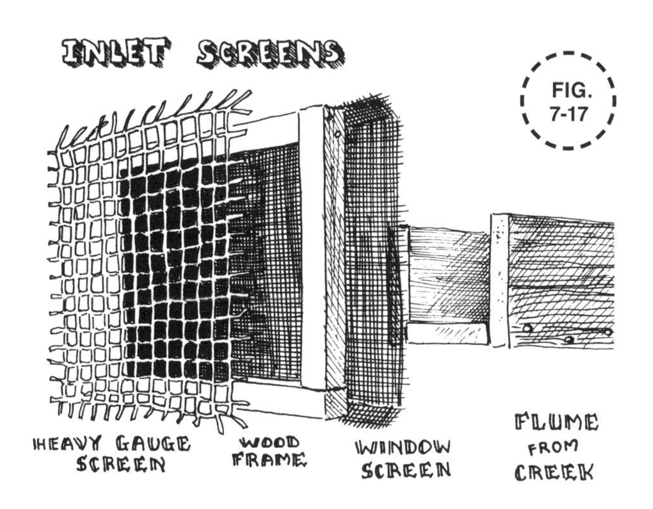

FIG. 7-17

The long, narrow opening is located at the bottom left of the intake, allowing the water table to reach that height and spill out without interfering with the inflowing water. If I were to redesign this intake, which I might do in the near future, I'd rig up a framed slot for a removable intake screen so that it could be easily replaced periodically.

The outlet is located at the lower left front corner, buried below the ground outside to protect it from freezing. It doesn't matter which side it is located on, as long as it is at the bottom and can be buried. It is a good idea to paint, tar, or otherwise protect this outlet area where it comes in contact with the ground, to preserve it from premature rot. The outlet has a 2" circumference to accommodate a galvanized fitting, which screws into it. The threads twist into the soft cedar, making threads in it to seal it in place. A galvanized fitting was used because a plastic fitting would never make a good tight fit. If the area around the fitting began to leak, this would cause a slight loss of pressure to the outlet and would eventually cause it to work loose from the box. A short piece of plastic pipe—just long enough to attach an elbow to—was clamped to it and another 2' length of pipe was attached, bringing the outlet line down so it could be buried sufficiently deep into the ground. Then another elbow was added. While it would have been wiser to have brought the whole front side deeper into the ground, we were confronted with a rock problem that didn't permit it.

The outlet is protected inside the box by a perforated milk can, which is suspended around the opening from a nail above (**FIG. 7-19**). The weight of the milk can holds it in place. Anything lighter, like a screened box, for example, should be weighted down from the inside with a good-size stone or something, or else it might float up, leaving the outlet opening unprotected. It is extremely necessary to protect this opening, otherwise large particles could pass through and collect at the outlet valves, clogging them. I once had a problem of wood particles collecting where pipe couplings reduced the line. I had to search for the connection, digging several holes into the ground until the problem area was located. That type of aggravating chore one could do without.

To control the flow of water to the newly installed collection box, I built a dam, using whatever branches, rocks, and logs I could find in the immediate area. This dam diverts a more than adequate flow into the flume (**FIG. 7-18**). A two-by-eight three-section, flat-bottomed flume comes from the dam at a slight downhill slope to carry the creek water into the intake. A flume such as this one is usually necessary when bringing water from a higher area to a lower intake that is not directly in the creek itself (**FIG. 7-20**). This flume should be kept free from accumulating rock and matter so it can continue to unobstructedly bring in a sufficient volume of water to the box. Try not to allow the water level in the collection box to go down past the outlet screen, otherwise air will get in your lines, sputtering and causing compression. This disturbance shortens the life

FIG. 7-19

FIG. 7-20

of your irrigation system. It is wise to periodically wander up to the collection box and check that leaves and twigs haven't clogged the screen.

Finally, a neighbor came down to help me shade the pipe in our trenches. This pipe should be filled with water prior to shading and backfilling, so as to lessen the chance of it collapsing. We checked all the connections to make certain they were very tight, because once the ditch is backfilled, there is no way, short of digging down 3', to tighten a loose connection. We filled in the trench 6" by hand to protect the pipe from large, sharp rocks that might cut or crush it when the machine backfills. The Cat came and did its job.

The water system has proven to be excellent, supplying us with all the irrigation and domestic water we need for our homestead. Without such an efficient system, many of our projects would not have been possible. The pastures are producing good hay, the garden is flourishing, and the orchard is filled with huge, succulent apples. To a homesteader cutting out his or her place in an undeveloped forest, the water system should be the first consideration. With it you can begin living on your land, developing the necessities, and irrigating the newly planted trees, which will someday provide shade for your future dwelling. Water is life.

CHAPTER 8

Another Gravity-Flow Water System

About 75 vertical feet above the flat where our houses are situated, the water from a tiny brook emerges for a final time on the mountain, then runs down a gully until it eventually sinks back into the ground. Technically, this flow is considered a brook since the water comes to surface at several places on the mountain, but for the purpose of taming the supply, it is treated as a spring and has been called such. And because we were the first people to settle on the surrounding land, the local water rights branch dubbed our source with our name.

At the point where the brook makes its final emergence, we constructed our water system. It consists of a dam to catch the flow and a box to store it for volume, thus allowing it to be pressure-released to the outlet pipe. The box, built first, holds approximately 800 gallons of water and is located at the first level spot below the spring, about 36' below the dam site. This location must have a substantial draining capacity and be large enough in size so that the stability of the box will not be threatened by eventual erosion. The size of the box was determined by our need for storage capacity and pressure since the spring reduces to a trickle in late summer. The dimensions of this box are 8' in length by 42½" in width and 46½" in height. It is constructed out of 3 × 6 inch double tongue-and-groove spruce lumber and sits on four creosoted railroad ties. The tongue-and-groove lumber was utility grade, purchased because we got a good price on it.

In hindsight, we should have been less miserly, because the imperfections in the wood made it difficult to pull the box tightly together. Consequently, leakage occurs. The ties were set on the ground, spaced evenly along the length of the proposed box, and were leveled. At another area, where it was easier to work, the long sides were each assembled and held in place with large bar clamps. Two lines were drawn 2½" apart near either end of the sides, then a channel was cut ¾" deep along those lines and the piece between the cuts was chiseled out to provide a snug notch for the shorter sides to seat in (**FIG. 8-1**).

Once the chiseling of the channels was completed, the walls were taken apart and were lugged up to the site. Nine floor boards were laid across the ties and were pulled tightly together, using the bar clamps underneath, between the ties, so they wouldn't interfere with the work (**FIG. 8-2**). The sides were constructed in courses, two long pieces set in place and the shorter pieces tapped into the channels. The courses were raised in this manner without the use of nails until all the side boards were raised.

Twelve four-by-four supports were cut into two lengths, six pieces being 53½" and six being 56". Holes were drilled as close to the sides as possible into the shorter supports which

FIG. 8-1

horizontally span the sides, top, and bottom, on either end and at the center. These holes accommodate the ½" threaded bar stock rods that hold the opposing supports together. Nuts and washers were fastened finger tight on all the ends of the rods to keep them in place. The longer upright supports were then positioned and dealt with in the same manner and each nut was slowly tightened, evenly and in succession with the others, going around the box as many times as necessary for a

FIG. 8-2

snug fit, stopping just prior to the walls deflecting (**FIG. 8-3 THROUGH 8-6**).

The interior of the box was divided into two chambers by a removable wall of one-by-tens fitted into grooves created by four upright two-by-threes that were nailed into the sides. The wall is neither nailed together nor to the box, for easy removal during cleaning. It could be made from any size lumber or waterproof plywood. When making any measurement for a water box,

remember that wood swells when it is wet. This works to your advantage when you want a tight fit to prevent leakage, but against you when you want to remove any portion after the box has been filled. The smaller chamber is 2' wide, designed as a space for depositing the incoming sediment, which passes through the intake screen instead of going on into the main chamber via the high 2 × 18 inch screened opening located at water level on the divider wall. (See **FIG. 8-8**).

There are three screens for filtration inside this box, all made from heavy-guage stainless steel screening. The primary screen, 24" × 24" is tacked to a frame of two-by-threes and placed at the point of entry into the box from the dam. The secondary screen is in the divider wall and the tertiary screen is shaped into a 10" square box and fitted over the outlet pipe (see **FIG. 8-7**).

So the progression of water, then, is: From the dam it goes through 2" plastic pipe to the intake screen and into the smaller chamber. There it gets filtered (not entirely, unfortunately) and progresses out of the secondary screen into the main chamber. We installed a V-shaped notch into the top board of a long side of the box to allow the overflow to spill out at that point and go down a wood trough into a natural gully instead of spilling out randomly and eroding the surrounding plateau. The water that remained in the box went through the outlet and down the pipe to irrigate the garden, fields, and dwellings. The outlet now consists of a 2" hole in the floor of the box with a galvanized fitting screwed into that hole. This fitting has a male end that accommodates 2" plastic pipe (**FIG. 8-7**).

The first winter glaringly exposed four mistakes in our design. Luckily, they were easily rectified the following spring. First, the water froze at the overflow trough because it was directly exposed to the weather. We substituted a 2"

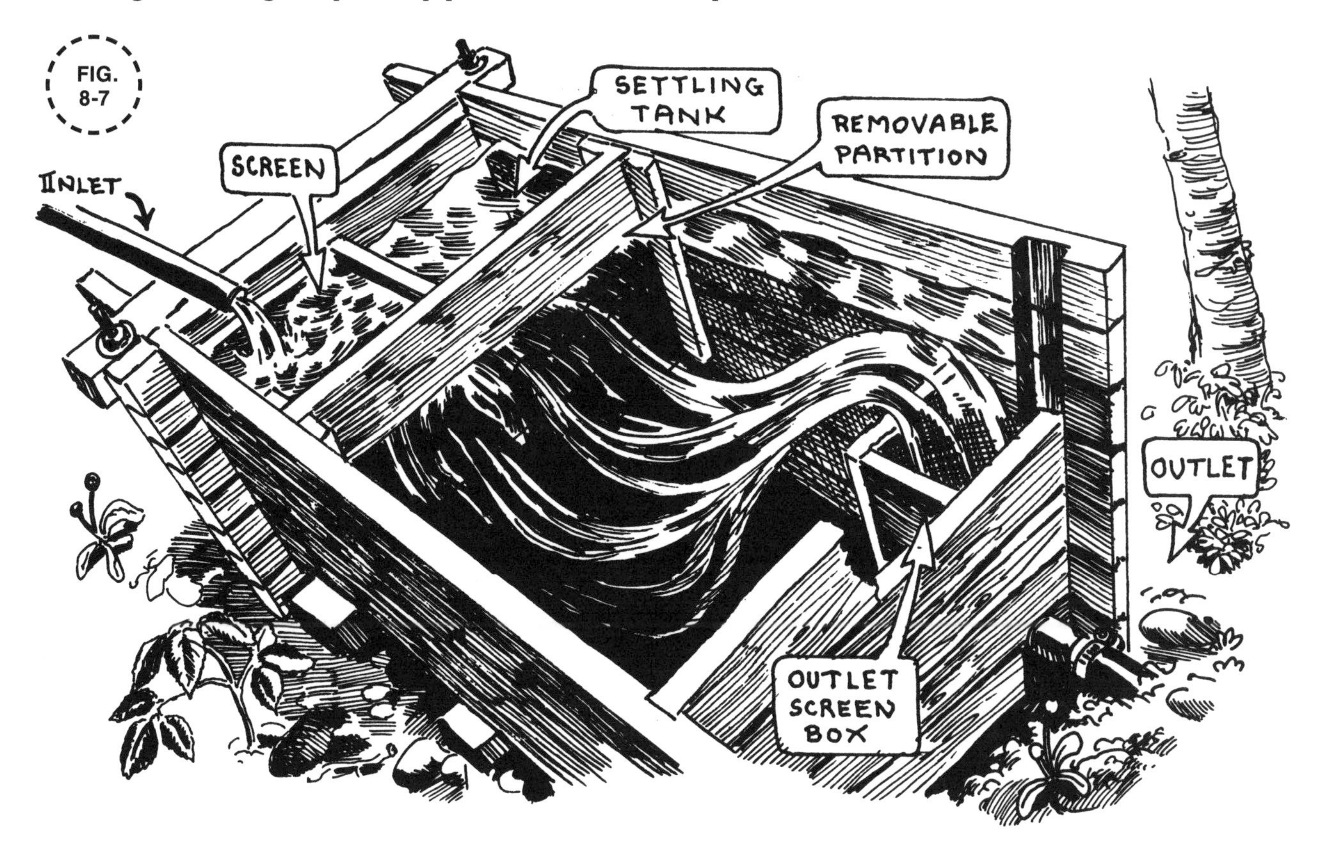

plastic pipe long enough to carry the water to the gully. The friction caused by the steady flow of water in the pipe kept it from freezing (**FIG. 8-8 AND 8-9**).

FIG. 8-8

FIG. 8-9

Second, we laid the plywood cover on the four-by-four supports, which protrudes 3½" above the box. This allowed the wind to whip under the plywood, leaving behind a "skating rink" on the surface of our water supply. Since we nailed two-by-fours around the top perimeter of the box, between the crossing supports, this problem has not recurred (**FIG. 8-10**)

FIG. 8-10

Our third mistake was to place the outlet hole on the side of the box, above ground, forcing the pipe to be exposed until it could be gradually bent down into the 18" deep trench where it is buried the rest of the distance to the houses. Even when we packed it with sawdust, it still froze at this point, almost leaving us without water for the remainder of the long cold winter. Once the snow melted, we emptied the box and cut a hole out of the bottom for the outlet pipe, then put dirt around the small portion of the pipe that goes above ground before entering the box. This alleviated our problem.

Fourth, we used a wooden V-trough of one-by-eights supported on crossing two-by-four braces to carry the water from the dam to the box. The water froze in it as well. We have since substituted a 2" pipe, which was laid in the trough and covered over in winter with one-by-four boards for added protection (**FIG. 8-11 AND 8-12**).

FIG. 8-11

FIG. 8-12

When the box was completed, we built the dam. The natural pool of the spring was excavated a bit to make it deeper and larger. At first we tried using the scrap pieces of tongue-and-groove lumber from the box to make the dam wall, but there was too much leakage under it. Consequently, we decided to reconstruct it using cement when we had more time to deal with it. A wooden wall is adequate if you are in a hurry and have enough of a flow so you don't have to be worried about every little drop when the usual late summer dry spell hits.

To make the cement dam, we dug a trench 18" deeper than the water level, across the ravine where the water flows. Forms were built for the 7" thick wall and a pipe was laid under the forms to drain the accumulating water while the forms were being secured and while the cement was setting up. Since the pipe could not be pulled out when the cement hardened (it became encased) it was threaded so it could be capped. To accommodate the outlet pipe that would take the water to the box, a hole was drilled in each form and the pipe was pushed through the holes prior to the pour. We also placed a small block of wood into the wet cement at the top of the wall to form a channel for the overflow during heavy spring runoff (**FIG. 8-13**). This pool was then entirely sheltered with an A-frame roof

FIG. 8-13

to keep out the debris, and a sheet of plastic was put over the exposed front to prevent freezing (FIG. 8-14). With this protection, there is no need to screen off the intake end of the pipe at the dam.

Now that the corrections have been made, this water system has been working great, supplying our domestic and homestead needs from a tiny spring that would otherwise just seep back into the ground unnoticed.

CHAPTER 9
A Recycled Hot Water System

Even though I choose to live deep within the forest, far from the madness of civilization, I still enjoy some of its conveniences and have given a few of its castaways new purpose in my primitive creations. I've gathered the elements of the hot water system and shower while shopping through society's greatest bargain centers—its trash dumps. I entered these establishments with little preconception of specific merchandise, and remained quite flexible in my design. I picked up an old radiator here, a discarded water heater there, and fittings and faucets and lengths of pipe wherever I could find them, then trucked the whole mess home to figure out what would come of it all.

With a few basic principles of hydraulic engineering surfacing in my consciousness as I pondered my latest acquisitions of useful junk, a design would begin to develop. Then would come the hard part—making all the pieces fit together to become what I intended. I'm sure most of the pieces would have rather remained where I found them, continuing their return to the earth. Most every fitting gave an unwilling struggle, voicing its disgruntled objection with each twist from my wrench. Finally, after what would have netted a paid plumber a goodly wage, the system was completed. The time spent on it was well enjoyed, unhurried, and thoroughly involving. For I had no other commitments—no bills to pay and no places to be except right here amongst my peaceful surroundings doing what I wanted to be doing.

The hot water system consists of a wall-type radiator, a 30-gallon hot water tank, a lot of piping, fittings, valves, and a showerhead. The radiator is located well outside the structure so that a fire could be built on it. It was set on its side and tilted slightly upward so the front outlet would be raised a bit higher than the rear one (**FIG. 9-1 AND 9-2**). This facilitates the convection

FIG. 9-1

FIG. 9-2

process. The radiator then acts as the convection generator. When it is heated up by the fire above, it initiates the circulating movement of the water by warming up the cold water that enters through the low inlet. The water fills the radiator, gets hot, and flows upward through the outlet pipe, returning to the 30-gallon collection tank as hot water. This cycle keeps repeating until all the water in the tank and pipes is as hot as the fire will allow.

FIG. 9-3

The collection tank is an old electric water heater tank stripped of its fancy enameled jacket. That jacket was replaced by protective layers of fiberglass, gunny sacks, and galvanized roofing, which act as insulation to keep the heat in (FIG. 9-3). It came complete with fittings in the proper places to facilitate the convection process. The water comes down the mountain from the source via plastic pipe to the galvanized tee

FIG. 9-4

at the bottom of the tank where it goes to the radiator and fills up the tank. Just prior to this union is another tee that goes off to the vertical cold water inlet at the left of the tank. At the top of that cold water inlet is a shut-off valve that controls the flow of cold water to the sink. This valve makes it possible to mix hot and cold water in the lines to get the desired water temperature for sink use (**SEE FIG. 9-3 AND 9-5**).

Six inches below the top of the hot water tank is the inlet from the radiator. At least 6" of space is required at the top of the tank to allow the hot water a chance to rise and create a convection area with that enclosure. The outlet comes out the top of the tank and elbows to the left to the shower and sink. Before it reaches the vertical cold water inlet, it is interrupted by a control valve that regulates the flow from it. From that valve, it either goes to the showerhead or the sink faucet, whichever is in use (**FIG. 9-4 AND 9-5**).

This hot water system heats up the water very nicely—too nicely. With the normal morning fire, it could heat up enough water to fill two 45-gallon tanks. Unfortunately, this causes the hot water to back up into the intake pipe coming from the source. In the present system, such a backing up causes the incoming cold water to heat up before it gets to its inlet pipe. Consequently, it temporarily converts the entire system into a hot water system, until enough cold water can run through the line to cool it off. The most logical solution to this problem would be to increase the size of the 30-gallon holding tank. This would provide enough hot water for a leisurely morning shower, the weekly wash, and all the stacked-up dishes in the sink.

FIG. 9-5

When I was a young boy, my father bought me a Lincoln Logs kit. That set of miniature logs captured my interest for hours at a time. Between chores, I would sit in the front yard and explore the many building possibilities the kit had to offer. One day when I was pulling firewood out of the bush with our old horse, I got an idea to use real logs, modeled from the Lincoln Logs, to make a larger fort, a fort which I could use as a hideout when I wanted to go off alone into the woods.

That fort had everything the smaller set had—only bigger. Even the window and door frames interlocked. But the only problem with my bigger fort was that it was too hard to disassemble and rebuild elsewhere when my hideout was discovered, so I had to construct another one every time this happened.

Designing log forts got to be my primary interest. Whenever I wasn't doing schoolwork or chores, I was designing and constructing new forts and figuring out new methods of skidding logs to my sites. I even figured out new ways of bringing in the firewood because I was tired of harnessing the old horse and having to hook him up to logs so he could pull them to our woodshed. It wasn't his fault, though—he was a nice old horse. I was just bored with doing that same chore every day. So I devised a pulley system, using an overshot waterwheel to skid the logs to the desired site.

I really consider myself fortunate to have grown up in a natural environment where I could learn how to take care of my own needs by utilizing readily available materials. I quickly learned to appreciate the woods. I still use whole timbers in my construction, to make use of most of the tree and to avoid having to buy expensive lumber and insulation materials. I admit building with timbers is a painstaking and sometimes tedious process, but it beats working long hours at a job to save enough money to purchase prefabricated materials. Also, I get a certain fulfillment from shaping each timber and making it fit snugly into place over the others. Each tree that was used was selectively thinned out from my own land and very little was wasted.

My wife and I live on an old homestead which had some poorly maintained buildings on it. They were usable until we could replace them with structures more suitable to our needs. Our homestead has an abundance of water, which rushes down from a large creek, irrigating our fields and providing us with enough electricity to power our lights and a few smaller appliances. It has plenty of timber for construction and heating purposes, and has excellent soil for our garden and hay crops. It is also isolated from neighbors and far enough away from town to be free of noise and air pollution. Since it is just what we'd been looking for, we plan to stay on it for a long time, and we are thoughtfully developing it into exactly what we want.

Our first building project was to replace an old root cellar which had rotted with age. We find it very important to have adequate food storage separate from the house, because of the possibility of fire. A root cellar also provides a constant cool temperature which keeps fresh fruits and vegetables very well.

The cellar was replaced with tamarack logs (western larch). The inner and outer walls were 4' apart and filled with planer shavings for added insulation. Its roof was constructed gambrel-style to allow for storage of empty jars, boxes, and other odds and ends. This smaller structure got us back in the practice of building and readied us for the next project, the house.

We chose for the site of the house a slope which overlooks many mountains in the area. This slope is part of a plateau in the center of our land. On one side of it were the hay fields, to the rear was the new barn site, the old barn, and the garden area. In front of the site was a terraced

Rebuilt root cellar

area which we later developed into a tree reserve. Here we've been pruning evergreens so they will grow as full as possible.

We began cutting the tamarack logs for the house in late winter. Most of them came from the mountainside across the field. After they were cut, we left them in place and peeled them with a square-nose shovel before all the sap was up. Peeling them before much of the spring sap flows helps keep them from mildewing and discoloring so readily. The logs remained on the hillside until the snow thawed enough for the horses to skid them to the building site. The logs were placed there on pole skids and left to dry for a couple more months.

We used tamarack logs mainly because they have little taper and are usually very straight. These timbers have few limb knots and are quite durable.

While the logs dried, we measured off a 28 × 30 foot area and outlined our foundation. The logs were 32 × 34 feet in length to allow for at least a 1' overhang on each side. We did not make any attempt to level the slope. Instead, we built our foundation with steps gradually going down the slope. The steps of this stabilizing wall went down at almost two-log intervals, allowing the adjoining logs to interlock with the flat rock and cement foundation for added structural support. Each of these adjoining logs was grooved

at the bottom so it fit snugly to the log beneath. They were trimmed on the underside instead of the overside to prevent moisture from collecting in the grooves.

One of the reasons for the combination of rock mortar and log in the foundation was that rock mortar had to be used where there was contact to ground. The logs would rot quickly if they were touching dampness. Logs—which we didn't have to buy and which are more pleasing to look at—were used where they were not on or below ground level.

The slant of the slope provides the house with a 14 x 30 foot basement with full headroom. Here, firewood can be stored and a Pelton wheel can be kept and protected from freezing. This area also houses an old wood furnace, which heats the upper sections as well as the basement. A cement floor was poured for this area and was slanted slightly downhill for drainage. The surrounding logs are interlocked with round notches as are the wall logs above. A long car axle was driven into these logs at each side to reinforce them where they join the rock and mortar.

The stabilizing walls on the 28' sides were 5" higher than those on the 30' sides to compensate for the stagger of the wall logs. They were constructed 5" higher because 5" is half the diameter of the 10" logs, which were used as the end logs.

The longer sill logs rested on the lower wall sections. The end logs, which would be notched into these sills, were raised by the higher wall so they could properly interlock with the sills and begin the necessary stagger. Anchor bolts, 12' apart, were sunk into the concrete for added support of the bottom logs.

We spaced the joists 2' on center and notched them into the sills with a lap joint. The joists were left round on the sides and bottom. Their tops were sized and hewed with a foot adze and a broadaxe to provide a level surface for the rough-cut 1" subflooring. Before I adzed the joists, I made vertical cuts into them every few inches with a very sharp axe. (It is important to keep your axe very sharp at all times.) Then I chopped out the pieces with a 9-pound-head broadaxe and went over the cuts with a foot adze to finish them off smoothly. If used correctly, a sharp foot adze will make a hewn surface look as if it had been planed at a mill.

We set up stringers or uprights every 6' as supports under the joists, where needed. Their lengths depended on the sloping of the hill. Their tops were notched and spiked into the spanning joists and their bottoms rested on large rock pillars.

We chose to notch all our corners with round notches because we've found from past experience that this type of notch looks best and lasts the longest. It allows the wall log to extend beyond the corners, making the notches both beautiful and strong. Our log ends are staggered, one long, one short, so the long one completely covers the shorter one, preventing moisture from collecting on it.

I've torn many an old log building apart, and, in most cases where dovetail notching was used, the corners were the first place to rot because they were not protected by any overhang.

In my opinion, the round notch is also the easiest of the notches to make. I started my walls by placing the first end log across the two sills, making sure there was an equal distance of overhang on either side. It was secured in place with log dogs. I then took a plumb bob and centered it

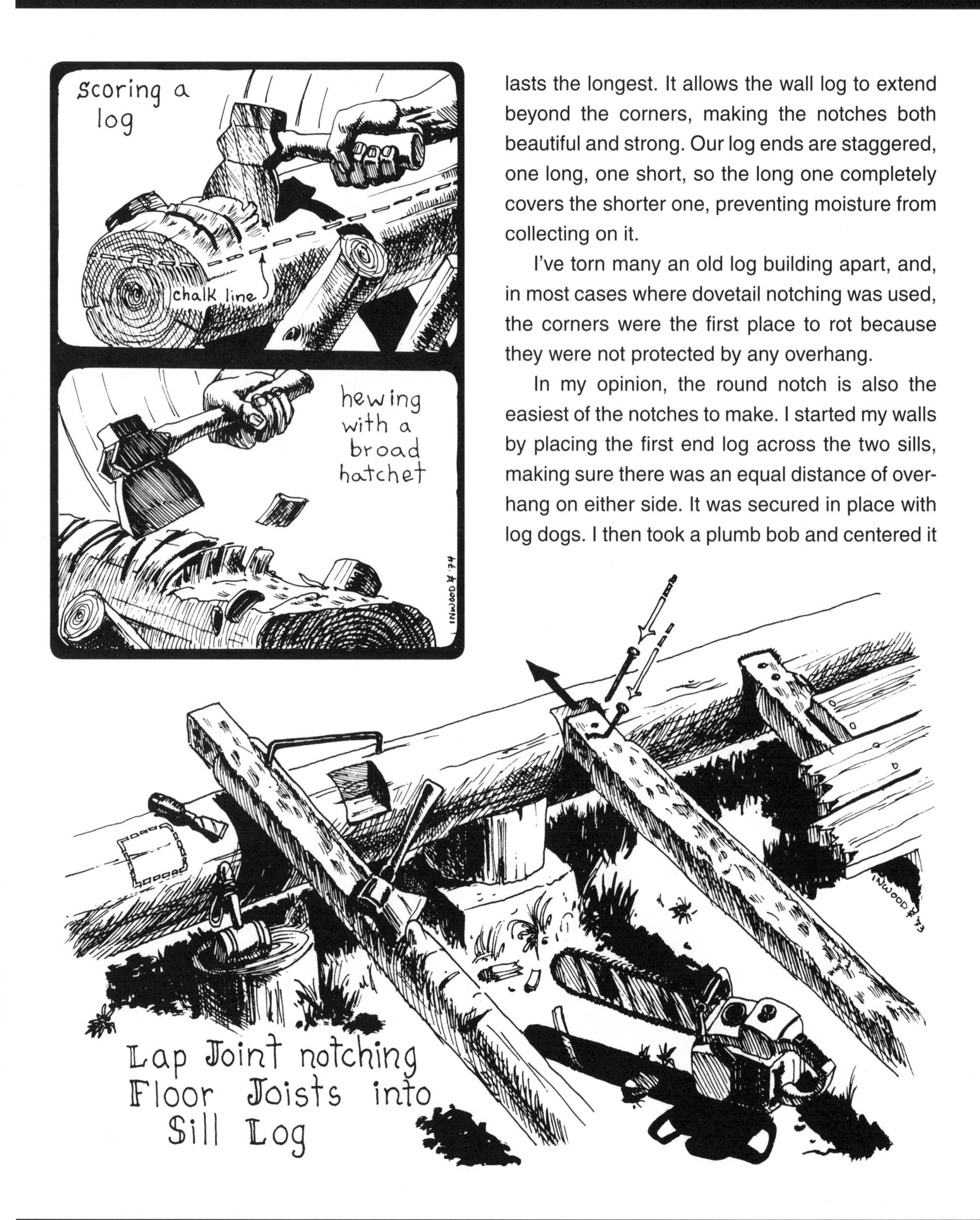

Round-notched log ends

on the highest point of the log end to line up the new log over the ones below. If you do not have a plumb, any pointed weight—a fishing sinker, for instance—will do just as well. I used this method to keep my walls straight.

To keep the wall logs uniform, I placed any log that was oval or unround so that its widest part was oriented up and down. Then I hewed it, to prevent obvious gain. First I used a pair of dividers to measure the gap between the stabilizing wall log below and the log which was to be notched. This gap would be equal to the depth of the notch prior to hewing. The dividers were opened to this size and were held vertical at all times as they were run along the contour of the crossing sill log's end. The upper point duplicated the sill log's contour onto the log above, a gap's width higher than the lower span. The outline of the notch was then marked on both sides of the upper log with a thick lumber crayon.

The log was turned over with a Pee-Vee and again dogged into place. Then, with a very sharp double-bitted axe, we scooped the notches out

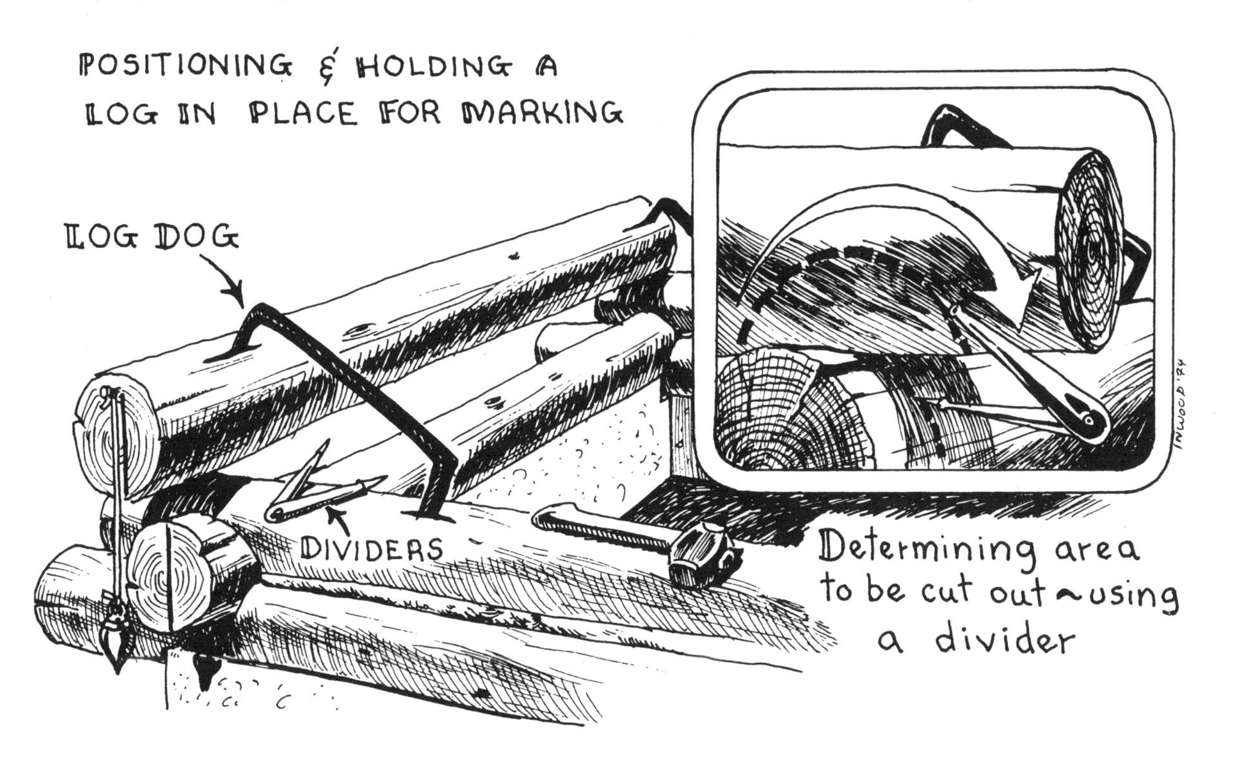

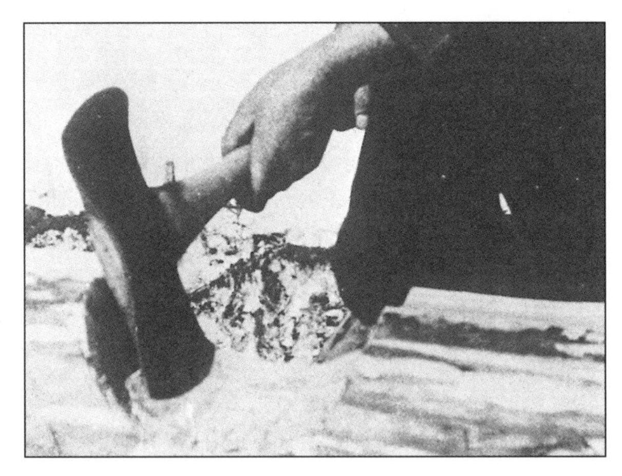

Axeing out a round notch

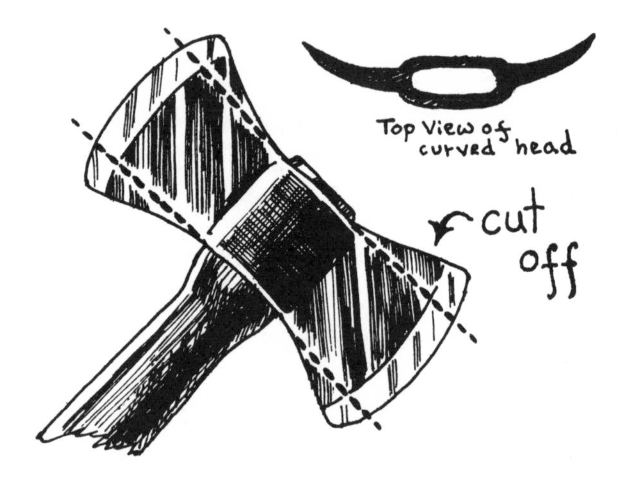

to just beyond the crayon's outline, to allow for hewing. The double-bitted axe I had been using was modified for this process. I heated the blades in a vise and slowly bent them into a curve. I then ground off the ends of the blades to round them, so they wouldn't gouge into the sides of the notches while cutting them out.

I learned many times over that it never pays to lift heavy logs into place without the aid of some animal or some simple machine. Too many people cause their bodies needless strain by trying to lift objects that are too heavy for them. Why chance wrecking your back when, with just a little thought and research, you can invent a tool to do your lifting? To lift heavy logs, I devised a swinging boom setup that is based on the principle of the log derrick, which at one time was used by professional loggers.

To make this swinging boom, I first set up a long vertical pole. This is called a stiff leg because it remains stationary. It can be supported by either a tripod brace or a stiff guy fastened to either a floor joist or a nearby solid structure. At the top of this pole, a block and tackle is connected to control a suspended, double-poled swinging

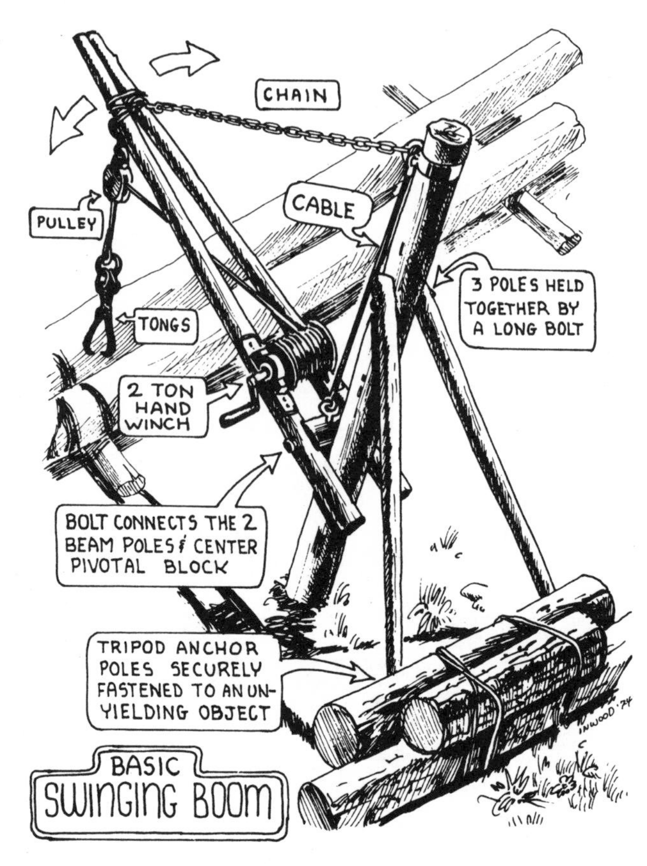

boom. At the bottom end, these boom poles straddle the stiff leg, and are held in place by a block of wood and a cable. This boom is free to swing around from the bottom and can lower logs onto any of the walls. A hand winch is bolted to the bottom of this boom and a pulley is attached to the top of it. The winch's cable goes over the pulley and is attached to a pair of log tongs, which grasp the center of the log, raising it and lowering it to the desired position. I found this device invaluable in my work.

The walls of the house went up relatively quickly, mainly because many of the logs I used were straight and had little taper. Many of them needed little more hewing than the 3" flat surface on top and bottom. These logs could be hewn easily enough by marking them with a scribe and scoring them with an axe or a chain saw, then chopping out the scored section with a foot adze. I made the scribe out of old scrap metal which I found in the dump, and I ground the marking tips out of old cobalt steel files. The tips were then welded onto the metal. One set of these tips is held shut with a piece of wire and the other can be adjusted to various gaps by a block of wood.

Some of the logs were very difficult to fit in places and had to be turned several times before they were straight enough to lay flat over the logs beneath them. These logs either had large tapers or serious bends in them. You'd be surprised how much a log can be corrected by hewing.

There are many purposes for hewing the tops and bottoms of the wall logs. One is to provide a shelf that insulation could be compressed into so it won't be exposed to the weather. I used a

Hewing with a foot adze

Homemade scribe

mixture of cotton and jute between the wall logs of the house. These materials were the stuffings of a couple of old mattresses we had lying around and a davenport I found at the dump. (When I ripped open the old davenport with my knife, people at the dump gave me strange looks. I wonder what they thought I was looking for!)

Since building the house, I've been using fiberglass insulation. I take a few rolls of the fiberglass and cut through each of them with a saw to get 3" strips. It is much easier to cut these strips while the wrapper is still on the roll, rather than stretching it out and cutting from long lengths of it. To use it, I just lay out a 3" strip and staple it to the top of a log that has been placed and hewn. It compacts very nicely between the logs and insulates extremely well.

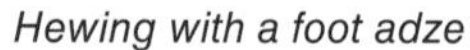

Another advantage of hewing the tops and bottoms of the wall logs—or any log for that matter—is to guard against random checking or cracking. Logs have a tendency to split where the sapwood dries out the quickest or where the sapwood is worked off. Some species, such as young 12"-wide cedar, have as much as 2" of sapwood all around them. If the log is left round or is hewn evenly on all sides, it will check the heaviest on the side exposed to either heat from indoor sources or heat from the sun, whichever is greater. If the sapwood is cut away from one or two sides, the larger checks will develop in those areas. If the logs were hewn on the tops and bottoms, the checking would be hidden between the logs.

Hewing the tops and bottoms of the logs is of primary interest to anyone who considers their log building a work of art and who wants to appreciate it for many years to come. Every log, no matter how uniform it may appear, has some taper and some bend to it. Unfortunately, most logs have a noticeable taper and some high spots which prevent them from seating snugly on the log underneath. To compensate for this taper, you must hew the gain accordingly. If a wall log has a 9" tip and an 11" butt, you should hew both top and bottom in such a manner that you gradually cut off an inch more on each side at the butt than you do at the tip. When dealing with a crooked or bowed log, you put the crook or bow up and down and hew the high points until the log is snugly seated. This sometimes takes many turns and much handling before the log cooperates.

It is also important to stagger your logs butt, tip, butt, tip, to prevent having to hew off larger portions to compensate for gain. This method makes the most use of each log.

A common mistake that is made when building log walls, and which can also be corrected by hewing, is the forming of mouse holes. These are spaces caused by notching too deeply. Mouse holes should be corrected

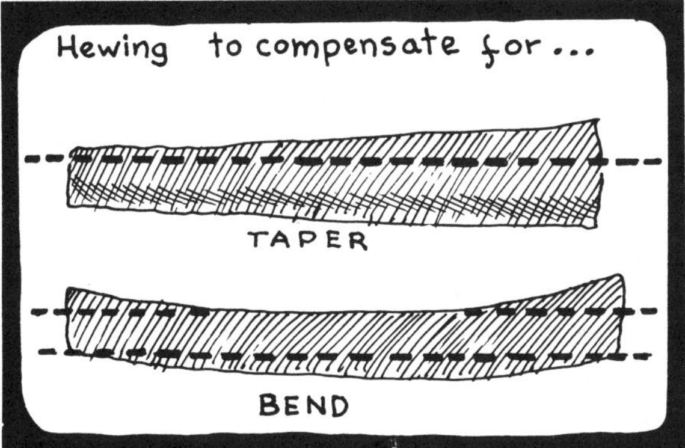

because they tend to collect moisture, which promotes rot.

For my house, each wall was made up of full logs, 32' or 34' in length; I used no pieces. After each log was notched and fitted in place, a 1" auger hole was drilled halfway through it on either side of the proposed door and window openings. A smaller-diameter hole was then drilled below the 2" auger hole to start the 12" spiral spikes.

These spikes were countersunk with a drift punch and mallet to secure the logs together. It is important to spike any future opening from log to log, to prevent the loose wall ends from collapsing when you chainsaw the doorways and windows.

The only disadvantage to using full logs seemed to be my wife's objection to working so long without knowing what the inside of our house would look like. Upon completion of the walls, I immediately chainsawed a doorway so she could inspect the interior. She was very pleased with the effect of being surrounded by golden tamarack logs, and insisted that we find some way to protect them from rot and prevent them from losing their color. We did, in fact, coat them with a plastic preservative, a brand of urethane. The house has been standing for more than eight years, and the interior looks just as new as it did the day we finished it.

After cutting out the first doorway, we tenoned the newly exposed wall ends with an axe and a homemade chisel. The end of the chisel was made from an old car-bumper support which I ground sharp on my grinder. The blade part fits into a cut piece of 1" galvanized pipe. I carved a piece of hardwood and wedged it into the pipe. This is the part hit with a mallet. I've used this tool for months at a time without having to sharpen it; the blade is so hard.

The ends of the logs at the door and window openings were tenoned so they could be sandwiched by the double four-by-four door frames. Two 2" holes were then augered into both tenon ends so the log walls could settle independent of the door frame. The double four-by-fours, which I hewed out of logs, were placed on either side of the tenoned ends, and two smaller holes, large enough for ⅜" bolts, were drilled into each of them at the bottoms of the 2" auger holes. The bolts fit through the entire frame, and the larger holes allowed the walls to settle without damaging the frames. Pieces of doweling were fitted over the bolts to conceal them.

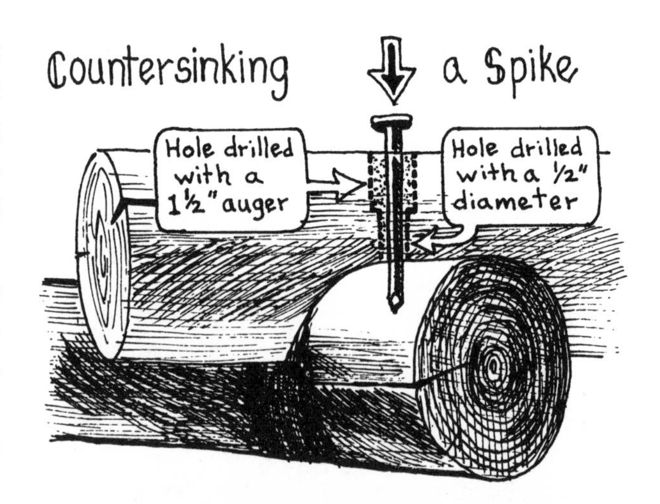

Living room corner

Window spaces were then cut and framed. Hand-hewn four-by-twelves were placed vertically against the wall ends and 3 slots, each 2"

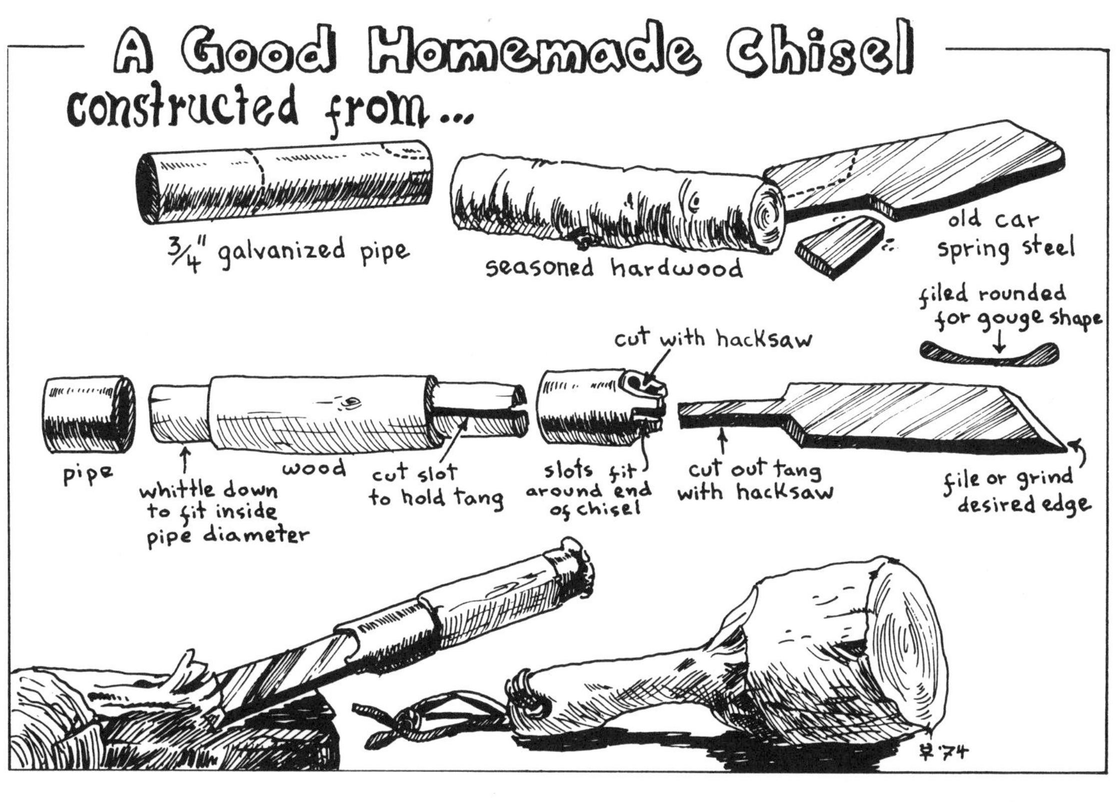

Concealing bolts

long, were drilled into each of them. Spikes were then hammered in at the bottoms of these narrow slots. A 2" space above the frame was left open, to allow for settling. This space was filled with cotton and jute (mattress and davenport stuffing). When the walls settled, they compressed the insulation, preventing damage to the windows and window frames. The windows were put in loosely and were held in place with ¾" lumber strips.

Allowing for setting

Before the top plate logs were placed above the 30' walls, several 4"-square laps were notched every 2' on center into the highest logs of these walls. Ceiling joists, which I hewed to a flat 4" top and bottom, were then spanned across from wall to wall, and their ends were fitted into the 4"-square laps.

The ceiling joists were supported in the center by a similarly hand-hewn crossbeam which spanned across the building to support these beams. The top plate logs were then notched into place.

Laying second floor joists

Except where the dormer was to be fitted in, 7" cedar rafters were round-notched over the top plate logs every 2' on center above the ceiling joists.

Then, the shorter rafter pieces were toenailed to a header above the dormer. To fit the rafters accurately, I used dividers to measure the width of the top plates, then cut my notches accordingly. The tops of the rafters were joined together at a final ridgepole, which spanned the building at the roof's pitch.

We had decided that we wanted extra living space in the roof area, so I designed an 8 × 10 foot log dormer which has 7' of headroom in the center. This area was supposed to be a sewing room for my wife, but it turned out to be used as a guest room for occasional visitors. To frame the log dormer, two-by-eight boards were first placed on the already hewn dormer-support rafters and were nailed into them. These boards extended from the roof's eaves to where the dormer-roof supports would intersect these support rafters. They provided a flat surface for the beveled ends of the cedar side logs.

The cedar side logs were round notched to the front dormer logs in the same manner as the wall logs, and their beveled ends were spiked into the two-by-eights. Where the top side logs intersected the support rafters, the bottoms of the valley rafters were nailed in. Their tops met at the ridgepole peak and marked the top center point of the dormer. At this point, the dormer's ridge log was connected to the ridgepole and the valley rafters. It extended horizontally out to the front

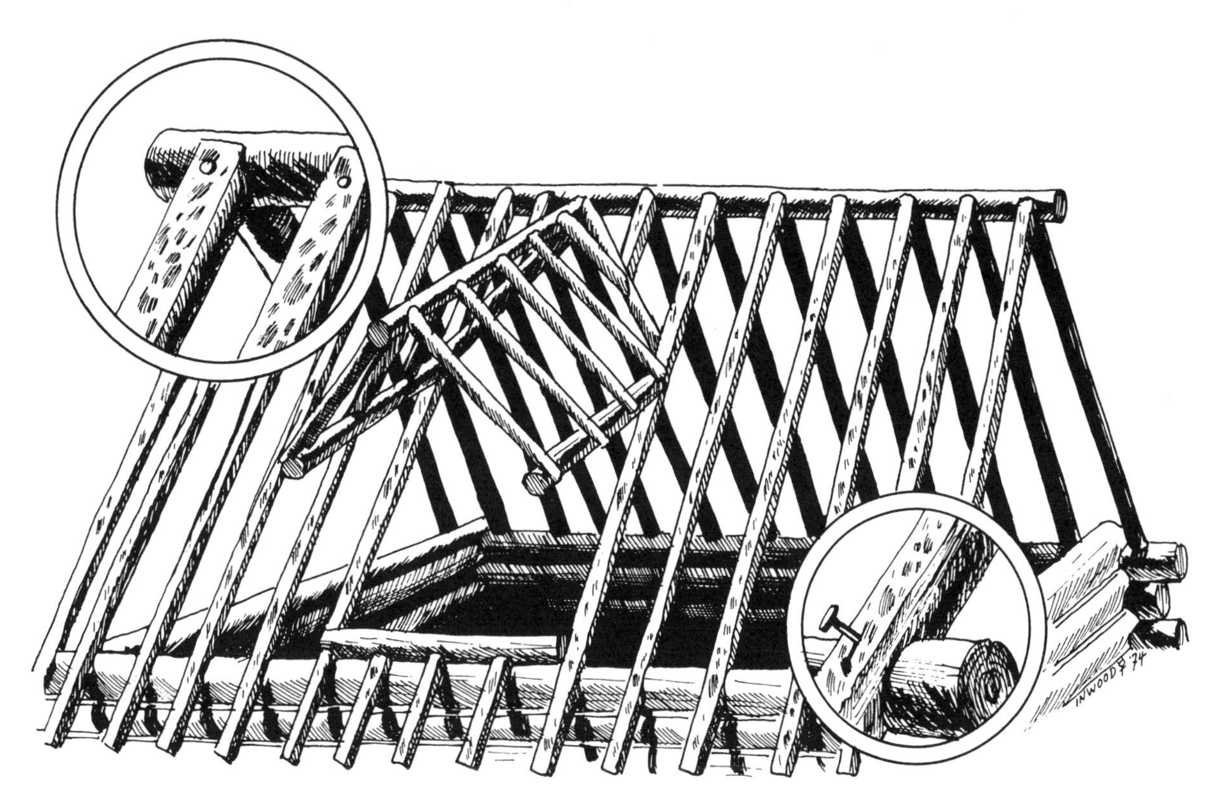

Pole framing of pitched roof and dormer

Cutaway view of log dormer construction

of the dormer. Dormer rafters were then round-notched to the top cedar side logs and their tops were connected at the dormer ridge. Jack rafters were nailed between the valley rafters and the roof ridgebeam for additional support. The front window space was later cut out and framed in the same manner as the wall windows.

After the dormer was finished, the gable ends were closed off with 1" lumber and later paneled over with cedar shakes. Each gable end has a 16"-square ventilator near the top for air circulation. I made nailers for the cedar roof shakes out of smaller poles which I hand-hewed, with a sharp axe, on two sides to a thickness of 2". Very few of these nailers were straight poles; in fact, a lot of them were quite snakey. But even with their bends, hewing them flat on two sides made them ideal nailers for shake roofing.

I split all the roofing shakes from cedar bolts with a shake froe. The job did not take long and

Log dormer

was very satisfying. After I split these shakes, I took a drawknife and went over them to smooth their surfaces. This, I later found, was a mistake. Shakes should have a rough surface for proper drainage. The rainwater goes into each of the rough crevices and rolls down off the roof. When the shake is smooth, the water tends to penetrate through instead of rolling off, causing leaks in the roof. So much for perfectionism.

The roof eaves are protected from ice buildup by an 18"-wide sheet of aluminum, which I nailed to them. The top of the aluminum is covered by the bottom shakes to prevent leaks.

With the main structural chores out of the way, it was time to add the finishing touches. We had been collecting rocks for quite a while. In fact, whenever we took a vacation, we brought back several beauties. Our collection consisted of various different types of rocks from Idaho, Washington, Montana, and all parts of western Canada. In my spare time, I used to practice splitting the rocks with a rock-splitting hammer. I tried to get them as flat as possible for a rock chimney. This chimney is now located at the northern wall, outside the living room. A 2'-square hole was cut out of that wall to allow for the chimney pipe of the antique Franklin stove with which we heat our living room.

Constructing the chimney was very interesting. I first gathered several possible rocks for it and hit them each with a blow of the hammer to check if they cracked easily. The ones that did were discarded because they were too brittle to work with and too dangerous to use. Such rocks have a tendency to explode when heated. We bought a commercial flue, which has an 8 × 16 inch draft opening, and began fitting rocks around it. Because it is very difficult to locate the

End wall with stone chimney

grain in most rocks, I found I had to use the trial and error method of facing them. I had good luck with many of them and was able to split them in half and get two flat faces from one rock. After facing them, I trimmed them so that when in position, they would be more inclined to tip backward, toward the flue. This way they stayed in place while the mortar was drying. I made other cuts on the end stones to square them off.

The mortar I used was a mixture of half dolomite grit sifted through a screen that had two hundred sections per inch (screen-fine, clean sand works just as well) and half Portland masonry cement. I mixed small amounts, only as much as could be used in a half hour. Before I mortared the rocks in place, I checked them once more to make sure they would tip inward instead of outward. When I was satisfied with the fit, I added the cement. Where there were small openings around the flue after the facing was in place, unshaped rocks were added to fill the gaps. An opening for cleaning out the creosote

Cleanout hole at base of chimney

TYPE OF HAMMER USED FOR BREAKING AND FACING ROCK
WIRE TWISTED AROUND FENCE STAPLES DRIVEN INTO LOGS BOND THE MORTAR TO WOOD SURFACE
FLUE
MORTAR
PLACE ROCKS WITH AN INWARD LEAN
TEMPORARY FORM MADE FROM SALVAGE 1"x MATERIAL
2"x4" BOARD NAILED TO LOGS
CLEANING HOLE
INWOOD '74
CONSTRUCTING THE CHIMNEY

was allowed at the bottom of the chimney. For safety, the chimney extends 5' above the roof.

I built forms around the chimney as I mortared the rocks, to help it maintain its tapered shape. Each form was a small section, no more than 24" high, to cover the portion that was being worked on. Each was propped up and held in place with poles that spanned from the top of the frame to the ground. The ends of the forms were also nailed to the walls, for added support. Later, when the mortar began setting, but before it dried, I removed the forms and swept the excess mortar from the rocks and joints.

After completing the chimney, I worked on the interior again. I framed in the room dividers with lumber, built a simple stairway to the basement, and constructed a spiral-type staircase to the upstairs bedrooms. We needed a spiral staircase

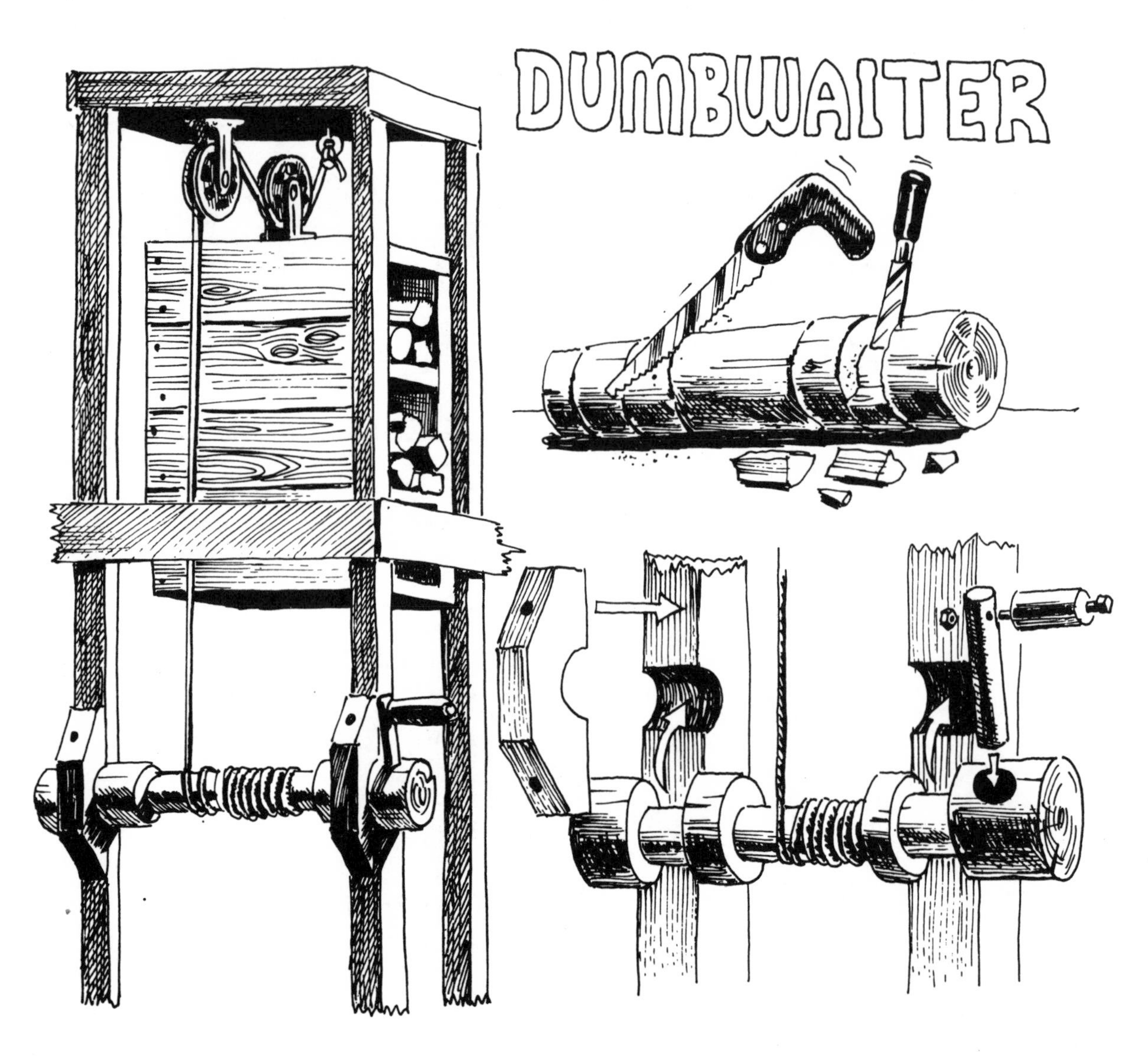

because of our lack of space. My wife and I agreed that we wanted a large living room, dining room, and kitchen, which would not be separated by closing doors. In the space left over, we wanted a good-size bedroom, an ample bathroom, and a staircase. With our design, the staircase had to suffer. Its width could only be about 36", which is convenient for everyday travel but impossible for transporting furniture up and down. All the larger bedroom supplies had to be brought in through the dormer window.

For finishing touches in the house, I designed a dumbwaiter-type wood lift between the basement and the kitchen to send up wood for the cookstove without having to haul it upstairs. Our dumbwaiter is a large rectangular box with an open face, which has two shelves in it—one for dry kindling to start the fire and one for larger material to keep it going. A small windlass with a hand crank is attached to the framework downstairs to control it. The rope which winds around its drum goes through two pulleys above the shaft and connects to the top of the dumbwaiter. A lever on the windlass's geared handle keeps the dumbwaiter in place until it is to be lowered and restocked.

I designed a kindling cutter to go in the kitchen, next to the dumbwaiter shaft. It has several notches for different lengths of kindling. The body of it is mounted on the wall, and the cutter,

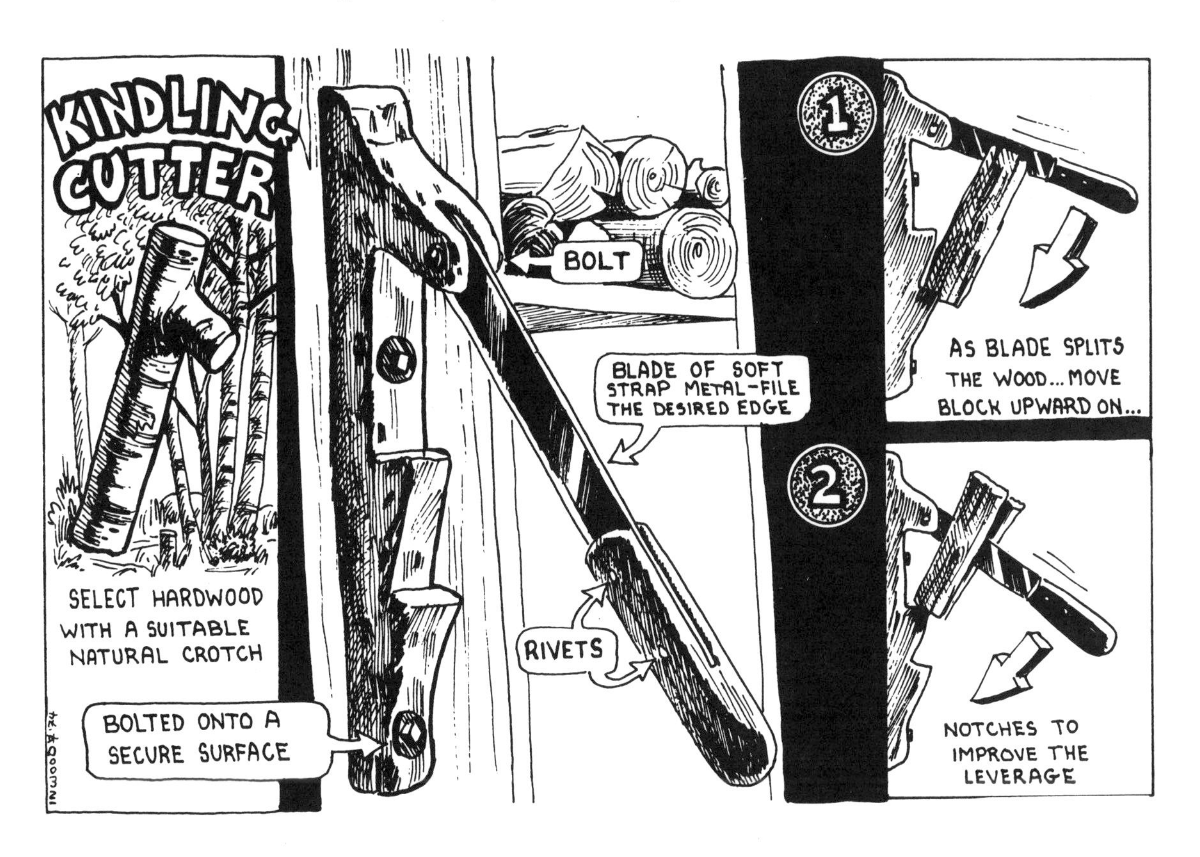

Log planter

which I made from an old car-bumper support and sharpened on the grinder, is attached to the top of it. The blade is free to swing down and chop off pieces with little effort. This device helps keep the kitchen clean by neatly cutting through the blocks without scattering splinters all over the place. It is mounted on the wall so my wife can stand while she is chopping. I also built outdoor planters out of logs. They attach to the front of the house above smaller supports. These logs were dug out with a chain saw and final cuts were made with an axe to clean them out.

My wife and I stood back and admired the completed house. We were both extremely satisfied with it. The one aspect of the sight we didn't like, though, was the rocky mound in front of the house. We wanted a proper lawn to accentuate our house's rustic beauty. We found it impossible to grow anything on that rock mound. We were frustrated with that area until the following season when I began plowing the hay field. At that time, I had a brainstorm. I carefully lifted large sections of turned sod and placed them in my pickup truck. I then hauled the grassy mounds to the front of our house and neatly fitted them together like a patchwork quilt. Before long the grass grew out and the sections grew together to make one big lawn for the front of our beautiful home.

The cost of constructing the house was very low since I made most everything myself. I spent a long time building it, and my time, like everyone else's, is valuable, but I can't set any price on it when I'm doing things for myself. I got pleasure every time I cleaned out a notch or put a log into place. I enjoyed it, not only when the notches were finished or when the logs were in place, but also while I worked on them. I don't appear to be a rich man, and I'm not, but I enjoy what I'm doing and I feel good.

Harry & June's Barn

With our primary shelter, the house, finished, it was time to think about the other buildings we needed. The goats and horses were comfortable in the small old barn, but there was not enough space in there to store all the hay that was needed for them. It looked like it was going to be a fine year for crops and we knew we would probably get a few tons of hay from our own fields. Since the weather can be unpredictable, I knew I could not just leave the freshly cut hay out in the fields to be sun-dried; more likely than not, an unannounced rain would come down and ruin it all. We needed to have a storage place ready by harvest time.

I considered different building designs. I wanted a large, open structure that would be both a storage and a drying area. It had to have an enclosed loft large enough to store at least ten tons of baled and loose hay. There were also a lot of tools and equipment around that had no definite places of their own. We needed a place to shelter all this miscellany from the weather.

I decided to build a huge barn-type shelter for everything that needed a "home." A 20 × 48 foot area was prepared and ten concrete piers were poured. We spaced these 12' apart on the 48' sides. Steel pins were then stuck into them

Post-and-beam barn structure

Spiked log piece

Shed area

to accommodate the hewn 12"- to 14"-diameter sill logs which were to stretch along those sides. These long sills would be the bottom beams for the 11' upright posts which would frame the first story. The posts were spaced 12' on center above each pier and were notched into the beams. A 48' top plate beam was then notched above the posts and the posts were braced in place.

Wherever beam logs came together over an upright post, that post was shortened and a 24"-long log piece was notched in above it to reinforce the joint. The log piece was then spiked to the post and each of the beam sections.

This area immediately seemed too small, so we widened it by building a 12' shed-roofed addition onto the north side, making the barn's width 32'. The new 8' outer wall was framed in the same manner as the other two long sides. Its rafters were spaced at 24' centers, spanning between the beams on either side. We spaced nailing strips across the rafters, and a finish roof of cedar shakes was put on. Thin poles were later run horizontally across the 8' upright posts and scrap one-by-four sheathing was nailed to them to enclose the shed area. This sheathing shielded the area from direct moisture but did not close it off from the crosscurrents of wind, which allow for good air circulation. This section became ideal for hay drying.

To make the racks for the hay, I simply ran smaller 7' rails along each 11' post of the inner wall and attached them at the top and bottom with 4" spacer blocks. Long spikes, spaced 18" apart, were driven through the rails and into the posts. These spikes act as rungs for the horizontal poles which cross the posts. I then extended five tiers of poles between the 8' and 11' side walls. Their ends rested on the crossing horizontal poles of each wall.

During harvest time, we spread out the freshly cut hay on each tier. When one tier was filled, the poles for the next were put in place. Fresh green hay was then spread out on that tier, and so on, until about an acre of hay was drying over the five racks. After the hay was dry enough to store, the poles were taken down and this 12' x 48' section

was turned into a winter storehouse for supplies and equipment.

We then constructed the loft area for storing the hay. A 48' log beam was placed over the ends of the shed roof rafters at the 11' inside wall. The 8" hewn-log ceiling joists spanned from above that beam to 2' beyond the top beam on the south wall. These 22' joists were spaced at 24" centers above the side walls. The 2' that went beyond the beam allowed for a roof eave, which would prevent rain and snow from getting into the open area below. The north side was protected by the shed roof that would be overlapped by the north loft roof eave.

I wanted the loft roof to have steep sides for better drainage. The better the drainage, the less chance of rot, because the moisture would not collect on the roof's surface. If the roof has a steep pitch, snow slides off before the load

Hay drying on tier

becomes too heavy. I also wanted the roof to be wide enough to accommodate the several tons of hay our animals would need throughout the long winter months. The common gambrel barn roof interested me, but I desired a different design, one that would give even more storage space over the 20 × 48 foot floor area. I remembered the old navy Quonset huts I used to be so familiar with. Surely there must be a way to utilize that design when working with logs.

I constructed a framework of log roof braces over the rough, one-by-four loft floor. The first section was 48' long, 14' wide, and 8' high. The sides were supported by uprights spaced every 12', directly over crossing joists. Also at every 12', a crossbeam was notched above the uprights to keep the long framework from spreading.

When I was done with this framework, I came down from the loft and examined the framed area. I decided to have the peak of the roof another 4' above it, making the whole barn 24' tall. In order to do this, I raised a sectioned ridgepole that extended 3' past either end of the frame. It was braced at 12' intervals by double 12' uprights, which spanned between the joists and the peak. These uprights straddled the crossbeams at their centers. They were doubled because each was bracing one end of the four poles that made up the length of the ridgepole.

The 48'-long upper purlins were then put in place halfway between the ridgepole and the lower purlins (the lower purlins are the 48' sides of the rectangular roof framework). These purlins also followed the proposed arch of the roof. They were

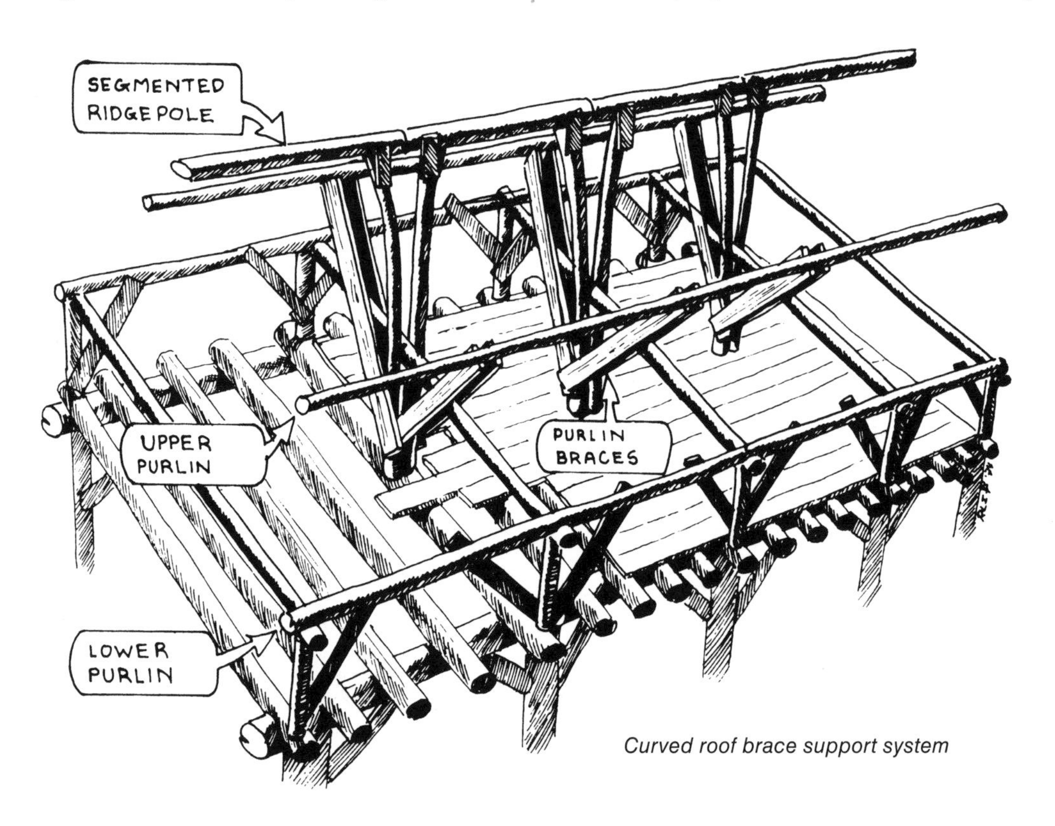

Curved roof brace support system

Interior view of roof construction

Interior view of roof construction

braced by short, diagonal boards which straddled each of the crossbeams.

With the roof-support frame finished, it was time to put my experiment into action. I found a thick stand of young fir trees and thinned out over fifty saplings, which were about 3" in diameter at

Anchoring the bottom end of the roof rafter saplings

the butts. We cut 16' sections with no large knots from them and these were immediately peeled. The saplings were then put into water to soak for a few days. While they were soaking, I augered a 2" hole near each end of every other log joist.

When the saplings had soaked until they were flexible enough to bend easily, I took one of them out of the water, trimmed its bottom end to 2", and stuck it into an augered hole. That end was spiked into the joist. The sapling was then bent over the roof frame and a piece of rope was tied to its top end. That end was pulled down above the ridgepole and the rope was tied to the joist end on the opposite side. The sapling was positioned and nailed to the purlins and the ridgepole.

I repeated this process until the rafters were spaced 48", or two joists, apart on either side of the frame. A pair of longer rafters, at either end of the roof, were attached to the top wall beams and were bent forward over the ridgepole to brace the overhangs.

For added strength, I decided to put another set of rafters over the ones already in place. To insure a snug fit, the undersides of the top rafters

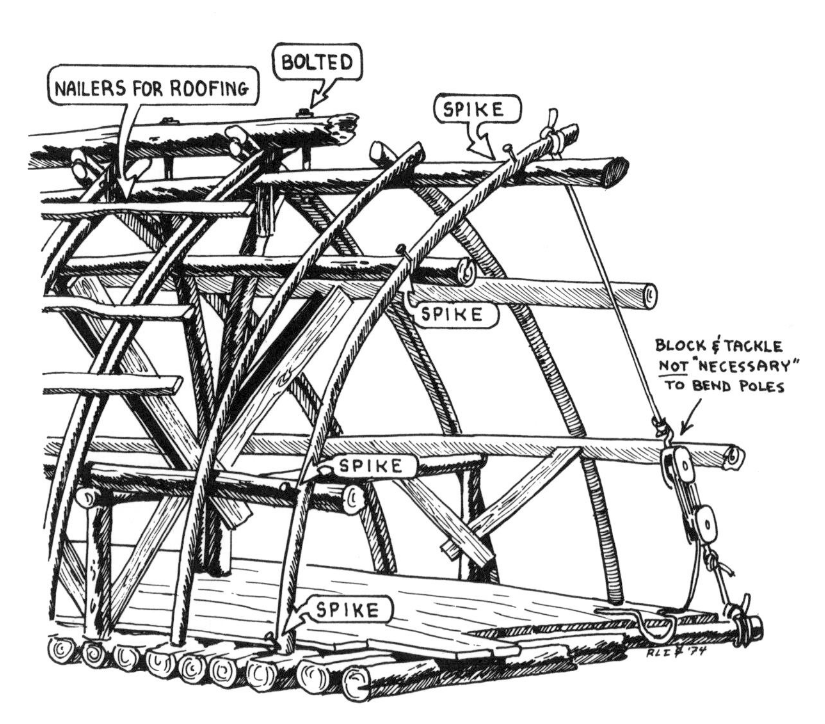

Bend the saplings over the brace

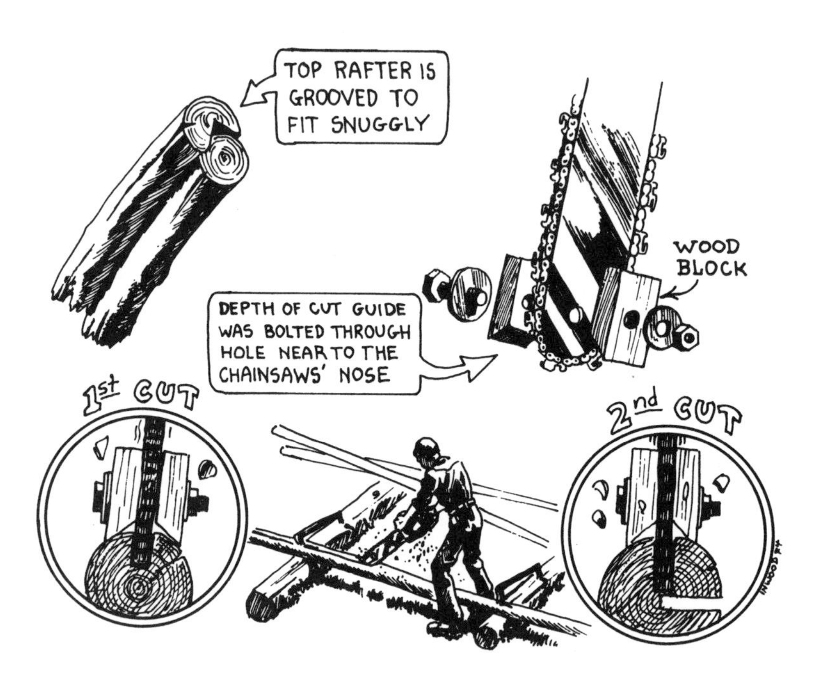

were grooved out about 1" deep so they cupped over the bottom ones. The grooves were made by a chain saw with guides on either side of its roller-tip bar. The guides were attached to each side of the bar by a bolt, which came through the roller-tip opening. The diagonal guides straddled either side of the pole, allowing the chain saw to cut a 1" groove in it.

To make the groove, I dogged a pole across two logs, stood over it, and cut along its length at a slight right angle. I then made a left-angle cut deep enough to meet the first cut, and removed the piece in between.

These top rafters were nailed above the bottom ones and the rafter ends were cut a few inches after they met the ridgepole, allowing each set to cross at the peak. After the rafters were all in place, a second ridgepole was put in over the crossing ends to sandwich them in. It was spiked to the rafter ends and bolted to the bottom ridgepole with ⅜" bolts at 6' intervals.

The ridgepole sections that stuck out beyond the end rafters were reinforced by a third piece that was wedged in between the top and bottom ridgepole ends. Before that third piece was bolted

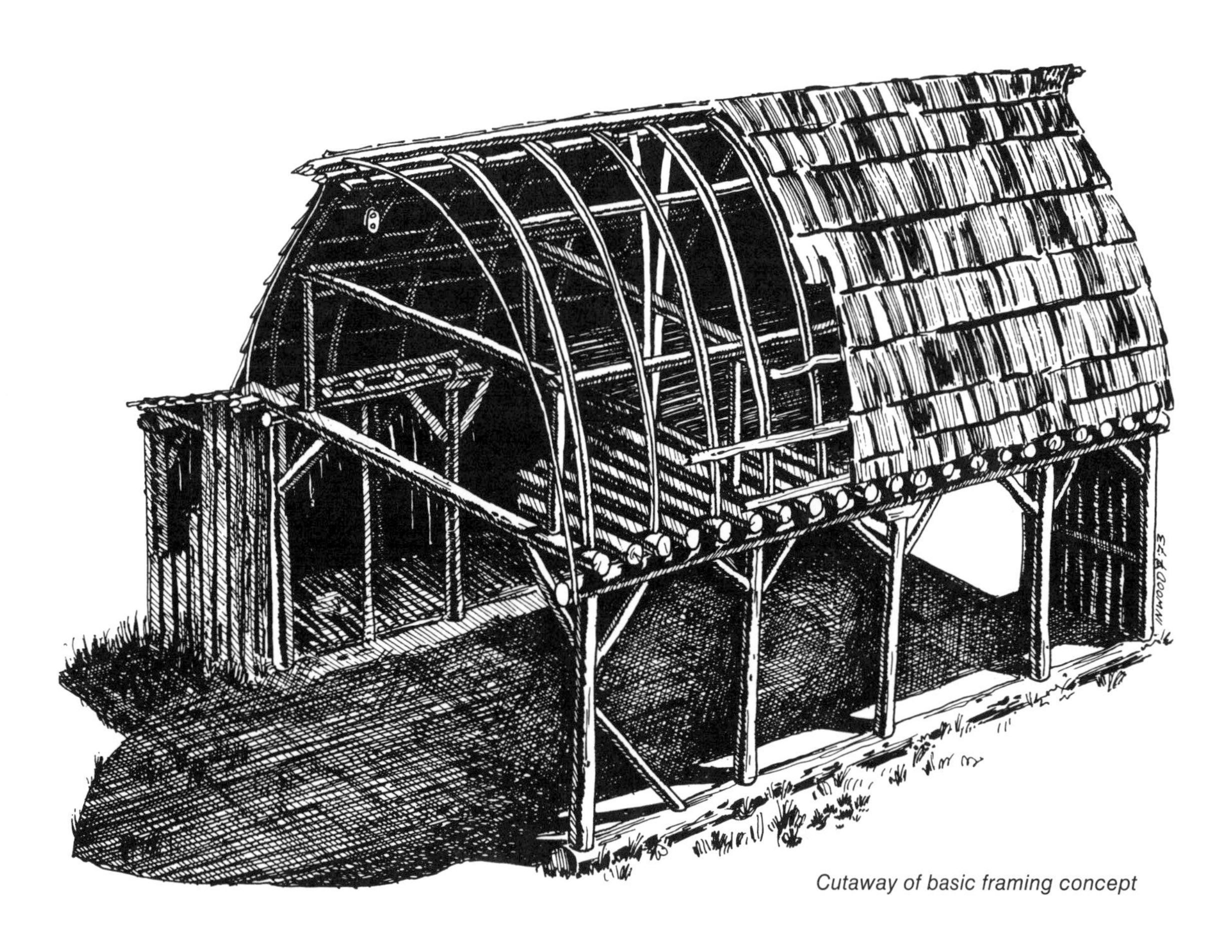

Cutaway of basic framing concept

in, two holes were drilled into either end of the bottom pole to accommodate two U-bolts from which large pulleys would hang.

Next came the nailers for the cedar-shake-finish roof. These nailers are 3"- to 4"-diameter poles which I hewed on two sides with a sharp broadaxe, making flat surfaces for the shakes to be nailed to. I prefer to do my hewing in the winter. For some reason, all types of wood split faster and cleaner when they are frozen. For example, try splitting a birch round in the middle of the winter, then try splitting that same birch round at spring thaw. You will notice an amazing difference.

I used the snakiest, most crooked poles for the nailers because it is not necessary for shakes to be nailed to a straight strip, as long as they have a nice flat surface. These nailers were spaced anywhere from 18" to 20" apart for the 26" cedar shakes. Because of the steep roof pitch, it was not necessary to put the nailers any closer, since the shakes do not have to overlap as much as they would on a roof with a less steep pitch. This spacing also saves on shakes.

The homemade shakes were attached to the nailers. The gable ends of the loft area were then sheathed in with 8'-long slabs that were being discarded by a local railroad tie mill.

Shake roof

Rear entry

Front entry

On the front gable, I framed double doors through which hay bales could be brought in. On each gable, a ventilation opening was left directly under the roof overhang to prevent the stored hay from mildewing. Just like the house, the barn was built slowly and carefully. I take great pride in my work, especially when I am building for myself. As I mentioned before, we had found a piece of land to our liking and we plan to live on it for a long time to come. So whatever we do on it has to be done right. Our buildings have to be beautiful as well as practical; our gardens and fields have to be neat and straight; our landscape has to be trimmed and well cared for; even our fences have to accentuate the natural beauty which surrounds us. Our homestead is a reflection of our own personalities; we respect it as we respect ourselves.

A Sheltering Arch

Section 3

ANIMAL SHELTERS

You wouldn't believe the miles we've traveled to gather firsthand information about efficient and practical animal shelters that have most of the ingredients necessary to house and maintain common homestead animals in the northern, four-season climate. The specifications and the measurements of the shelters are as subjective and unique as are the homesteaders themselves, although each dwelling had a few basic ingredients in common. And amazingly enough, no matter how radical the design, most of the well-constructed shelters were suitable within their own variables, even though most of the builders couldn't explain a lot of the reasons for specific sizes, shapes, and designs.

Among all the farm animals, goats are either the most notorious or the most complacent, depending on the circumstances and the management involved. Opinions concerning the animal, from the people I visited, ranged from "Never again will I attempt to deal with those stubborn, ungrateful beasts," to "They're the most gentle, loving critters we have." Why such a diversity? There are several reasons, I'm sure, but the single greatest reason has to do with the shelter; lack of genuine "heart-to-heart" contact is the second reason. From our own personal experience with goats, I realized that proper shelter and management could have prevented much anguish that we had with these sensitive creatures. Now that we've shared the experiences of many successful and unsuccessful goats-people, we're ready to try developing new relationships with that animal.

We bounce back and forth from a communal cow, which is shared by a few families, to two or three goats. We love the dairy products from the cow—the milk, yogurt, cheese, butter, and ice cream—but we are overburdened by the cost of maintenance. One cow eats about the same amount as sixteen goats and doesn't give as much in return. If you are on a small piece of land without much cleared acreage and have to depend on bought or traded hay and grain, a cow would be rather expensive, even more so than a small herd of goats. Though goat's milk is harder to deal with for butter and other cream products, since it is naturally homogenized (meaning that the cream doesn't easily separate), it is much easier to digest because it consists of smaller fat globules and is more in tune to the needs of the human body. Goat's milk is very versatile for cheese-making and other by-products. Just ask a few people who know about goats for some of their special recipes.

The main problem with goats in so far as shelter goes is that they are natural escape artists. They get out of places unanticipated by the inexperienced goat-keeper. This is especially true of the kids. They have the ability to spring high up or squeeze through small holes. If they

find a weak spot in their enclosure, goats will gnaw, bang against, or butt at that section until it gives way. They will get out and head straight for tender little fruit trees, colorful flowers, and tasty garden vegetables with no conscience whatever. They'll eat and eat until they are as wide as they are long. And when you return from that trip to town to see the fruits of your labor lying in ruin, no matter what you do to punish the offenders, they will feel no remorse and the first chance they get, they'll do it again. So now that you are a bit familiar with the "nature of the beast," you can plan accordingly.

Like any being, human or otherwise, the more contact and interaction you have with the goats, the more rapport and understanding will be developed. Be prepared and do your homework before starting out with any animals. Exercise your advantage of being able to obtain textbook knowledge about them even prior to beginning the relationship.

Since in our travels we could not locate any ideal goat shelters that incorporated more than a small percentage of labor-saving devices in their construction, we will describe one type of shelter, critically explaining the good points and shortcomings of its features. Then, we will share our own design, which has been checked over and approved by local experts.

The goat setup that most appealed to us, but was far from being error-free, was a system involving two base shelters and a lightweight portable pen and tethers for controlled grazing. The main shelter was built to house four kids and three adults in small communal-type pens. Communal pens such as this, especially with a larger herd, are much more in tune with the animal's gregarious nature. Such a space offers interaction, choice of area, exercise, liberty, and social amenity, as opposed to stall-type housing. Stalls isolate the animals and do not provide them with enough area to exercise in.

Generally, the only time it is really dangerous to put goats in such a communal environment is if some of them are horned and others aren't. The dehorned or unhorned ones do not stand a chance against their opponent's "weapons." Even in the most gentle goats, there is a pecking order, but in my experience, the competition never gets too serious because the order is soon established and usually all the members of the group accept their status.

The 20 × 16 foot main enclosure is separated into four sections. Two of the sections are covered with a shed roof for protection against the weather and the other two are open exercise yards. All are complete with feeders and water bowls (**FIG. 10-1 AND 10-2**).

FIG. 10-1

FIG. 10-2

The covered sections consist of three areas. The two 5 × 8 foot areas within the sheathing are indoor quarters for the kids and does. The area on the right is partitioned from the area on the left and is open to the front section. It is for the milking does. The upright pole and heavy-guage 2" mesh poultry netting partition prevents the kids from getting to the does, yet allows them some restrained contact (**FIG. 10-3**). Most breeders recommend taking the kids from the does immediately after birth, not even letting them get their first natural feeding. This protects the does' udders, which is said to extend their productive years. It also makes the transition of separating the mother and offspring much easier, especially if restrained contact can still be made.

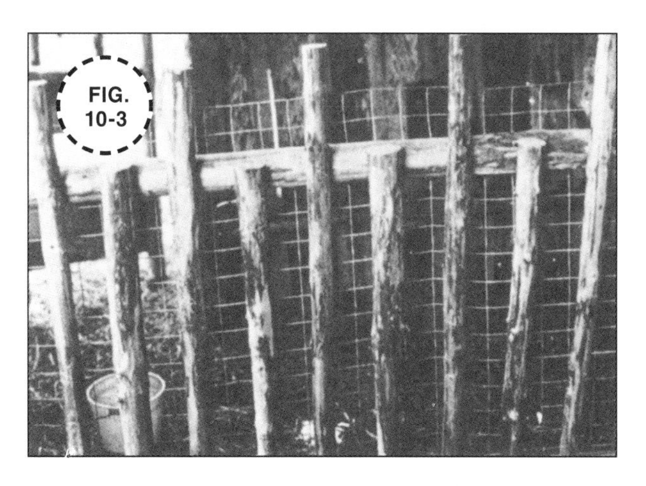
FIG. 10-3

The does' quarters are nothing more than a loose shelter. In the colder months, it is closed off by the door that separates it from the exercise yard (**FIG. 10-4**). Beside the door, on the outside wall, is a feeder box that is used in winter to feed the confined does.

FIG. 10-4

The hay is placed in the box from the outside. The does go to the wall from the inside and eat through small holes within the wall, which ration the hay but deter tany waste by only allowing a small amount to come out at a time. Certainly this wouldn't prevent the goats from slowly working it all into the shelter and using the spoiled hay for bedding, but it does make it more difficult than if they were using an open feeding system. The best control against food wastage is the American feeder, which I will describe later (see **FIG. 10-33**).

The same type of controlled feeder is used on the wall that divides the kids' shelter from their small outside exercise yard. This feeder box can be used at either side—from slots through the inside wall or through openings on the front of the covered box (**FIG. 10-5 AND 10-6**).

FIG. 10-5

FIG. 10-6

A larger feeder for the does is located on the inside of the lattice-fence enclosure. This supplies them with their food in the warmer months when they do not need to be enclosed (FIG. 10-7). Though goats should be separated while being fed grain, there appears to be no advantage to isolating them at other times except during kidding and disease control.

The flooring both inside and out follows the principle of underfloor heating, which goats themselves originated. Goats' natural bedding is of accumulated droppings mixed with whatever grass and other roughage they can find. This provides a warm and comfortable bed, which absorbs and evaporates the urine, making its odors inoffensive. It is the most economical in labor and materials and seldom needs to be changed. In fact, it should be left to accumulate all winter to provide sufficient heat and prevent floor drafts. It can later be put on the garden.

As any experienced homesteader will attest, goat manure is one of the best nitrogen-rich manures for the garden and is also the easiest to deal with. I personally prefer wood shavings that are free from turpentine—such as cottonwood, birch, or poplar—for bedding and flooring. Wood shavings or sawdust are easier to lift off the floor when ready to be changed because these

FIG. 10-7

materials consist of small particles that don't intertwine and blanket together as does hay or straw. They are also cheaper and readily available wherever there is a sawmill nearby.

The fences around the enclosure are of latticework, designed for beauty and not so much for efficiency, though they have proven to work well for containing the adults (**FIG. 10-8**). They did not work so well for kids, on the other hand, because these nimble little creatures could easily squeeze and push through most of the few inches of space between the cross members, and once they did so, they proudly pranced over to the nearest food source. When using a wooden fence of any type to contain kids, the rails or members should have no more than 3" of space between them. Any more than this and the kids will either get their heads stuck when attempting to escape or succeed in their attempt. The designer constructed a removable upright partition similar to the one he built between the kids and does in the shed to separate their exercise yards, but failed to use poultry netting of sufficient gauge to reinforce it. This resulted in an eventual break out, as can be witnessed in the photograph. Diagonal slots were sawed out of the supporting uprights for the cross members of the partition to fit down into, for easy removal of the partition when the kids are old enough to run with the adults (**FIG. 10-9**).

Since the main criteria for this structure was to provide a small, comfortable, lightweight and portable shelter that could be easily moved when the occasion arose, its most definite shortcoming was that it did not provide adequate protection against the cold winter. Barn weather should be regulated at around 55° F. If does are not provided with proper shelter from the extreme weather, they will utilize their energy to keep warm instead of to produce milk. This problem can be easily remedied with the addition of insulation and the

FIG. 10-8

FIG. 10-9

FIG. 10-10

closing off of all openings, but that would make the structure less portable, unless all the walls and fences are built as separate units that could easily be taken apart when deemed necessary.

The secondary shelter is a combination milking-doe quarters, storage shed, stanchion area, and open-horse stall, which incorporates some very intricate and interesting notch work and other unique construction designs. The enclosed area consists of two stalls and one outside area. The stall to the rear is a smaller shelter for a couple of the heavy-producing milking does. It has a feeder attached to the front outside wall that allows them controlled hay through two slots in that wall. It is conveniently placed near the raised milking stanchion in front of it and the feed shed beside it. In the feed shed, the grains and other perishables are stored in large galvanized trash cans to prevent rodents or other pests from eating and spoiling them (**FIG. 10-10 THROUGH 10-13**).

FIG. 10-11

FIG. 10-12

FIG. 10-13

The milking stanchion is raised high enough for the milker to stand during this operation. This particular milker works best in this position. The platform is held up from the underside with diagonal braces. The neck-locking device is attached to the front outer wall. It consists of a frame of two uprights straddled at top and bottom by one-by-four crosspieces. Another center upright is held at the bottom by a dowel that loosely fits through a hole in it. It is free to swing from side to side at this point. At the top, it is sandwiched by the upper crosspieces and positioned by a peg that goes through those supports. When the movable upright is on the outside of the peg, the goat can be led up the steps and put into position, her head going in between the far uprights. When she is in position, the milker removes the peg and swings the top of the upright toward the goat, locking her neck in place, then replaces the peg. Here the goat stands patiently until milked out (**FIG. 10-14**).

A milking doe will usually happily go straight to the milking stanchion without any trouble since her udder is full and getting quite uncomfortable. Such a stanchion is useful for several reasons. It should contain a platform that raises the udder to a level that is comfortable to the milker whether a standing or sitting position is preferred (**FIG. 10-15**). It should also have a neck-locking device to prevent the doe from squirming around or backing up while being milked, and perhaps upsetting the bucket after she has finished her grain. Grain is usually fed while the doe is being milked, to distract her and keep her content throughout the operation.

There are some goat people who prefer not to feed the does when they milk. Instead, they keep the does' heads in a controlled locking mechanism so they can contentedly chew their cud and peacefully meditate while being milked. This method discourages the nervous gulping down of grain that is usually the habit of a goat that is fed while being milked. She races through her portion, seemingly competing with the milker, seeing who can finish first. If she finishes first, she usually fidgets, cries for more feed, and sometimes holds her milk back because of the distraction.

The milking stanchion should be away from the main pens or stalls and removed from possible sources of contamination. It should be kept spotlessly clean because milk is quick to pick up surrounding odors and bacteria.

FIG. 10-14

FIG. 10-15

FIG. 10-16

FIG. 10-17

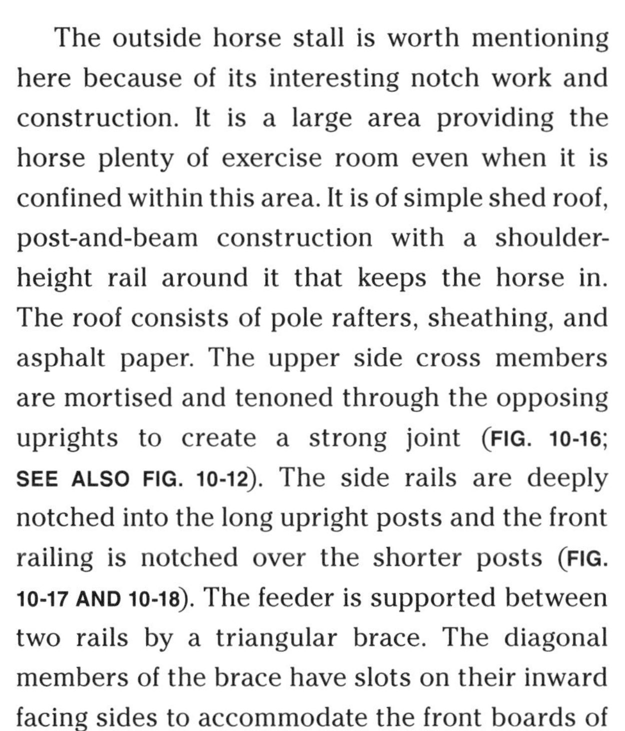

The outside horse stall is worth mentioning here because of its interesting notch work and construction. It is a large area providing the horse plenty of exercise room even when it is confined within this area. It is of simple shed roof, post-and-beam construction with a shoulder-height rail around it that keeps the horse in. The roof consists of pole rafters, sheathing, and asphalt paper. The upper side cross members are mortised and tenoned through the opposing uprights to create a strong joint (FIG. 10-16; SEE ALSO FIG. 10-12). The side rails are deeply notched into the long upright posts and the front railing is notched over the shorter posts (FIG. 10-17 AND 10-18). The feeder is supported between two rails by a triangular brace. The diagonal members of the brace have slots on their inward facing sides to accommodate the front boards of

FIG. 10-18

the feeder. The side and back boards are nailed to the bracing (**FIG. 10-19 AND 10-20**).

One of the most laborious methods of controlling animals—and one which is definitely against the inherent nature of goats—is tethering. Its only saving grace is that it allows them the controlled grazing of various portions of a pasture or a forage area without letting them get near gardens, fruit trees, or other off-limits areas. The disadvantages of this system are that the tethers have to be frequently changed, they only allow a circular area to be foraged, and they are difficult to maintain and easily broken away from if they are poorly designed. The proper post tether should consist of a rod that is at least 24" long and that can be pounded deeply into the

FIG. 10-19

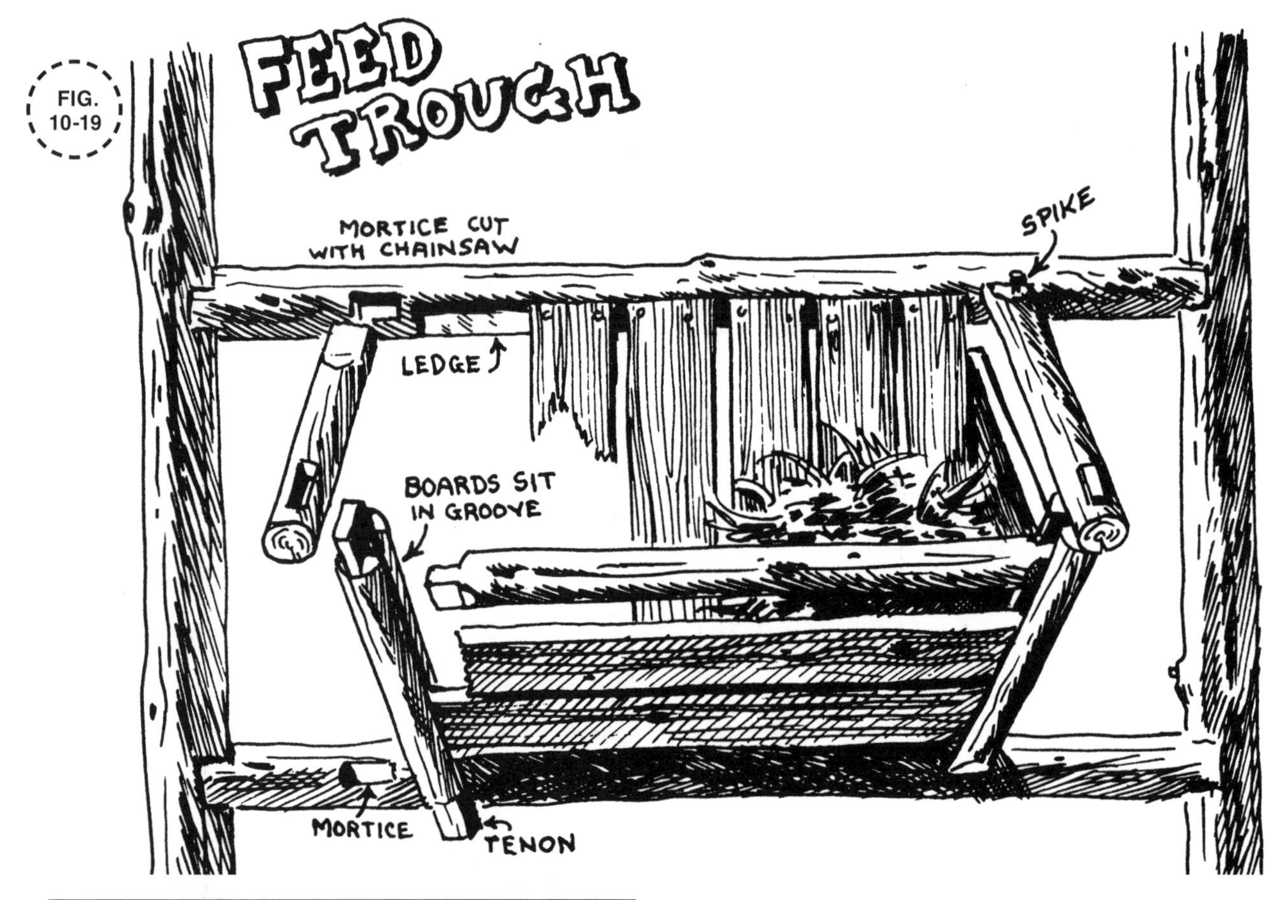

FIG. 10-20

ground until only its top ring is exposed above surface. Attached to this ring should be a swivel ring and a strong lead chain. The chain should be no longer than the radius of a daily portion of forage, 10' at the most, and should have a swivel ring attachment where it connects to the goat's collar to help prevent the animal from getting tangled on the lead. Even with all these precautions the animal usually finds some way to make this method inefficient by either getting tangled up or pulling out the post.

An alternative to this primitive method is the running tether, which allows the goat to cover more lateral ground and more efficiently forage a given area. It is also less demanding of maintenance. Since it covers a larger area, it only has to be changed every few days, and even then, only one post has to be moved. Using the other post as a stationary point, the one being moved could be positioned at various points and eventually encircle it, or each post could be moved forward every other time until the running tether is staggered along the forage area. The running tether consists of a short length of chain or rope, no longer than 3'. It is attached at one end to a swivel collar and at the other to a swivel ring, which runs on a wire or rope that is stretched taut along the ground between two posts. The posts are pounded into the ground, so all that is exposed are their top loops. These posts could be placed to an excess of 100' apart if the terrain allows (**FIG. 10-21**).

One goat alone on the tether, out of the sight of the others, is an unhappy animal. A number of goats tethered within sight of each other will be much more contented and quiet. Their productivity and general attitudes will also be considerably better.

Another alternative to the tether is the movable pen, of which this eight-sided enclosure is a good example (**FIG. 10-22**). It is made up of eight equal 10' sides that are 4½' in height. Each side is composed of two two-by-four end posts with four

one-by-four horizontal boards approximately 1' on center. The spaces between the boards are filled to a height of 30" with heavy-duty 2" mesh poultry netting, which is stapled to the boards and stretched from the top with a spiraling rope. The sides are held together with two heavy-duty hook-and-eyes on either end. Each piece can be easily moved by one person. And the structure can be added to or sides can be taken away, depending on the shape and size you want the enclosure to be. This enclosure is excellent for adult goats, but is not recommended for kids. The older a goat gets, the less spring it has. Young kids have plenty of spring—too much. The only problem with this enclosure is a simple one to remedy: The hooks are just not long enough to clear the bottom of the eyes by more than ½". This makes it quite easy for an energetic, rambunctious goat to buck at the side highest on a slope until the hooks come out of the eyes on one end. The remedy would be to have longer hooks that extend at least 2" beyond the bottoms of the eyes (**FIG. 10-23, 10-24, 10-25**).

A large number of variables determine the design best suited for a goat shelter. These are found around the immediate exterior of that shelter. Is there plenty of surrounding browse and forage? Is there water near? Is there shelter from direct sunlight? Is there room for an outdoor exercising pen? Are there a lot of immediate temptations such as a garden, fruit trees, and flowers? Each question has to be considered before locating the optimum site for the goats' main living quarters. It should be close to forage, browse, and water so the goats don't have to be led across long distances to be taken to the desired areas. It should be in a place that has some shelter from direct sun so they can go out to a good-size exercise yard and still get shelter from the heat when they need it. It should be

FIG. 10-23

FIG. 10-24

FIG. 10-25

out of the way of unprotected fruit trees, flowers, and gardens. Actually, it would be beneficial to have the garden nearby if it were properly fenced, to facilitate the transferring of manure from barn to garden.

The building itself should be well insulated and draft-free, containing a large enough loft above the living quarters to hold all the hay and bedding the animals will need for the long winter. Our goat living quarters design takes into consideration the housing of four to six milking or dry does, four to six kids, and one or two functioning billies. Included in the design is as much pole construction as possible to give the place a warm, natural feeling.

One enters the gambrel roof, 24'-square (outside diameter) barn from the west Dutch door entryway. To the immediate right of that entrance is a supply cabinet containing milk pails, rags, medicines, detergents, tools such as nail clippers and knives, and disinfectant for the milking stanchion. The cabinet has a double door and a latch to prevent it from being opened by a curious goat who has managed to escape its confines. To the right of the cabinet is the 3 × 5 foot hay drop chute, which is directly under the lidded hole entrance to the loft. Beside the chute is a stairway and a pulley rope for easy opening of the lid. On the far side of the ladder is a deep-bowl galvanized sink for washing (**FIG. 10-26**).

As you continue facing the aisleway down the center of the barn, there is a does' pen and milking stanchion to the south, and kid, billy, and isolation pen to the north. The 9 × 6 foot milking stanchion has a 4½' high horizontal pole partition separating it from the entrance area and another separating it from the adjacent doe pen to the east. It is located to the far south, facing west, away from the doe pen, so the doe that is being milked will not be distracted by the pen. There is a long double window of salvaged, shatterproof windshield glass above it to allow in the southern

FIG. 10-26

light. The milker goes into the stanchion area, through the gate at the east, and readies the area for milking. Then the milker goes into the doe pen through the gate within the stanchion area and gets the usually willing doe. The doe climbs up on the 3 × 5 foot stanchion platform, which is about 1' off the floor. This height makes the udders easily accessible to the seated milker. The doe puts her head inside the neck-locking unit and busily nibbles at her grain while the milker milks.

The feeder unit for the milking stanchion consists of two 36" uprights spaced 4" apart. These uprights were held together at the top with a crossing one-by-four and were braced to the front end of the platform. Below the crossing one-by-four, a 12" long by 1½" wide piece was cut from the insides of the uprights to allow the goat's head to freely enter and slip down the narrower 4 channel to the raised 8 × 8 × 6 inch deep feeder box. After her head goes down the channel, the applewood bolt is drawn across the branch intersection latches. Her head is then locked in the narrow opening until after the milking (**FIG. 10-27**).

The 10 × 14 foot doe pen is plenty sufficient for the growing herd. At least 12 to 15 square feet of floor space should be provided for every adult goat. This area is well lighted for solar heat in winter by three windshield-glass double windows along the south wall and one on the east wall, making the pen a cheery environment instead of a dark, depressing one. The floor is dirt, covered with a thick layer of fine poplar, birch, or cottonwood woodchips, which are the quickest species to decompose and are the best for garden compost. The droppings and urine mix and soak up well in this porous material, and it is all easily transported into the garden once it is the proper consistency.

FIG. 10-27

There are two 3 × 6 foot portable raised platforms, approximately 1' off the floor along the south wall for the goats to sleep on. They are set 1' in from that wall to avoid wall drafts. Don't use slats for this platform, because goats do not like them. Two rather large stones were brought into the pen for winter exercising. Goats naturally crave some contact with hard, abrasive surfaces such as stones. This contact helps keep the hooves trimmed and cuts down on unsightly surplus growth in which foot rot develops. This growth will not accumulate if the goat is regularly exercised on hard ground for even a short period each day.

Contrary to popular belief, goats are quite fussy about what they take into their bodies. They won't drink soiled or contaminated water and they won't eat food that has been even nosed over by other members of the pen. But they do need plenty of water to produce milk. For this reason, I prefer a self-regulating, running water system with a steadily trickling intake and a high outlet to control the overflow. This system consists of a 5-gallon can raised on a 1' platform. It is located just inside the east wall. The intake pipe tees up from the underground water line and elbows into the bottom of the can. The outlet nipple is also brazed to the bottom of the can. It has a fitting for a length of plastic pipe, which extends to about 2" from the top. The water fills up in the can and the overflow spills down this pipe to the runoff pipe below (**FIG. 10-28**). This runoff is buried under woodchips and sufficiently protected against freezing in the winter. It empties into a subterranean boulder pit. In the summer, it feeds the outside waterer.

The water is kept warm through the winter by a tube heater, which is suspended near the water level by a float attachment. Half the can top is covered by a lid to lessen the chance of the water being contaminated. The salt lick is placed beside the waterer to encourage the goats to drink.

Goats are wasteful feeders. If their food is dropped on the bedding, they refuse to eat it, yet if allowed to, they will pull out large quantities at a time, spreading it out over the floor and eventually spoiling it. Therefore a well-designed feeding system is a major consideration for an economical, efficient goat operation. It should be a system that enables you to feed them without having to enter the pen.

Along the center aisle, opposite the kid and billy feeders, is a row of partitioned American feeder–type racks, which act as a center enclosure for the doe pen and separate it from the outside feed trough. This system has proven to work extremely well because it prevents the goats from pulling the hay back into the pen and wasting it. It also allows the does to eat their hay ration at will when the locking latch is not secured and keeps them at the feeder eating their grain when it is secured. The partitions discourage the more aggressive, long-necked ladies from slithering beyond their own areas to their neighbors' portions.

The feeder racks of the various pens are designed around the basic dimensions of the American feeder type, but they deviate from the original design in many ways. The row of uprights across the front of the doe pen, which ends at the east wall gate, are 48" in height with crossing one-by-fours on either side holding them all in place at top and bottom. This rack is supported by upright poles that extend from the ceiling to the floor at 30" intervals. The feed trough partitions are attached to the rack uprights with metal brackets, separating the feeders. They can be readily removed for easy cleaning of the long feeder trough.

The fourteen one-by-four feeder rack uprights are spaced in such a way so there is 18" of distance between the centers of each neck slot and 4" of space allowed for each. They are 48" long from top to bottom. Six inches down from the top, a 1½" wide by 12" long strip was cut from the insides of the uprights, making a 7" head opening between the boards. The doe's head goes into the opening and her neck comes down into the neck slot as she feeds. This allows her to feed comfortably and not spill her hay into the pen. When grain rations are served, the handy apple-wood bolt is drawn across the latch below the head opening to prevent the more aggressive goats from gulping down their portion, pulling out, and bullying the weaker goats into giving up theirs (**FIG. 10-29 AND 10-30**).

The gate at the east end of the feeder rack opens the doe pen to the sliding back door. This door acts as a dual entranceway to the two outside areas. If the door is slid halfway to the north, it opens the pen to the forage grounds. If it is opened halfway to the south, it leads to the small exercise yard. The doe pen gate and the kid pen gate both open outward, preventing the animals from going any direction but out the sliding door when it is opened to them (**FIG. 10-31**).

The kid pen, billy pen, and isolation pen are designed similarly to the doe pen. Each contain

FIG. 10-29

FIG. 10-30

FIG. 10-31

feeder, waterer, sleeping platform, rock, and salt licks. The main difference is the sizes. The kid pen is 8 × 10 foot and the billy servicing pen is 6 × 10 feet including feeders. The waterer is built into their partition railing so it can be shared by animals in either pen. The head opening in the kids' feeder is 6" and the neck slot opening is only 3". The head opening in the billy's feeder is 9" and the neck slot is 6" because of the animals' larger features.

The isolation pen at the northwest corner is 5 × 10 feet including feeder. It is reserved for does who are preparing for labor, for sick animals, and for newborn kids. The enclosure is reinforced with heavy-gauged, 2" square poultry netting to hold the kids in. This stall only has a bowl for water since it is not in use throughout the year (**FIG. 10-32**).

Well, that is my design. I don't expect anybody to copy it, but I do hope that it inspires some thought and creativity in designing your own goatherd shelter. And take into consideration that the oversize dimensions of this shelter were designed for a growing herd. Goats do not need as much room as I have allowed them.

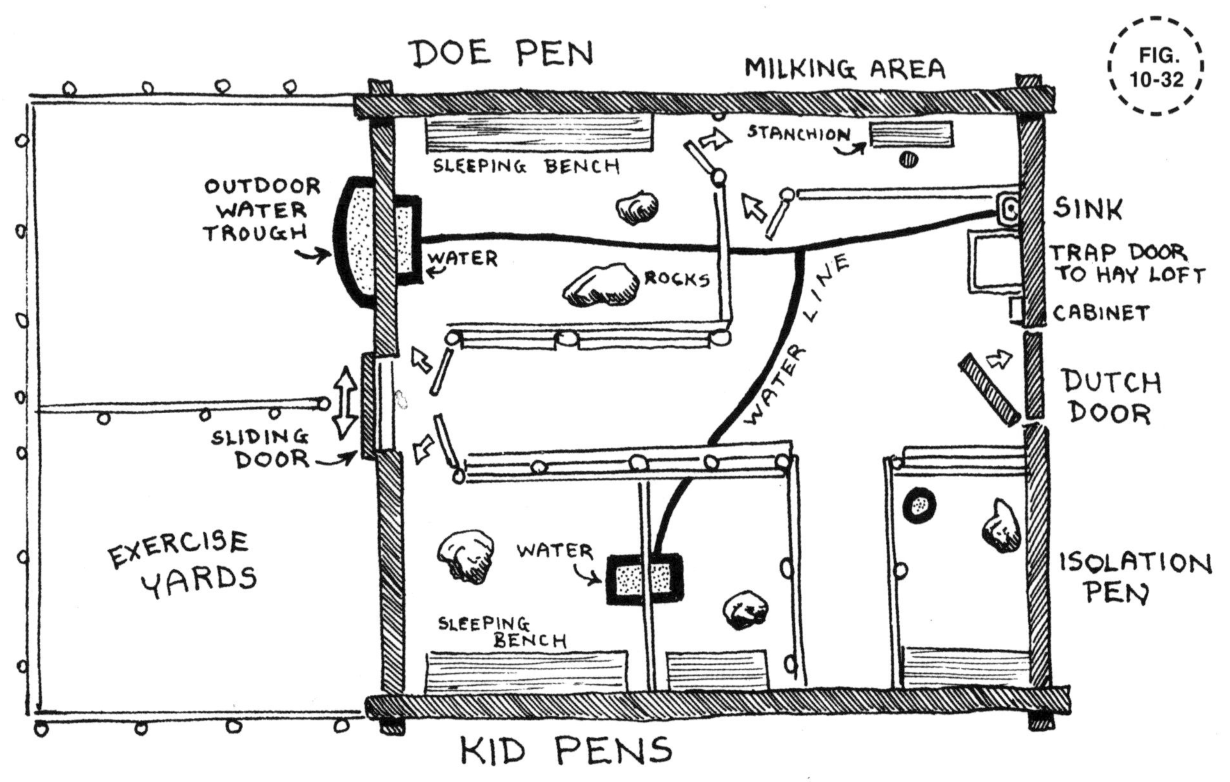

Chickens. No homestead or farm is complete without these multipurpose little noise makers who welcome us to each new morning with vibrant glee, usually way before we're ready to start the day. From my travels, I've noticed that the potential of these necessary and versatile beasts is grossly underestimated. They are usually confined in areas that are too small and are not allowed to forage for their own food, but instead are fed with expensive mixtures of grain and mash. The nutrients they get from these purchased products could be easily supplied from garden trimmings and wild greens.

Some people forget that all our presently domesticated animals, in the not-so-distant past made it on their own, gathering their own food, finding their own shelter, and producing their own offspring. Sure, I grant that egg production of wilder birds isn't as efficient in the first two seasons as that of chickens in the larger poultry farms, but compare the egg production after that initial chemically stimulated period, and the quality of eggs throughout the laying period. If the worn out, bedraggled commercial birds were still laying after the conditions they suffer during their first two seasons, it would be a miracle. The wilder birds, I'm sure, would just be reaching their prime, still healthy and proud, laying thick-shelled eggs with bright orange-yellow yolks and firm whites. That is what we've witnessed, anyway, with our own birds and with others that are allowed to roam around large open areas to find their own nourishment. And, come to think of it, when ours would return to the coop in the evenings, full of natural, green, plump little bugs, and sandy grit for digestion, they would take their place on the roost and the next morning, when I would come to collect the eggs, each nest would be full, meaning that most of the birds were laying the bright orange-yellow-yolked eggs daily. What more could we ask for? But, of course, our system had to be worked out with the variables we had to deal with. We had to supplement their natural forage with some grain and we still use laying mash in small quantities. In fact, the laying hens barely touch the commercial mix when the weather is good enough for them to go out (and when I don't forget to open the gate for them).

The main problem with our system is that someone has to make sure to let the hens out in the late morning, after they have had a chance to lay their eggs in the familiar nests, and close them in after dusk when they return to the roost on their own. If they are allowed to go out in the early morning before they lay their eggs, they will search out hidden nesting places, making it difficult to collect the eggs. Be sure to fence in your garden well, because that's the first place they'll head. And protect new flower bulbs and seeds. Some friends of ours actually go as far as setting up chicken paths through compost heaps, outhouse holes, and areas for future garden additions, so the omnivorous birds can peck through the heaps, helping break down the decaying compost while they supplement it and the future garden spot with their own nitrogen-rich droppings.

Other people we've talked to don't use commercial laying mash at all. Instead, they just purchase the individual grains and mix their own feed—by far the cheaper way to go. And this way, you know what your animals are eating instead of having to trust vague labels that don't explain how all the ingredients are proportioned. A good home mixture would be include some corn or other whole grain for scratch, plus some cracked egg shells to replenish the calcium lost in egg-shell making. We save up all the old egg shells, bake them until they are crisp and dry, then crush them and return them to the hens. The protein supplement is necessary unless the chickens have a large area in which to forage.

Here are examples of two efficient poultry coops which we found in our travels. Both are complete for the care of adult hens; the second explains some necessary hints in introducing new hens to a coop of seasoned layers.

In my opinion, the design of this first barn is by far the most practical multi-animal structure I've come across. It is made up of three sheltered areas covered by a huge 18 × 50 foot modified gambrel hayloft with long, steep sides that extend 18" beyond the walls of the 12 × 50 foot lower structure for optimum hay storage. A spacious 15 × 50 foot chicken yard is located behind the barn to provide the birds with plenty of foraging space, yet it keeps them enclosed and protected from the wild beasts that prowl the area. This building is constructed mostly of log post and beam with slab sheathing for the walls and lower sections of the loft roof. The upper sections are cedar shakes to prevent leakage through that low-pitched area. The loft has space for much more loose hay than could possibly be needed for the farm's own use. Such a large space comes in handy, though, when the hay has to be brought in early to dry before it gets rained on. It is just spread out on the crossing pole floor so the air can circulate through it and dry it out.

The 15' square front section of the lower area is the quarters for the cows when they need to come in out of the weather or when a calf has to be separated from the milking mother. It's simply a comfortable open area that protects them from direct precipitation and chilling winter winds. That's all the shelter cows really need. It is complete with a slatted feed trough, which is fed from a convenient hole in the loft floor directly above it (**FIG. 11-1 THROUGH 11-5**).

The center section is split into two areas, a narrow 5'-wide milking area just inside the doorway and a larger storage area for feed, grain, and other supplies. The milking area is a long corridor which leads to the milking stanchion in the rear. It is fine when you have a cow with a good "reverse gear," but would be a difficult space when dealing with one who insists on turning around to head out when finished milking. The stanchion is supported at top and bottom by horizontal two-by-fours braced by three uprights. It is composed of six boards nailed and bolted together and is suspended on a chain to make it mobile instead of rigid. Two 18" long horizontal pieces sandwich in the 4' uprights at the bottom. The upright to the left is stationary to these crosspieces, the one on the right is loosely bolted between them. There are two 24" long horizontal pieces also on either side of the uprights at the top.

A loose bolt with double nuts holds the movable upper crossing boards to the left upright. This is the pivot point when the crossing boards

FIG.
11-1

FIG.
11-2

FIG.
11-3

FIG.
11-4
HAY SHOOT

MILKING
STANCHION
FIG.
11-5
STALL
MILKING
AREA
FEED
STORAGE
CHICKEN
COOP

are raised and lowered over the right upright. This right upright then can be adjusted to either side of a bolt near the center of the crossing boards. When it is on the inside of the bolt, there is a parallel space of 7" between the two uprights. This is just wide enough to hold the cow's neck and keep her from fidgeting and possibly kicking over the bucket (FIG. 11-6 AND 11-7). For many cows, such a restraining apparatus is unnecessary, but their part Black Angus cow needs it. Black Anguses are on the whole more rambunctious than most breeds (they are usually raised as meat) and require such precautions. I've had dealings with a milking black Angus and sympathize with the problem.

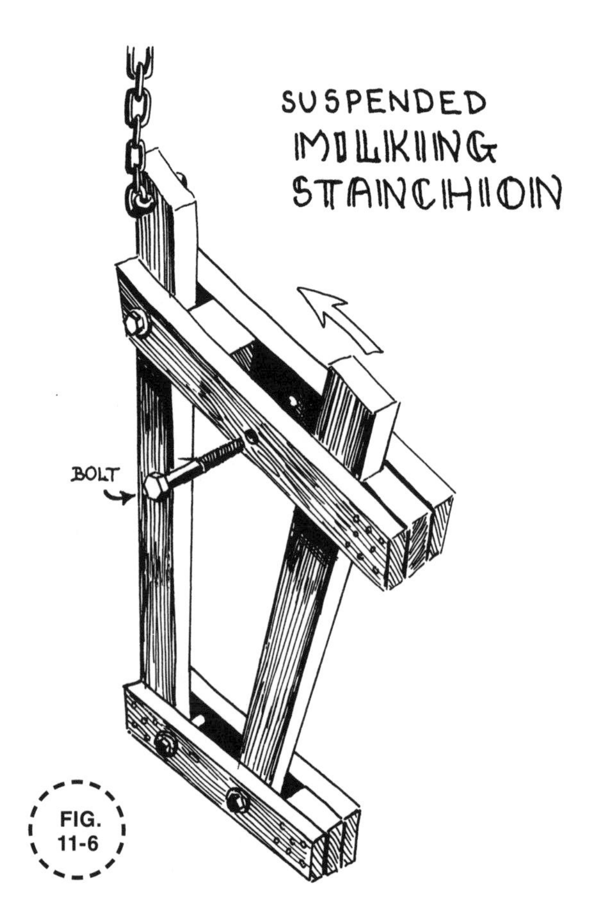

FIG. 11-6

FIG. 11-7

The third section is the 15 × 15 foot chicken coop. This builder and I agree on the point that the first and most important consideration in proper animal care is housing. With adequate housing, you can curtail disease and anxiety/tension caused by overcrowding. Sufficient space for the birds is very important since there are so many of them sharing the same dwelling. Allow at least 4 square feet per bird; less than this amount causes havoc—remember what it was like back in the city? It promotes cannibalism and pushes the weaker birds off the roosts. The so-called pecking order can be easily witnessed in an overcrowded henhouse. It is easy to spot the weakest, most shy hen. She cowers away from the others, in her own corner. I've seen a few of the more aggressive hens peck at a "wimp" until they drew blood from her, then all the others charged in to make a meal out of her until I pulled them off and gave her a separate cage so she could strengthen up.

This chicken coop is more than adequate size for as many chickens as the average smallholder would care to deal with. It is laid out very efficiently and care has been taken in the design of all the essentials. The builder feels that a chicken coop should be as spacious as one can afford to construct and be properly insulated from the cold. Chickens like a lot of room and they crave a lot of light. The builder put in a high-watt unfrosted bulb to compensate for the short days of winter. The south wall has several large windows of opaque plastic, screened with poultry netting so the birds won't peck through the plastic. Chickens should have adequate lighting so they can see their food, because they identify their food by sight. They do best in a well-lit coop with good ventilation. A dull, gloomy coop makes the birds lethargic and nonproductive, but too much artificial lighting can cause a false molt, which would result in the premature ending of that season's egg production. About twelve to thirteen hours of light a day seems to give the best results.

Chickens thrive in a warm, dry environment where they have plenty of indoor scratching room and sufficient roost for the meditative evening hours. The floor should be covered with some type of absorbent litter that can be kept as dry as possible. A 4" layer of sawdust is used on this floor. It is turned quite frequently to mix the droppings in and provide a clean, dry surface. The hens even help in keeping their floor dry by scratching through layers of sawdust. This material is one of the best for litters because it is so absorbent and can be thrown on the garden after it has soaked all it can take. Only hemlock, birch, poplar, and cottonwood were used on this floor, because these species break down the fastest and do not contain turpentine, which could be harmful to gardens in large quantities.

The two roosts go the length of the coop between the north and south walls. They are 18" off the ground, both of equal height. Roosts should be on the same level when there are a lot of chickens in the coop, to prevent excessive bickering over who gets the highest roosts—the old pecking order again (**FIG. 11-8**).

Controlled feeding and watering with minimal maintenance are other highlights of this chicken coop. The waterer provides an endless supply of

the vital liquid without any maintenance. In fact, it fills and regulates itself without demanding time and energy as one of the multitude of daily chores. There's so much routine that has to be contended with when dealing with animals, that you want to get a break from it whenever possible. The watering system consists of a 5-gallon can with a ¾" galvanized elbow intake, which is soldered to the bottom center of the can, and a length of 1" plastic pipe, which extends from a soldered nipple at the bottom to about 1" from the top of the can as a drain. It is capped with a perforated metal plate to prevent the birds from pecking at it. The upper edge of the can is lined with a split heater hose to keep the "dumb" chickens from cutting their necks when watering (**FIG. 11-9 AND 11-10**). The water slowly trickles up from an insulated pipe under the coop, entering the can through the elbow. The water then fills the can to the top of the plastic pipe, and the unused water drains down the pipe, running out below the coop. This simple system works fine, doesn't freeze in the winter, and provides a constant, dependable flow of water year-round (**FIG. 11-11**).

The feeder consists of a lidded bin and a feed trough. The bin is 12" wide and 42" long with a sloped roof that is 38" tall at the high point. It is large enough to hold 200 pounds of feed and has a lifting lid to prevent the birds from entering it from above. It is enclosed with one-by-four boards except for the 4" at the bottom along the length. An 8" high feed-trough rim, extending 3" out beyond the open bottom, surrounds the bin, providing a lip to keep the feed from uncontrollably spilling out. At first we had the trough enclosure only 4" high but soon doubled that height because the chickens could get in and peck out all the corn, leaving the oats and laying pellets. With the addition of another 4" board, it is impossible for them to get down far enough to choose their favorite food, leaving the rest. Chickens should have a continuous food supply, which should never run out. This is a dandy feeder for the job, which requires filling only once every several days (**FIG. 11-12 AND 11-13**).

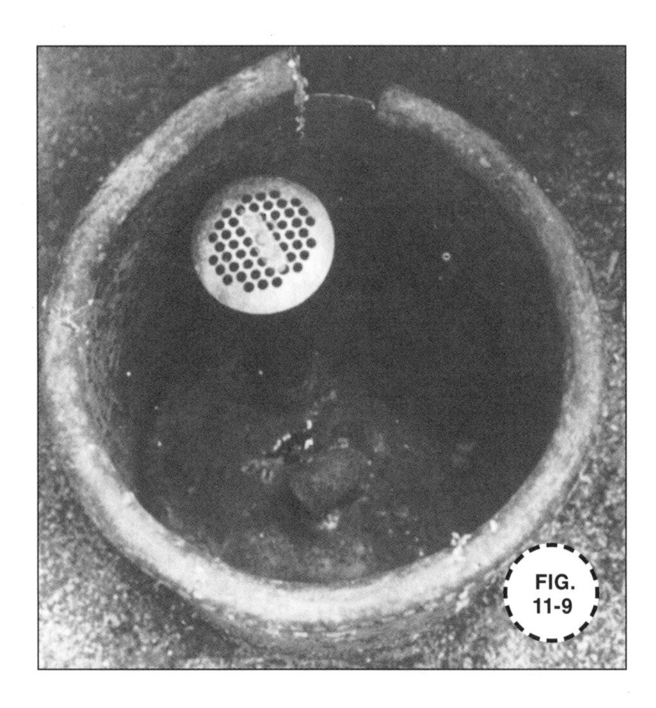

FIG. 11-9

FIG. 11-10

FIG. 11-11
SPLIT HEATER HOSE FORMS RIM GUARD
FRESH WATER
DRAIN
LID
LEATHER HINGES
FEED
FIG. 11-12

Another interesting and efficient feature of this coop is the covered group-laying box, complete with large hinged door, ramp, and small entranceway. The hinged door is kept shut and is opened only when the eggs are being collected

or when the box is being cleaned out. This keeps the laying hens in the dark, preventing them from eating their eggs or pecking at the others. All the eggs that are collected are whole and unbroken, proving that this method works well (**FIG. 11-14 AND 11-15**).

An improvement to this design would be to include an outside collection setup so the eggs can be gathered without disturbing the temperature of the coop in the cold of winter (**FIG. 11-16**).

And, of course, then there's the chicken yard. At the south wall there is a small doorway (see **FIG. 11-8**) that leads out to the enormous chicken run. The larger the run, the better it is for exercise and the greater the variety of food the birds will forage for themselves, saving you the expense of being sole provider. This southern-exposed yard is shaded by a forest of surrounding trees, keeping the direct midday sun off the area. Chickens become dull and inactive in the heat. The run is fenced with 7'-high stakes split from large cedars. They not only keep the chickens in, but deter the prowling wildcats, bears, and coyotes, which still claim their rights to this land that was once their domain.

Give chickens as much room as possible to grub up worms and scratch the dirt for other edibles, and you'll save plenty on the food bill and get the richest yellow-orange yolks you've ever seen. And throw in your weeds, garden trimmings, and compost. They'll love you for it all (**FIG. 11-17**).

The second, quite interesting chicken setup is this double yard henhouse and post-brooder quarters combination. The baby chicks are transferred to the post-brooder quarters when they can withstand the natural temperatures and do not have to be warmed by artificial heat. This length of time varies with different breeds and existing weather conditions. The chicks are housed in a small, tight, well-built dwelling that insulates them from the cold and protects them from extreme weather conditions. It is small and cozy and is adequate for them during this growing stage when they want to spend a great

FIG. 11-17

FIG. 11-18

FIG. 11-19

deal of time outside, curiously venturing around, exploring their exciting new world. This yard is a good size for them to get the exercise and exposure they need (**FIG. 11-18 AND 11-19**).

The pullet yard is adjacent to the henhouse, allowing the developing hens to become familiar with their future environment and also giving the grown hens a chance to have some preliminary contact before they share their coop with them. Just prior to the amalgamation, both doors to the yards are left open so the hens can share a

FIG. 11-21

FIG. 11-20

common foraging ground (**FIG. 11-20**). This cushions the shock of the move and acquaints them with each other. If the newcomers were just thrown in with the established residents there might be quite a battle. It's usually easier to share one's territory with those somewhat familiar rather than with a complete stranger, especially when it is a home one is sharing.

The existing hen coop is rather small but adequate for up to fifteen layers. It is a shed-roof dwelling with approximately 8 × 8 foot interior dimensions. The two large southern windows and the diamond-shaped west window allow in all the light that the coop needs and provides it with enough sources will be required. The roosts are two crossing lengths of slab on upright braces. Behind them on the west wall is an entrance to the yard. Just in front of the entrance is a small portable feeder (**FIG. 11-21**).

The nest area is divided into four partitioned laying boxes, 1' wide by about 14" deep, sufficient for about four chickens each, one at a time. The high partitions prevent the neighboring layers from looking over and pecking at a nearby egg while they are laying their own (**FIG. 11-22**).

Such a chicken setup would be fine as a temporary shelter in the warmer months, but the buildings should be more insulated, to be warm and cozy during the long northern months. If the dwellings are not properly insulated to retain the birds' own body heat and the heat of their droppings, the egg production will go down considerably and the hens will be listless and disgruntled. Keep them warm and dry and they'll return the favor by laying efficiently all winter long, greeting you with healthy activity.

FIG. 11-22

CHAPTER 12

The All-Weather Stock Waterer

After one has been sufficiently exposed to the time-consuming chores that are omnipresent on the homestead, one begins developing the ingenuity it takes to design labor-saving devices such as this handy tool. "Let them water themselves," the homesteader stated as he explained his creation. He needed a stock waterer that was maintenance free, which supplied water to his cattle throughout the year, even in winter without freezing. He designed just that, consisting mostly of old items that were salvaged instead of being discarded after their initial use. Never throw anything away if there is a remote possibility that it could ever be used again for something.

The stock waterer consists of an insulated tub with an intake and outlet system and a floating electric heater. The intake is a ¾" galvanized pipe that tees up from the existing water line to just above an old drain hole in a junk washing machine tub. The intake pipe was cemented into the center of this hole to prevent leakage. The pipe is reduced in diameter shortly after it enters the tub. An elbow is connected to the top of this nipple to accommodate a float valve.

The float valve regulates the amount of water that enters the tub. As the cow drinks, the float valve opens, allowing water in until it automatically shuts off the intake (**FIG. 12-1**). In case the float valve system fails or a large amount of rain enters the tub, the excess water can escape

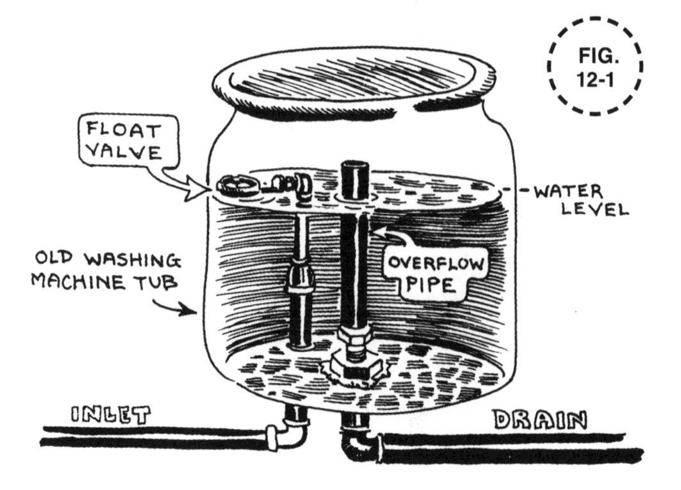

through the 1½" plastic pipe outlet tube that extends to just above the regulated water level. This outlet tube prevents the water from overflowing and possibly causing the whole system to freeze. It is fastened over a plastic fitting, which is screwed into a larger galvanized hex reducer. The galvanized reducer is brazed into the center of the tub bottom where the agitator mechanism used to come through (**FIG. 12-2**).

Fastened to the 2" male end of the reducer, on the underside of the tub bottom, is a coupling that takes the male end of a plastic pipe fitting for the drain pipe. The drain pipe extends to well beyond the waterer so the overflow won't collect around it. The tub is mounted about 12" above the ground on two two-by-fours, which are notched into the slab sides of a surrounding box. This is a convenient height for the animals. The 32 × 32 inch outer box protects the system from direct

FIG. 12-2

FIG. 12-3

contact with the animals, and shelters it from the elements. The inner system is also protected by sawdust insulation to prevent it from freezing. A plywood lid keeps the whole box covered. Only an 8½"-diameter hole is allowed in this lid for the cows to extract the water. This is sufficient for them (**FIG. 12-3 AND 12-4**).

The water inside the tub is heated with a small electric tube heater that is attached to a floater. This electric heater is submerged just below the water level, heating it up to the temperature cows like most, between 60° and 70°. It maintains this water temperature, helping keep the whole system from freezing, even in below-zero weather.

The cows love the waterer. They go to it and just slurp until they get their fill, anytime they want, instead of having to wait for someone to haul it. The average milking cow requires at least 15 gallons a day—that's a lot of water to be hauling for them in the middle of winter when most outdoor water systems are frozen solid (if they haven't been drained).

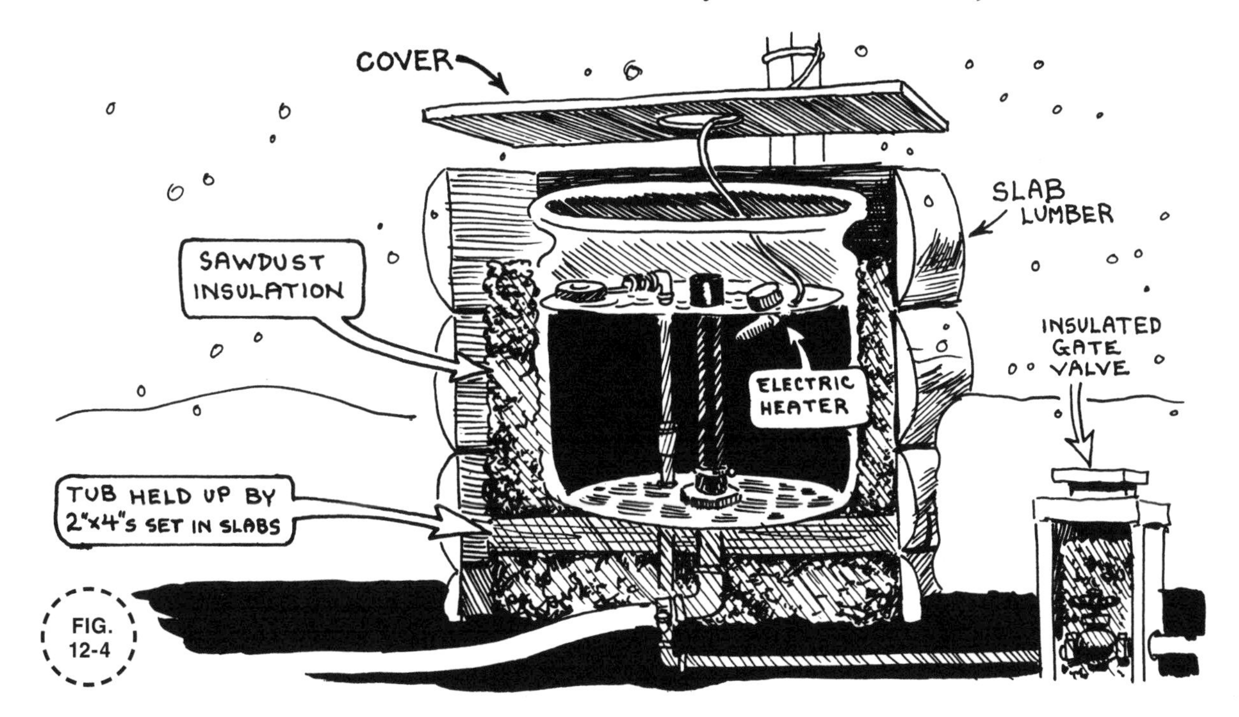

FIG. 12-4

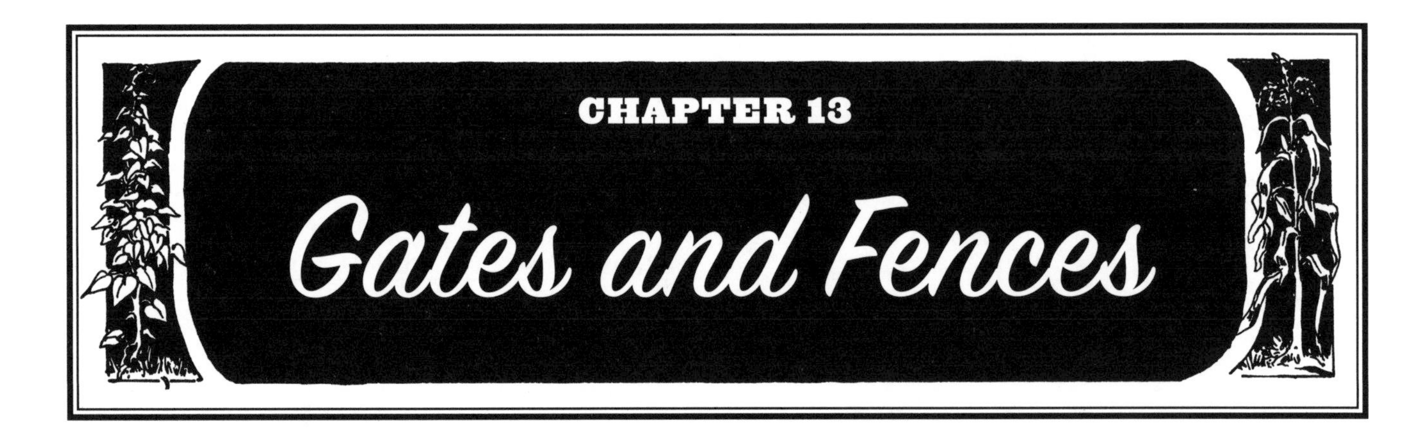

CHAPTER 13

Gates and Fences

Gates and fences, like shelters, dwellings, and other homestead buildings, can exemplify the uniqueness and creativeness of the builder. There are no rigid design rules, except to become familiar with "the nature" of the animals you are enclosing or keeping out. Wisdom of their individual capabilities, habits, and personalities will prevent hostile actions, which usually result after an offender has butted, gnawed through, banged against and finally collapsed or otherwise negated the exhausting efforts of the designer.

Here are a few gates and fences that are outstanding examples of how detailed some builders are in creating an efficient and aesthetically pleasing environment, expressing their sensitivity to the surrounding elements. Working with care and beauty inspires one to remain careful and beautiful.

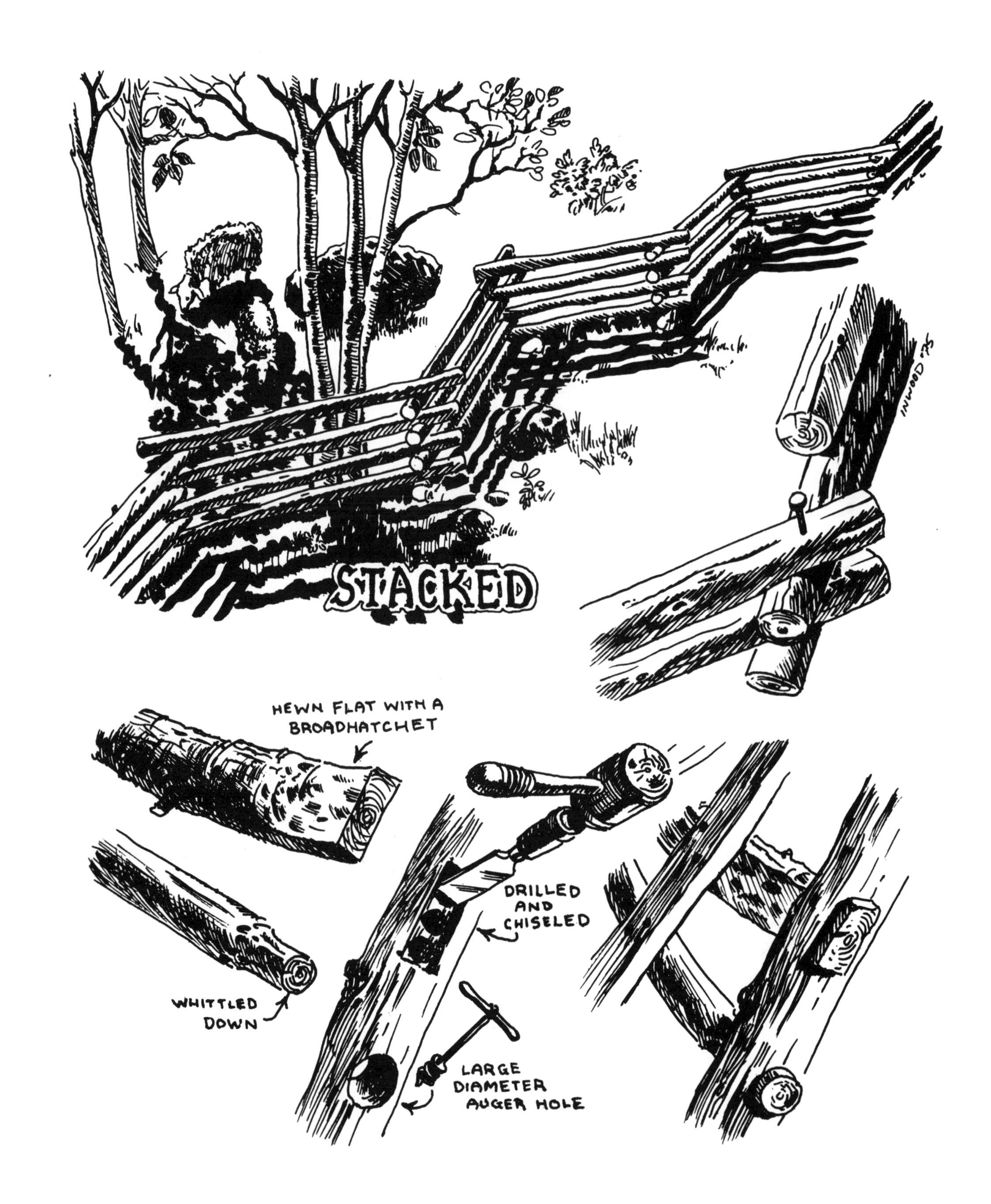
STACKED
INWOOD '95
HEWN FLAT WITH A BROADHATCHET
WHITTLED DOWN
DRILLED AND CHISELED
LARGE DIAMETER AUGER HOLE

HORSE-PROOF FENCE OPENING

EASY CHAIR

FROM AN UNUSUAL TREE

NAILESS FENCING

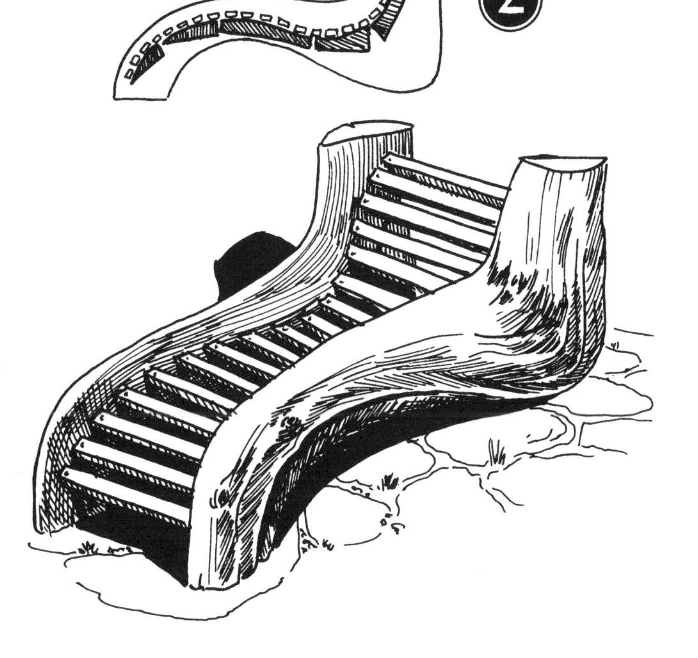

In the spring of 1969, Shelley and I left our native California in search of land. It didn't really matter where we found it, as long as it satisfied some of the basic criteria we had set for "our place." Being somewhat private people, we knew that the land must be relatively secluded—we had come from ten acres of country property in California that housed five families, and found this type of situation most unsatisfactory for who we are. The land we were looking for had to have water, plenty of trees, a low price per acre, and be situated in an area where there was a small community of people we could relate to. We need our own territory, as well as social contact. We felt lucky to find a most beautiful valley, some like-minded neighbors, and sixty acres of land we could afford.

Our land begins about three hundred feet from the highway and then climbs gradually up the mountain through a series of benches. It had been logged twenty years ago and had been able to reproduce a thick second growth of fir, pine, and cedar. Logging roads met at several points on the land, which made immediate travel by car quite easy. We decided to build in one of the logging turnarounds, on a very small clearing—at best, a quarter of an acre. We built on an upper bench of the land, in a place that would be inaccessible by car in the winter. The spot was quiet, a bit remote, and had a nearby spring from which we could get water.

We soon set up camp in the clearing and began making lists of priority jobs. A road from the main highway to our bottom property line had to be put in. It just happened that the road was designed to cut through a group of large cedar, fir, and pine trees. As we began the job of felling trees, limbing, and peeling them, our original plans of building an A-frame house gave way to the obvious and overwhelming urge to utilize these trees in a log cabin. We were not only making a road through the bush, but also felling the trees for our home.

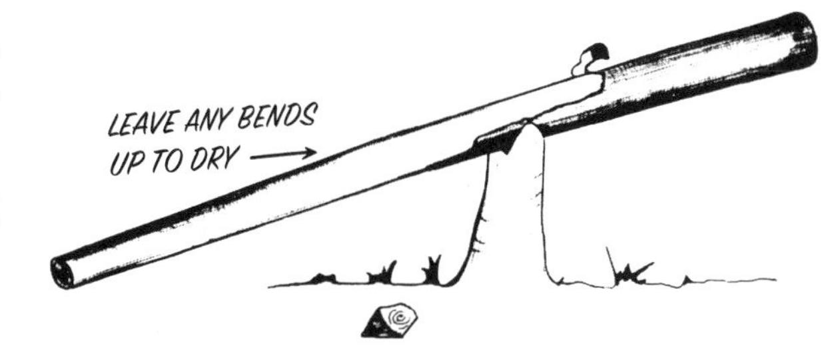

Propping the fallen logs

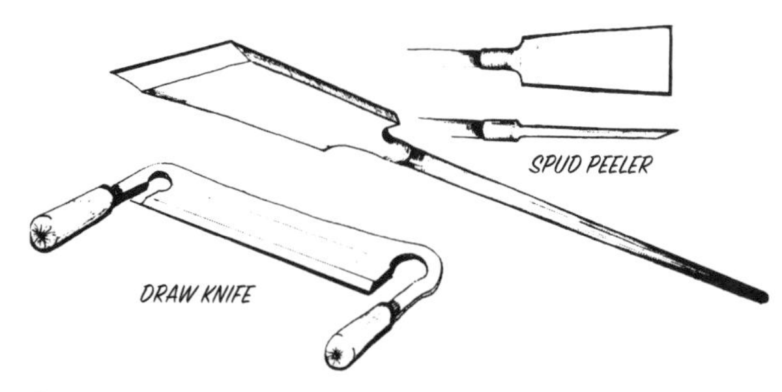

Peeling tools

I can't count the number of times Shelley and I sat around the campfire at night, surrounded by a forest we were just getting to know and feel comfortable in, wondering if just the two of us could possibly assemble our home, expand our clearing, bring new life to this area, and claim a portion of territory. Our talk was exciting and exhilarating, our bodies dirty and tired.

Each morning we would go out and choose the trees to cut. I would fell them with my chain saw, then cut a "V" into the top of each of their stumps. This way, I could prop the fallen logs up on something, instead of letting them lie on the snow-covered ground until I got around to skidding them out.

I peeled them in that position with either a drawknife or a peeling spud. This method worked well and enabled me to do that tedious task without having to bend down to the ground; I just lifted up whichever end I was working on until I was done peeling it. When I finished that job, I left the logs to dry in that position until I needed them.

The logs which had serious bows or bends were set up on skids with their bows up so that the weight of a drying log helped straighten it.

Note: Logs should always be peeled no later than a couple of weeks after they are cut. This will ensure clean, easy peeling, because the sap under the bark will still be wet.

We needed approximately forty good-size logs, with an average diameter of 12", for the walls. We also cut twelve pine poles, 20' long with 7" to 8" butts, for floor joists, and 30 small white-pine poles, 20' long with 6" butts, for roof rafters.

Another good reason for raising the peeled logs off the ground was to speed up the drying process and avoid their becoming mildewed from excess moisture. We allowed them to dry on the skids and stumps for four months. That summer we searched around for a good house site while we were waiting for our materials to age properly. We thought about the sun and wind and their relationships to the house. We wanted the living room to face west so it would get the afternoon sun and we wanted the bedroom to be exposed to the morning sun. Because we would be surrounded by windbreaking trees no matter where we picked our site, we knew wind exposure would not be a problem. We chose to have our house on a bedrock mound at the high end of the clearing. This spot was ideal because bedrock does not shift with the frost. In fact, it does not move at all. Nothing can grow on bedrock, so we did not have to waste good, usable land for our site.

To locate where to put our foundation piers, we just drew a 20' line and eyeballed a right angle of 24', then staked both ends and the corner. We repeated this for the other corners and checked to make sure they were square by first measuring a line 8' long on one side, and another line 6' long on its perpendicular. If the diagonal of these points turned out to be 10', the corner would be a true right angle.

After the logs were sufficiently dried, we hired a neighbor with a tractor to pull the large wall logs to our site. Logs for the 24' sides were placed in one pile and logs for the 20' sides were placed in another.

We were able to handle the floor joists and the roof rafters with our own pickup truck. With most of the materials at the building site, we started the actual construction. There it was, summer already, the hottest time of the year, and we did not even have a foundation set up. We really wondered if we were going to have a roof over our heads in time for winter.

The first step, of course, was the foundation. We used cedar piers, dug 3' into bedrock. We painted them with creosote, then coated the part of the pier to be buried with tar, and wrapped the tarred area in strips of black plastic. The treated piers were set in the foundation holes and packed in the sand. Thirteen of them were spaced throughout the house site.*

Once these piers were in place, we chose three of our largest logs for the sills. They extended 10' in front of the house as supports for the front porch. They were notched with chain saw and chisel, then were set lengthwise on the foundation piers, spaced 8' on center apart from each other.

*EDITOR'S NOTE: This is not an advisable practice. Concrete should be used below ground, as even-treated wood will eventually rot.

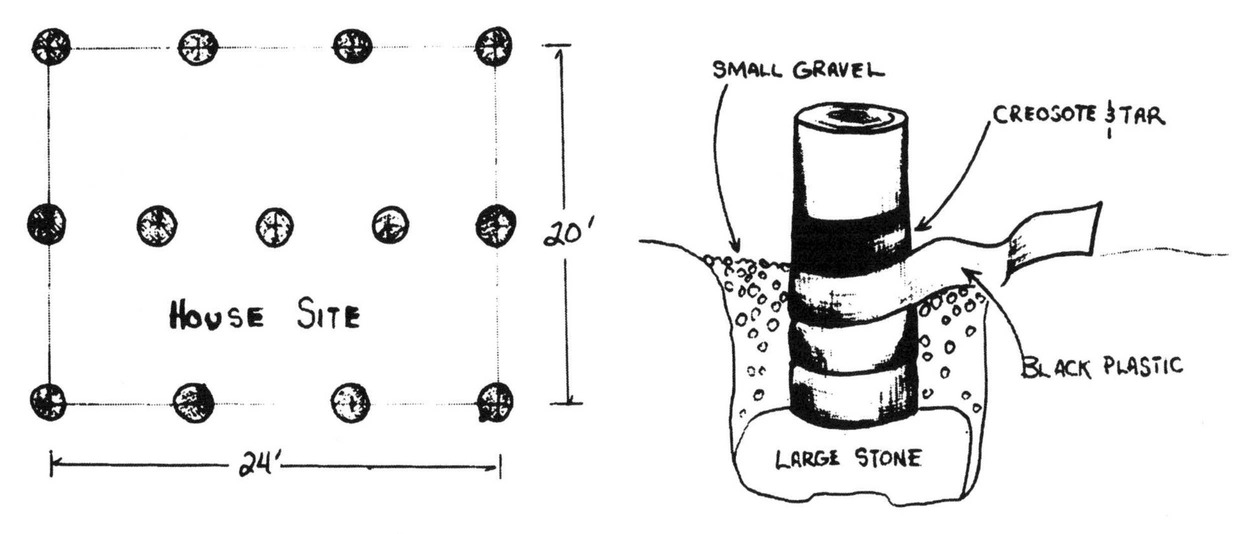

FOUNDATION PIERS

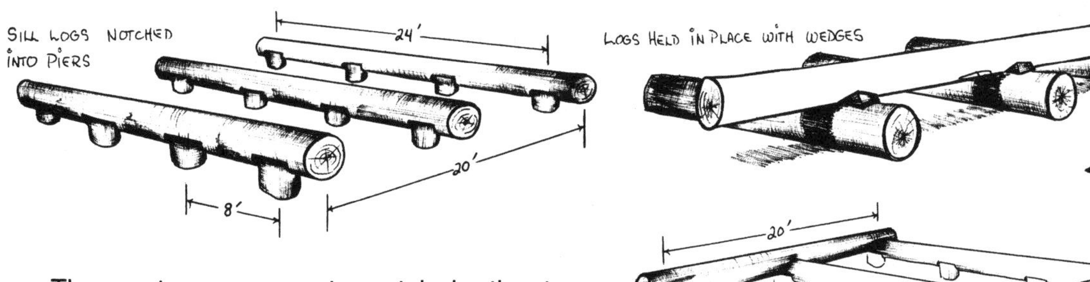

The next move was to notch in the two cross-logs at either end of the 24' sills; thus the walls began to rise. We used simple saddle notches on our wall logs because we wanted the log ends to extend past the corners. They are easy notches to make once you learn some of the tricks to them.

The first cross-log was rolled into place above the sill log it would be notched to. It was kept from rolling too far with small pieces of wood acting as wedges. We then determined how deeply this log was to be notched. To measure the notches, divide the diameter of the end of the log being notched in half, and mark that distance on each side of the log exactly where you want the high point of the notch to be. If a log is 14" at the butt and 12" at the tip, the notches will be 7" and 6" deep, respectively. Then, with a T square, mark the width of the bottom log on the side of the log that is to be notched. Remember to always mark both sides of the log, and make sure that each tier contains logs as close to the same size as possible, the larger ones at the bottom of the wall and the smaller ones at the top. This ensures that each staggered tier will interlock correctly to the one below it.

After you have marked the widths of the bottom log on the log you are working on, take a scribe, or whatever you can find to estimate with, and mark the contour of your bottom log's end on the upper log. This can be done by eye or with a measuring tape.

When the logs have been marked, roll them over so the area to be cut faces up. Hold the logs in place again with wedges. Then make vertical cuts, about an inch apart, with the chain saw. Cut each as deep as the lines scribed on both sides of the log end, and, with a 1" chisel and an applewood mallet, knock the pieces out.

When the notch was completed on both ends, I rolled the log in place to be fitted. If the notch needed to be larger, I rolled the log back and chiseled more out of it. This was usually necessary. Sometimes a log was rolled back and forth several times before I made a proper fit.

After the first wall logs were notched over the sills, we put in the floor joists. The sill logs were notched with a chain saw and chisel, and the floor joists were set into the sills at intervals of 2' on center. They were then spiked to each sill. Any humps or high points in the joists were made level with a foot adze. The joist tops were not hewed because we thought it unnecessary to go through all that work just to have a flat surface. A round surface was just as good, as long as it was reasonably level. We had to make sure that we butted the ends of the floorboards exactly at the center of the crossing joists to keep everything level.

We used two-by-six fir boards for the subfloor. We found that laying an early subfloor gave us a nice surface to work on when we set in the wall

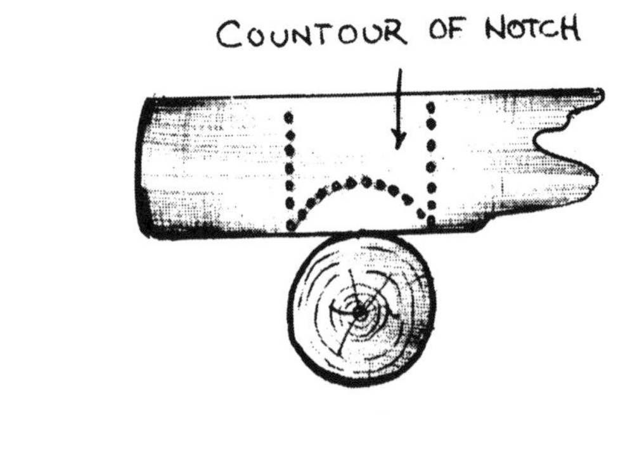

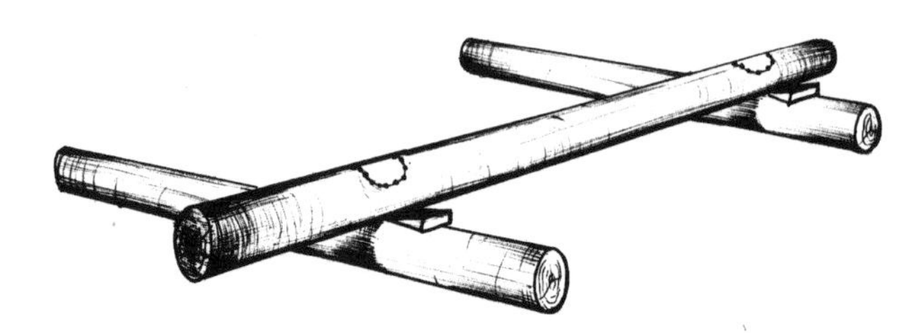

Marked logs ready for cutting

logs. It was better than hopping from one joist to another.

With the subfloor laid, we concentrated on the walls. Each wall log was saddle-notched in place as described and fitted to the log beneath it. The logs rarely spanned from end

Chiseling out marked notch

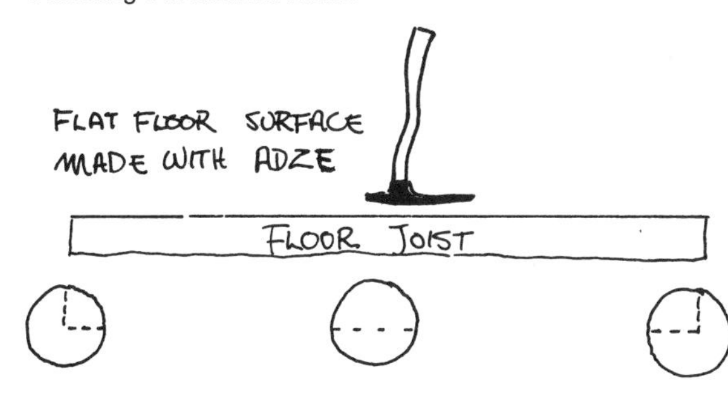

Sill logs notched for floor joists

to end without having some high points that prevented the two notches from resting snugly on their cross-logs. To make them fit correctly, we placed wedges in any large gaps and ran a chain saw between the horizontal logs. The wedges prevented the top log from settling as the cut was being made. If either of the logs still had high points, we repeated this action until we achieved a tight fit.

We then rolled the top log back and stapled a 2" strip of fiberglass above the log beneath it. The log was rolled into position one more time and was spiked in place at the ends.

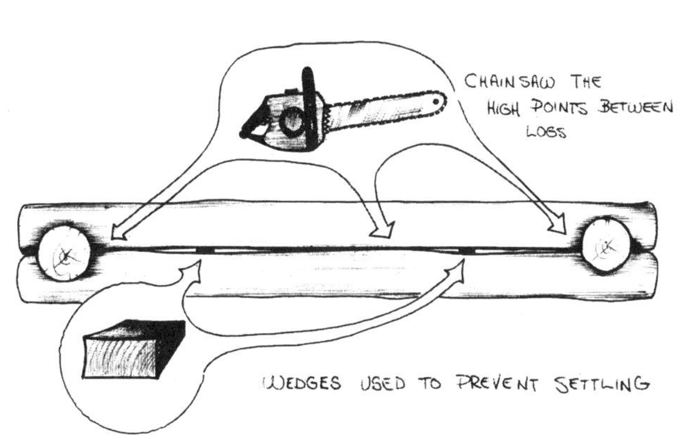

Round saddle-notch joints (1)

Round saddle-notch joints (2)

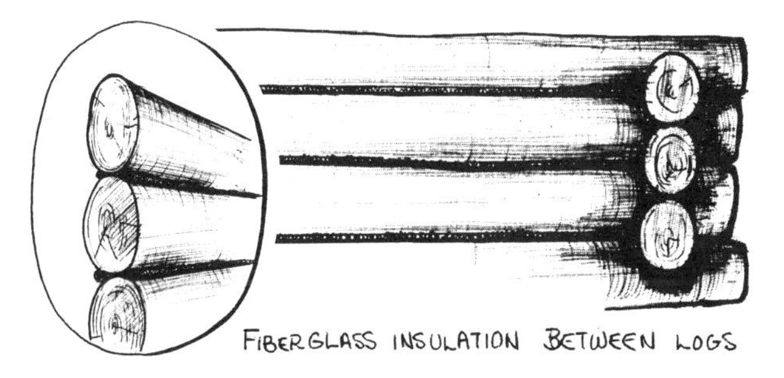

Shelley and I could lay eight logs on a good working day when it was not too hot to work. It sure is a good feeling to see your house grow two-logs-per-side higher by evening. It gives you something to celebrate. As our walls grew, we devised a system of using two poles and a rope to roll the logs up from the ground. The poles were used as a ramp, and the ropes, one on each end of the log, were tied to the inside of the building and were run up over the walls. The ropes extended from the ramps, encircled the new log, and again returned to the inside of the building. We then rolled the new log up the ramp.

Friends sometimes stopped by just in time to help us with the heaviest logs—or, should I say, we worked on other projects instead of hassling with the "heavies," until friends came by to visit. We raised the walls to a height of 9', then cut out a door space in the front wall.

The top side logs and the center beam extend 10' beyond the front of the building to support a front porch on the upper level added later. This area, since it faces south and gets the full arc of the daily sunlight, would be the site of our small greenhouse. Along with full sunlight, it will also get the heat rising from the house to keep it warm during the winter months.

We set in a vertical center pole and notched it to the center beam above. We then laid 20'-long joists across the front half of the roof to support the loft area. We left the rear 12 × 20 foot section open so it would have a high ceiling and give us a feeling of space.

When we were in Oregon a couple of years ago, we saw an old log barn with a very high, peaked roof. That building stuck in my mind because it was so unlike any others I had seen before. I wanted our house to have a roof with a very high peak. I experimented with various-

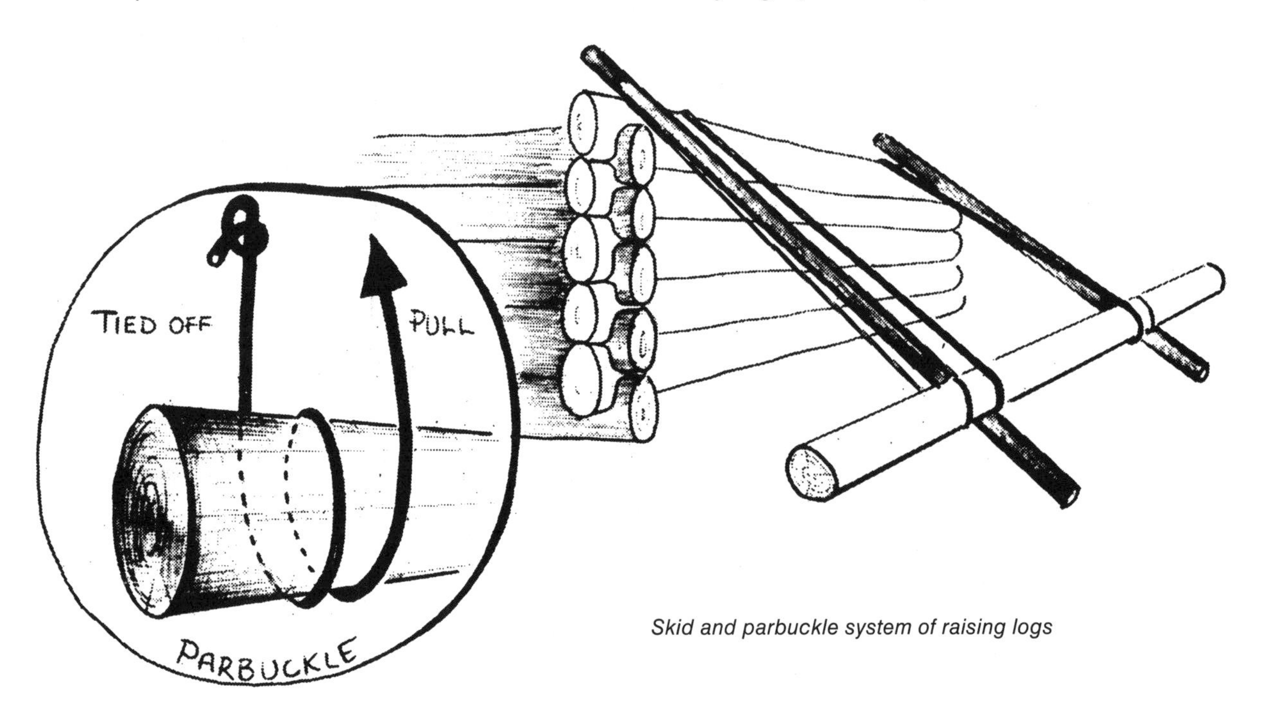

Skid and parbuckle system of raising logs

size upright poles. The higher the uprights, the more severe the roof pitch. We put up three 16' uprights, one at either end of the center beam and one above the center pole. They were temporarily held up by two-by-six braces. I envisioned the roof rafters meeting a ridgepole at the top of those uprights. The peak was incredibly high. I had to whack 2' off the tops of those poles. The peak still seemed too high. I ended up cutting two more feet off the tops, making the uprights 12' in length. A ridgepole was notched in across the tops of the uprights, then the rafters were set in place.

The 22 rafters were spaced at 2' on center. Each spanned from above the ridgepole to 18" beyond the walls, giving the house long eaves to protect the walls from direct precipitation. Afterward, the top side logs on each side were notched at right angles where the rafters met them. Each rafter was notched the same way to set into the top logs. They were then spiked into place at those joints.

The tops of the rafters extended enough past the ridgepole to cross with the rafter coming from the opposite side wall. Though they were all cut the same size, 19', some were longer than others when they met the ridgepole. We sawed off the tops of each rafter at an angle so they could face each other where they rested. They were then spiked in place.

The next job was to nail on the double roof. We used one-by-eight fir boards for the inner roof. They were nailed across the rafters and covered with tar paper. Then two-by-four spacers were placed 2' apart above the fir boards. Then came 2½" of fiberglass insulation, which fit between the two-by-fours. The roof was finished with sheets of thin, galvanized steel that were nailed to the spacers. We chose this material because it sheds snow well, which means we

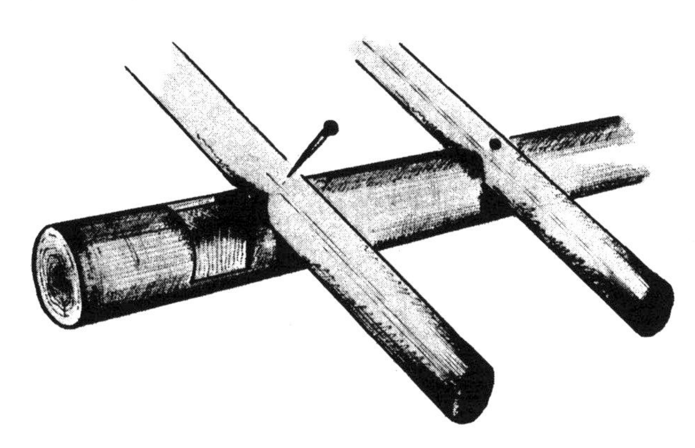

Notching rafters into top wall-log

Positioning and spiking the rafter poles

High-pitched roof

do not have to climb up on this roof every winter to shovel it off. Galvanized sheeting also reflects heat, keeping the house cooler in summer and warmer in winter. I must admit, it's not the most natural roofing we could have had, but it sure does its job well.

The roof has so high a pitch that we could suspend a clothesline under it, and the clothes hanging from the line would be far enough above the main level that they would not be in anyone's way.

The main floor is finished with one-by-six hemlock. It radiates out in four triangles from the center pole and spreads to the ends of the walls. After the floor was laid, we built a staircase to the loft area so we could put in the flooring up there.

The loft area is approximately 10 × 12 feet. It has a double floor of fir one-by-four boards. Our bedroom area is located here. At the east wall of this area, we put in a 5'-wide dormer with a peak that extends 7' out past the roof line, directly over the log wall. The morning sun comes in from this small study area to get us up bright and early.

The gable end, south of the loft, has two windows and a door, which allow the midday sun to brighten this area. The window that opens was placed directly across from another opening window at the north wall for good cross-ventilation. They release the warm rising air and aid circulation.

The door to the porch terrace is a "Junky John" special. "Junky John" is the local dealer of any

Log staircase

Floorboard pattern

Outside view of bedroom dormer

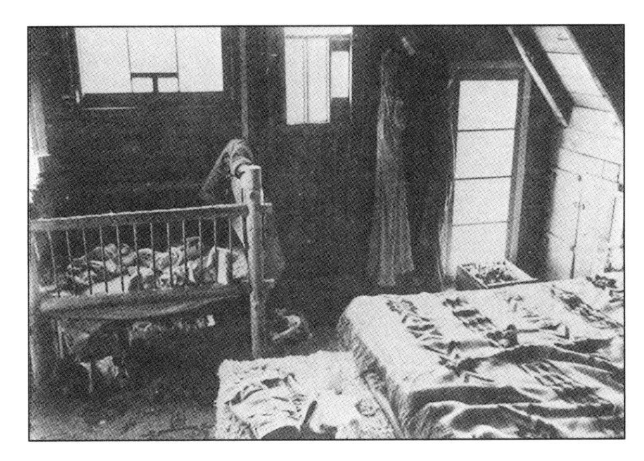

Bedroom area

South wall exterior

used item from toilet bowls to electrical supplies. Sometimes, if the offer is right, he even sells you things from his own house as his wife stands in the background cursing in Russian.

The end gables were paneled inside and out with one-by-eight cedar boards, which were nailed to two-by-four frames. They were insulated with fiberglass.

One of the main advantages of using full-length logs instead of pieces for the walls is that you do not have to figure out where you want the windows until you have lived in the house awhile. They also make the walls more structurally sound and more aesthetically pleasing. We definitely wanted several large windows in the west wall so we could get full afternoon sunlight and so we could see the sun set over the mountains. The kitchen and living areas face that wall. From the large windows, we can watch everything grow and fade with the changing seasons in our landscaped clearing and our rock-lined garden.

Once most of the structural work had long been finished and we were living comfortably in a well-insulated, cozy house, we had more time to enjoy and learn about our surroundings. Already, in some of the places where I cut the logs and thinned out the evergreens, huckleberry and elderberry bushes were sprouting up abundantly. Every once in a while I took time from my daily chores to thin out more areas so other small plants could get the needed sunlight and could feed from the limited nutrients in the soil. I used a lot of the wood I took out for making furniture and other useful objects.

I made several pieces of furniture that fit in with the rustic style of our house. A couple of my favorite pieces were made for the newest member of our family, who has inspired a great

The way our land looked when we first saw it after it had been logged 20 years earlier—and our house today.

deal of creativity (and other reactions, as well!) in both Shelley and me. Using nature's own bending and spiraling designs, I made our daughter a high chair out of vine maple and pine. I also made a crib with trimmed rail ends that interlocked with the augered holes of the upright posts.

I am learning much about working with wood as a hobby, as a necessity, and someday, I hope, as a source of income. Wood is very satisfying to work with and will always be plentiful if people will learn to be selective about their cutting. If a forest area is thinned out instead of wiped out, removing many crowded trees but leaving one tree standing every 10' or so, the growth of that area will actually be promoted. The trees remaining will be able to grow straight and tall, and will receive all the sunlight and nourishment they need. We will continue to preserve our woodlot and make it a more beautiful area by thinning it out in this manner. We realize how important wood is for our future survival and for yours.

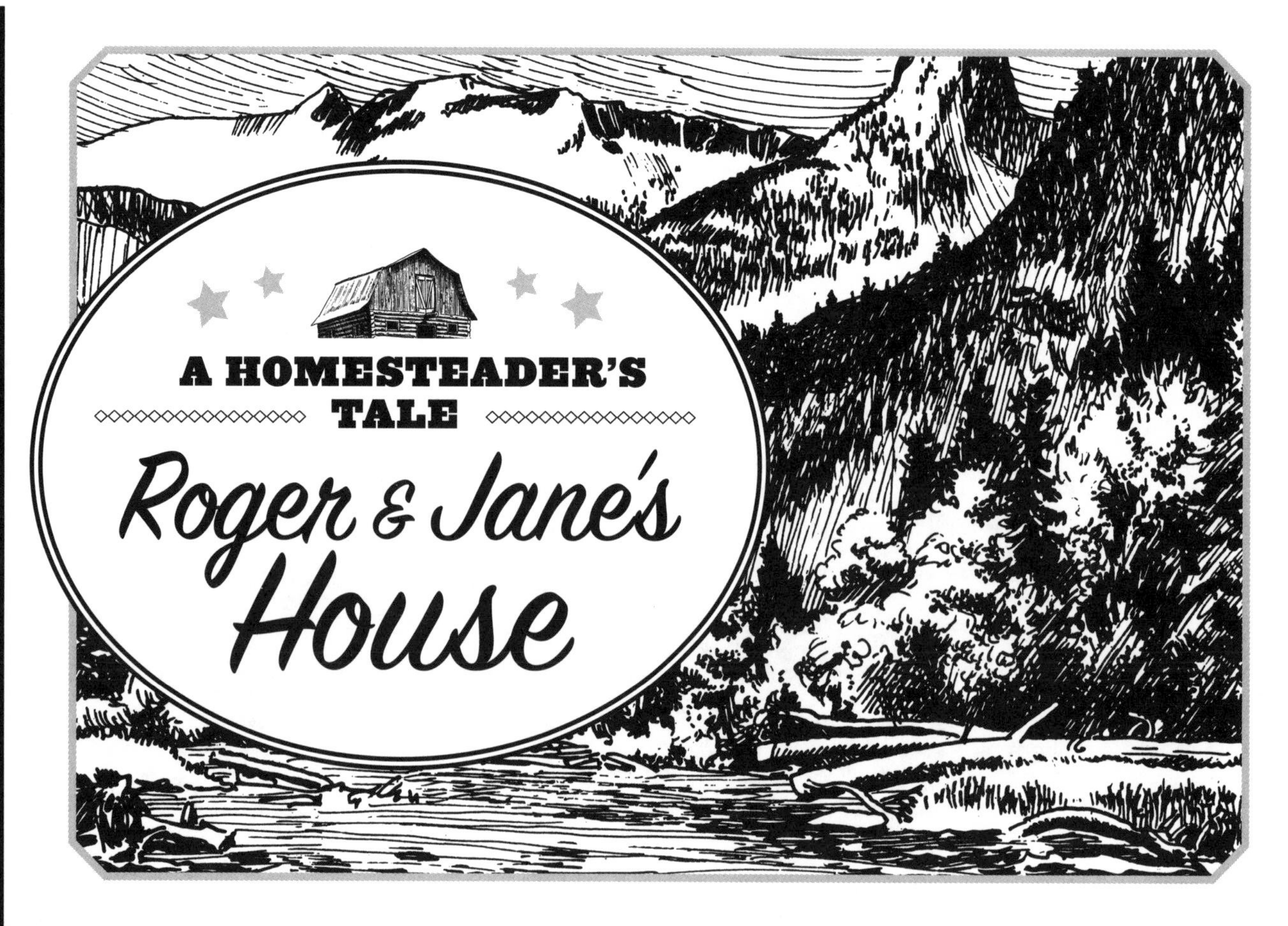

Since my childhood, I've had a longing in my heart to go back to nature and, once there, to become self-reliant. Like many people, for whatever primitive motive, I wanted to step back into a wooded wilderness with nothing more than a trusty axe (and, in these modern times, a chain saw), a rifle, and just a few other belongings. I wanted to live off the land, build myself a little log cabin, and become one with the natural order of things. But because of immediate necessities and a learned societal craving for financial security, I had to compromise these yearnings and enjoy a creative outlet within the environment I was familiar with—the environment from which I was receiving my "daily bread." This creative outlet was painting.

I spent several years in colleges and universities, and, at one time, had a strong ambition to become a university art instructor. I was among those many qualified potential university instructors who received advanced degrees in art but who were not successful in finding suitable teaching positions because of a trend against the support of higher-learning institutions.

So I was left without a job. Because of the order of things, I could begin thinking about those things I had always wanted to do. In effect, I was forced to find alternative goals and to do the two things I wanted to do most: start my own art school in order to share some important ideas I had about art, and go out into the woods and hack out an existence for myself and my family. I founded an arts-and-crafts school and taught at it for 22 months. I realized that the local community was too small to support such an endeavor, so I gave it up and devoted my energy to developing our little homestead. In the process, I have discovered the importance of working together with my family toward a common goal.

The first few months that we lived on the land seemed like an extended vacation from our former rat race. We found solace in nature and in hard physical work. We cleared areas and set up a temporary camp. We spent very little money and kept ourselves busy doing what needed to be done. We made rough plans for a woodshed, an outhouse, and corrals. The building site that we picked for our house was relatively level. We agreed not to employ any heavy equipment on the land, and we even parked our car near the road to keep from disturbing the natural setting.

Though I had previously worked as a laborer on a couple of ranches and did log work while employed by the U.S. Forest Service, we were somewhat inexperienced and idealistic in our approaches to the new land. I had once made a giant monkey-bar set out of logs, but I had never built anything like a house. Still, I had a pretty good idea of where to begin, from having read many books that taught some basic techniques and having a memory full of building ideas that went all the way back to my childhood. I wanted to build a house of my own design and imagination with only a few tools and without any conventional materials except for nails and wood. I knew that with practice, I could eventually develop my own building techniques and be satisfied with my achievements.

It didn't take very long for my plans to begin materializing. I remembered the way the Forest Service used to peel logs for the quick structures they built. They would cut the log to size and, with a sharp axe, cut off thin strips of bark on three sides of the log. This allowed the moisture to escape from the peeled areas and also controlled checking. The remaining three strips of bark would easily peel off after the logs dried out and shrank, which would take anywhere from one to two years. Then, when the seasoned logs got a good soaking from a heavy rain, the bark could easily be removed. This method also produced a nice color effect. As the peeled areas aged, they turned grey. The protected areas stayed a golden to rust color until the bark loosened. The log, when fully peeled, looks wonderfully rustic.

I found the chain saw to be my most valuable tool. As I knew from previous experience, nothing could surpass its efficiency and speed in cutting and bucking firewood or felling and limbing trees. While it is a machine and, like most machines, is sometimes temperamental, with experience you begin to learn what it can do for you. It also costs a bit to maintain, but every time I rip through logs to make my own boards and cut poles to the size I want, I realize that I could not do these jobs by myself any faster or cheaper with any other tool. However, just as a matter of personal taste, I use the axe for limbing

Peeling logs

and notching. It may take longer, but there is a certain satisfaction in using an axe and it doesn't cost anything in gasoline and oil.

Since the construction job was primarily my own, and because we could not afford hired help, I had to design a method of building our house in which I could handle all the materials by myself. Though my wife was on the project with me, she was busy with housekeeping chores, gardening, animal maintenance, and civic activities. I didn't want to be calling on her every few minutes to help me fit something in place or assist me with other awkward jobs. The building had to be a one-man effort with lightweight and easy-to-handle materials.

The site we chose for the house was on a slight slope overlooking our garden and the pasture in which our two-year-old colt was tethered. I compensated for the slope with fir pillars of varying lengths, which were set on stone foundations. Fir was chosen because of its strength, weather resistance, and ability to hold nails.

The pillars were very short at the top of the slope and became gradually longer toward the bottom. Each pillar was peeled on three sides, and each rested on a pile of rocks, which went roughly 2½' down into the porous ground. For added drainage and strength, I placed larger foundation rocks in the bottom of the hole, and smaller ones going toward the top.

The rear kitchen and dining room area were built into the slope on the same level. As the terrain dropped away, storage space was provided for under the front unit.

The many-sided rear section gave us the room we needed for cooking, dishwashing, and dining. It lacked the monotony that a traditional four-sided room can have. I originally planned

Stone foundation posts

Crawl space storage opening

to have seven floor joists of various lengths fanning out from a 9' beam in front, to a 19' beam at the far rear, going into the slope. The distance between the two beams was 12'. These beams were notched over the five vertical posts beneath and they were each hewn level. Even at that point, I could tell that the space wasn't going to be adequate, so I added two 10½' beams, which were almost at right angles with the 19' beam. Another floor joist was added to either side to secure the beams and lengthen the floor support. On the east side of the area, an 8' beam was notched to the front of the 10½' beam and was connected to the front beam. On the west side, I added two other beams to include space for a doorway and a stairwell. I then added another 2' area into the slope behind the 19' beam for seating space, storage, cupboards, and counters.

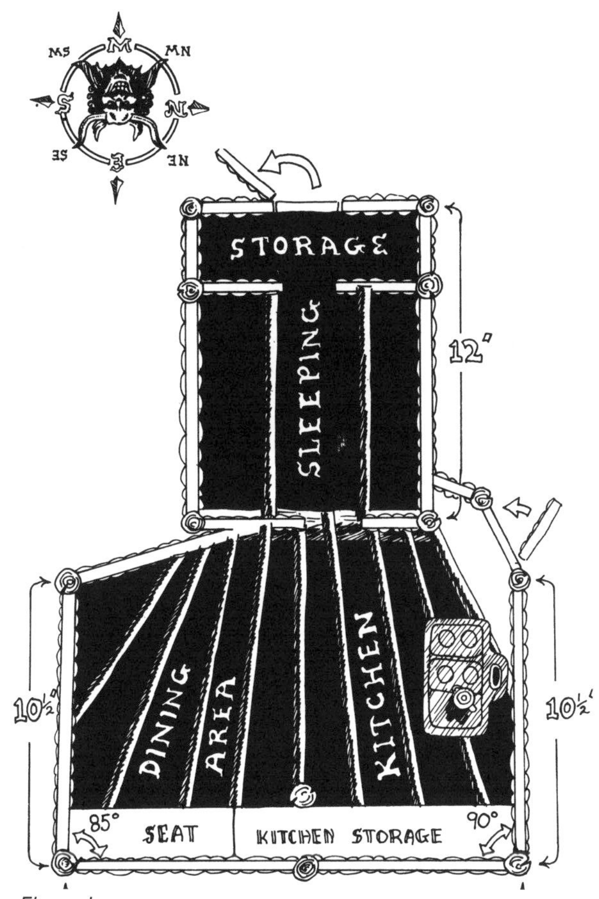

Floor plan

The 9 x 12 foot front section was a bit simpler. It was raised on six upright supports. The two uprights in the rear also support the 9' beam in the kitchen area. In front of these are two other uprights, forming a square. Two more are in front of the others. They border the 3 × 9 foot closet area. This section was framed with log floor beams, and three joists were added for more support—two running lengthwise on either side of the doorways in the 9 × 9 foot section and one going across to frame that section.

With the pillars and joists in place, it was time to work on the subfloor. I went around to several of the local mills and priced some low-grade two-by-tens and two-by-twleves, but the cost was beyond my means everywhere I went. So I decided to make my own boards.*

First, I felled a tree. I left it where it was while limbing off the branches from the butt up to the length I wanted the boards to be. The tree was dropped across a stump or blocks, so that the butt of the tree stuck out parallel to the ground. The lower branches were then trimmed off, and the upper branches were left intact to prevent the tree from rolling while I chainsawed. I stood along the side of the tree, started my chain saw, and began cutting off one of the outer slabs. I used a downward motion, as used for bucking cordwood. I didn't use any device for measuring the thickness of the cuts—I just eyeballed them, and before long the boards were coming out reasonably straight.

*EDITOR'S NOTE: We advise against making your own boards.

After each board was made, it was detached from the log. You wouldn't believe how bad the first bunch of the boards I made looked. They were so warped and their widths varied so much from end to end that I really didn't think they could be used, not even for the subfloor. By doing some hewing, though, I managed to make them usable.

With practice, I developed the technique of ripping my own boards. I learned to keep my chain saw very sharp and to file the rakers of the chain lower than usual. That enabled each tooth to take bigger bites out of the wood. If the rakers are very low, though, the chain grabs the wood too much instead of sawing through it.

Most of my boards were about 1¼" to 1½" thick. They were of white pine and cedar because these are softer woods, which are easier to cut through. White pines in our area don't have much taper and are very straight, so long boards can be cut from them. They would produce boards which were the same width from end to end if my cuts were straight.

Since we live in an area that has cold winters, we wanted to make sure we had a well-insulated floor. We covered the subfloor with a vapor barrier layer of black building paper and spaced 1½" pine saplings every 16" across the paper. Then, 1¼" boards were placed over the pine saplings as a temporary finish floor. I later spaced more pine saplings over it, and added a final board floor made of fiber. Needless to say, with all these layers, we have no problem with cold or dampness coming through the floor.

When I was a young boy growing up in Idaho and Montana, I saw many old lambing and pioneer

Making lumber with chain saw.

Ripping first board

Trimming off outer slab wood

Cutting boards off log

Built-up floor

sheds. They had post-and-beam walls, which were sheathed or paneled with rough lumber or vertical slabs. They sure seemed to have been around for a long time. I liked that method of building, so I incorporated it into my house, using half-rounds for siding. I further modified this style to give the house proper insulation by adding a few more layers of boards. This added strength and helped to keep the heat in.

I set up and nailed in five posts on either side of the front area and a post on either side of the doorways of the 9' sides. Beams were notched and spiked across the tops of these poles to brace them and tie them together. Diagonal poles were then added for more support.

The sections without windows had diagonals spanning from the bottom of one corner to the top of the opposite corner. The sections with large windows didn't have any diagonals. The headers and sills of those windows provided horizontal support because they were notched solid into the uprights. The sections in the far front had smaller windows, which are also framed into the uprights. There were additional braces below the bottom sills of those windows to make sure the walls wouldn't go anywhere.

Style of pole framing

The same basic plan was followed for the kitchen/dining-area walls. What differed was the placement of the diagonal bracing. In sections without windows, instead of the braces going from the bottom corner to the opposite top corner, they only went from the bottom corner to the middle of the opposite pole. Though this method was sufficient and used less material, the method used in the front section was structurally far superior. It could support more tension without giving way because the diagonals were wedged into opposite corners and were braced by them. The larger windows in the rear section have V-braces under them. These consist of two poles that meet in the center of the floor beam and span to opposite posts under the bottom windowsill, providing strong support.

I was next faced with the problem of how to make the walls thick enough so I would not need store-bought insulation and yet would still give them a finished

Built-up wall diagram

look. I came up with the idea of using peeled half-rounds. I ripped 6" to 10" logs, peeled them, and spaced them upright against the outside of the post-and-beam frame, then toenailed them into place. Small quarter-rounds were then ripped and placed between the half-rounds to fill in the cracks.

Many 5"- to 6"-diameter cedar saplings were then ripped in half for an inner layer of sheathing. These half-rounds were nailed side by side, and quarter-rounds were later added between them to seal off the walls. This inner wall decorates and adds to the insulative qualities.

Each season the quarter-rounds can be wedged deeper into the spaces between the half-rounds, as both layers dry with age. The tighter they are driven in, the greater the wall's ability to seal the dead-air spaces between the slabs. The insulative quality of the wall is excellent when it is properly sealed, and this method is not costly. To finish off most of the walls, many more poles were ripped and their half-rounds were set upright side by side along the inside of the post-and-beam frames. They were then nailed to those frames.

I made sure to cut the ceiling joists 6' longer than the end of the walls to allow for 3' of eave overhang on each side. These joists were then notched into the upper log of the double top plate. They act as floor supports for the loft and prevent the walls and roof from spreading. The longer the eaves, the more the walls are protected from the weather. But to have long eaves, you must make sure to brace them properly because there is a great amount of snow and ice that builds up on this area. I not only extended the ceiling joists to the end of the eaves, I also notched in diagonal braces from the upright posts to the ends of the joists, for additional support.

Long eaves are a disadvantage when the house they are on is surrounded by a forest. In colder seasons, the long eaves will tend to prevent the warm sunlight from shining on the walls and entering the windows. In hot weather, though, this is an advantage.

The end ceiling joists also acted as end logs to begin the loft area. The loft was the small area above the piano and the children's room, which we used for our sleeping quarters. It was 5' high at the peak and four logs and one rafter high on either end. The loft gables were made up of three horizontal logs, which were round-notched to one another, and several smaller stockade rounds, which followed the roof's slope. At each end, there was a window for light and ventilation. These windows were designed to be large enough for us to crawl through in case of fire.

To make the roof's slope, I simply extended two poles over the gable ends and met them at the peak in the center. This gave the 5' peak. We

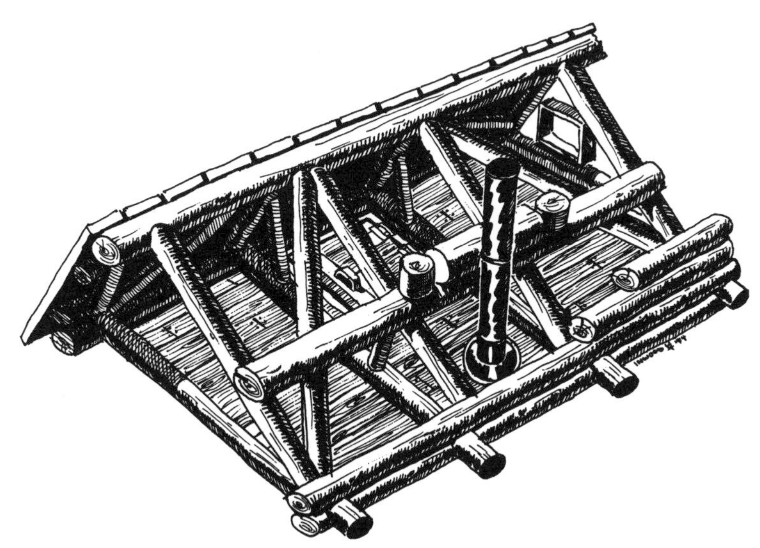
Roof frame

End view showing supports over extension

didn't need a higher space because, after all, we were just going to sleep in this area. To raise the peak would have given us more space to heat and a greater chance of heat loss. We wanted the ends of the loft to be a bit higher, so we raised them by adding a 5"-diameter log on either side above the eave supports.

A second set of roof rafters was put up at the rear of the loft and a 24' ridgepole was nailed to their joined tops. Two 24' purlins were notched in across the center of the rafters. This framework was then cross-braced in many places so it could support the heavy shake roof and an extra heavy snow load.

Included in the roof supports was a set of two angle braces, which hold up the 8' front extension. Roofs are often the first section of a structure to go. They are exposed to all the weather conditions and have the greatest tendency to rot and weaken; therefore, they should be well made.

Instead of laying rafters for the roof, I decided to rip extra-thick boards and span them from the ridgepole to the eaves. These boards take the place of rafters and provide the finish roofing with superstrong supporting nailers. These nailers are made of fir, because it is almost as weather resistant as cear and is as strong and durable as any of the other evergreens. It also holds nails much better than cedar.

To make your own roof shakes, you need several log bolts. Each should be between 18" and 32" long, depending on how long you prefer your shakes; it's all a matter of personal taste. I like to use 21"-long shakes because they cover a substantial area and they can be overlapped nicely to prevent leakage. For me, this length is easier to make than the usual longer ones, because our shake trees are young and small. They have twisted grains and many knots.

To find ideal shake material, you should go where a logging company has logged an area rich in first growth cedars. There you will find many high stumps that were left after the "slaughter." Such stumps, which are over 18" in diameter and have at least 6" of radius before the center rot, are an excellent source of shake bolt material.

To make these shakes, I take a mallet and froe and split the bolt into several sections. A section is placed on a stump or log-round, then the bark and outer layer of soft sapwood is trimmed off. I cut out half of the top of the working stump or round to make a safety ledge. The half that remains acts as a backstop to provide better control of the shake sections as they are being worked on. The next step in making the shake is to trim off any portion of heartwood that is dry-rotted. With a froe and a hardwood mallet, the shakes are then split off the bolt section. The

Cedar-shake block

Trimming layer of sapwood

Trimming off dry rot

Splitting with mallet and froe

Splitting of a shake

Tapered shake

bolt section should be turned end for end each time a shake is split off. This develops the taper effect of the shakes.

Because of the poor quality of our shake trees, I had to make thick, 1"- to 1½"-wide shakes. I'm beginning to like heavy shakes, though. Most homemade or commercial shakes are half that thickness. Heavier ones are more durable and are better insulators. Shakes are cheap if you make them yourself and they are, to my eye, the most beautiful of all roofing materials.

To lay the shakes on a roof: You start from the eaves and nail down the top ends of a full row of shakes to the accommodating nailer. To do the second row, you start a third of a shake's length up until you complete that whole row, then repeat this procedure. Note that each row covers a large portion of the row below it. This ensures

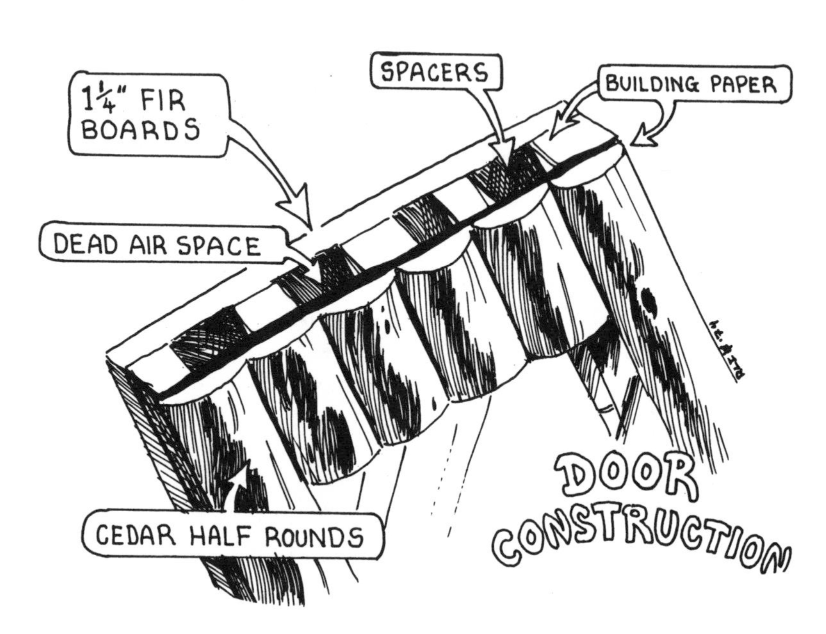

a 100 percent leakproof roof. This method is continued until the roof is finished. At the top of the pitch, a ridge cap of aluminum or some other material is placed over the highest shakes. It ensures against leakage at the peak of the ridge.

My next job was to hang a door. Our massive door was made out of fir boards on the outside and cedar half-rounds on the inside.

Between the two layers is a vapor barrier of building paper and a layer of one-by-fours spaced a few inches apart to create a dead-air space for insulation. The door was finished off with a locking bolt on the inside, which was secured in place with 8" lag bolts. With that, the house was sealed off from the weather and was ready for us to move in.

I really enjoyed the interior work that followed. From leftover cedar and other wooden half-rounds that I ripped for the purpose, I made cabinet doors. Using small birches, I designed dining-room chairs and a dinner table.

I wanted everything in our house, except for stoves and cooking facilities, to be handmade, nothing store-bought. I had little choice in the matter, actually,

Rounded cabinet made of half-rounds

Kitchen cabinets of half-round legs

Handmade table

because of our financial situation.

After we had a suitable roof over our heads and warm living quarters, I started working on other necessary projects. One such project was to make a bordering fence around the horse's pasture. Because our ground is very rocky and is difficult to dig fence posts into, I built a jack fence instead of the typical post-and-rail fence.

To begin this fence, I first cut a few hundred poles to length and brought them to the site. I took two short poles, about 6' or 7' in length, and notched them together near their tops to create an inverted V-jack. These jacks were spaced every 16' or 17' apart to receive the horizontal poles, which were notched and joined to them. The legs of the jack prevent the fence from falling forward or backward. A diagonal brace, running from the crotch of one jack to the bottom of the leg of the next jack, was put in at every fifth section of the fence to prevent any part of the whole structure from collapsing lengthwise. The braces were placed on the opposite side from the horizontal poles to not interfere with them.

A jack fence also has one line of poles running horizontally at the bottom of the jack legs on the back side of the fence. This prevents the leg of the jack from being accidentally forced sideways, thus undercutting the support. The horizontal poles on the front of the jack are alternated—butt, tip, butt, tip—to balance out the weight on each jack. Also, the jack leg on the uphill side is shorter than the one on the downhill side to maintain balance.

The front of the fence—the side with most of the horizontal poles—should be facing the livestock. This side can withstand the most pressure.

Freestanding pole fencing around horse pasture

Goat corral

The poles, otherwise, are subject to being forced off the jacks when the fence is rammed by a frisky animal. This design, of course, is reversed when the fence is being used to contain goats or sheep, because these nimble creatures are apt to climb the progression of horizontal poles.

Perhaps something should also be said about my well. I dug a 10'-deep hole, going below the water level, down to a bedrock base. This hole is 6' in diameter and is lined with a hefty boulder cribbing which encircles the well opening. The boulders for this cribbing were carefully chosen so they would remain secure when stacked. This kind of cribbing will never deteriorate and will never pollute the water.

The well is bordered above the ground by a 2½'-high rock wall to minimize the danger of animals and humans falling into it. Boards should cover the well until a hinged cover is made for it. The boards are a precaution against children climbing over the wall. The well was then enshrined with a distinct shake roof to shade it in summer, keep leaves from failing into it in autumn, and prevent snow from dropping into it in winter. This roof is substantial and has a certain aesthetic appeal.

My idea, though not entirely clear at first, was to incorporate a shake pattern into the well's roof. From previous experience, I found that shakes can have a somewhat sloppy look, especially around the eaves. I wanted to compensate for the "loppy-eared" look that I had been getting with the heavy shakes I was using. By building up the roof's eave support a little higher than normal, the first shakes went on in a near-horizontal position, but not horizontal enough to cause leakage. I found to my great surprise and satisfaction that the lower end of the second row of shakes came to rest, not on the upper part of the previous row as was usually the case, but near the center of those first shakes.

In using this design, I realized that I had discovered something about the way pagodas were made. I studied how I could develop the idea throughout the roof. It looked simple and proved to be. It was all a matter of common sense, but it took an eye for shaping to make the roof to my own taste. I couldn't wait to get each row of shakes on and see the effect. In order to fit the inner logic that was unfolding before me, the shake nailers had to become steeper as they went up. There was a functional reason for this design as well. The more pointed the roof peak, the better it would shed the water, like the bow of a ship. The final step was to place an inverted, hollowed-out log over the crack formed at the apex of the roof.

Creating curved-roof effect

Pagoda-style shake roof

After a few years of hard but gratifying work, with the blessing of a strong back and the necessary assistance from my wife and young children, I had come a long way toward accomplishing my primary "back to nature" objective. I have built all the necessary outbuildings, including a 5-ton capacity hay shed, a woodshed, a laundry house, a bunkhouse, and a chicken coop. I constructed a footbridge which spans our early spring runoff bed, and made most of the furniture we need. I hand-cleared over an acre and planted pasture grasses and legumes for the animals to graze on. We raised animals and fowl for meat, milk, eggs, and hides; planted a full garden each

Hauling the newly made lumber

Log pole bridge

year for vegetables and fruit; cut our own firewood and cooking fuel; and did a bit of seasonal hunting. In short, we developed for ourselves a new way of life, a life of hard, steady, year-in and year-out satisfying physical work. And someday we hope to achieve an even greater degree of self-sufficiency and self-reliance. Growing and processing our own clothing would be one of the next, very important steps in our efforts to achieve these aims.

What we have been doing on our land, we feel must be done—not only for our own good but, if only in a small way, perhaps for the ultimate good. The results of our labors are clearcut and tangible. As anyone who has made something more or less original and usable with their own hands knows, such results give an incomparable sense of personal satisfaction. There is a New Testament saying which I sometimes catch myself quietly quoting as I am working: "Ye shall know them by their works."

Section 4

STONE STRUCTURES & BLACKSMITHING

CHAPTER 14
Stoned-Stone Masonry Walls

It isn't necessary to waste thousands of dollars on a shelter when most of the materials are right out in the open if you take the time to look for them. Utilizing timber and local stones, one could bypass the enormous expense of lumber and concrete in mass. After working with logs for a while, I couldn't even consider dimensional lumber as a main material for a shelter. It is too limiting and expensive, and is a huge drain on several natural resources. There have to be fallers to fell the trees, skidders to take them to the trucks, the trucks to drive them to the mill where the timbers are sawed into right-angled boards, wasting much of what remains. This process involves hundreds of thousands of people, huge machinery, and an enormous expenditure of the energies we are quickly running out of.

An alternative to this would be to find a piece of wooded land, thin out a few of the crowded trees, and build with what you have. Sure, there is a lot of labor involved, but at least you'd be working for yourself with materials that don't tend to suppress creativity as does dimensional lumber. I am against working with this material because of its rigidity. With it, you are dealing with straight lines and right angles; you are not allowed to flow with the incredible array of shapes and contours that you can with natural timber.

And rock: so few people even notice the beauty and potential of this extraordinary and omnipresent material. And few realize how permanent and maintenance-free a stone foundation or wall can be if done right. But then again, stonework isn't for everyone, the same as finding a wooded lot and working with logs and natural timber isn't for everyone. It requires a dedication. One must really get into working with it and not worry about how long it takes to accomplish one's goal. But we should enjoy the act of creating beauty. People should be able to sit back and appreciate their efforts and not worry about spending too much time on a project. I helped a friend build the stone foundation for his home. It took us two years of working on it whenever we felt like it, and we never regret the time spent (**FIG. 14-1**).

I apprenticed with limestone, but give me the strength and finish of granite any day. It is the least affected by the weather, while limestone and

FIG. 14-1

sandstone are the most affected because they are so porous. Granite is not at all porous; in fact, it has irregular seams and grains, and is very difficult to split properly. So when working with it, take this into consideration and always try to find the most square and angular rocks you can.

The durability of any stone wall will be determined by the quality of the joints between the stones. Incas, Egyptians, Greeks, and people of other ancient civilizations used dry-fit mason work of incredible precision. Many of their structures still stand. Depending on the climate, a variety of elements can attack stone. In colder climates, water can seep between the joints or into cracks, freeze, then expand and cause even huge rocks to shift or break. This is the most important factor to be conscious of in stone-wall building. If you do not prevent this, your wall will not last through the first winter. Wind can also erode stone, although it is a much slower process than with running or dripping water.

Before we attempted to figure out the design and materials needed for my friend's house, we chose a building site. Actually, the first site chosen was in a small clearing of usable soil. We soon decided against it since there was so little cleared area on his land. Why waste the part that has some topsoil when there was a huge hunk of bedrock nearby, which couldn't be used. Besides, with bedrock, we are guaranteed against shifting and erosion. What more natural complement to the bedrock could we offer but a foundation of granite stone? Our decision was simple. We trucked in several loads of beautiful, multicolored granite rocks of all sizes and shapes. As we began working with them, we soon realized the solidness of this material. We attempted shaping and splitting possibilities with a two-pound mason's hammer. First, we chipped a line across the section we wanted to cut and went back over that line several times, patiently, with the sharp end of the hammer, then we tried to knock off the unwanted piece with the blunt end. The more you chip at the line, the better chance your rock will split at that point. This is a process that works well with softer stone but was just too unpredictable with granite. We ended up breaking several pretty rocks into small chunks, which were later used only to fill in behind the facing. But we did have a few rocks with small protruding knobs, which were usually quite easy to knock off in this manner (**FIG. 14-2**).

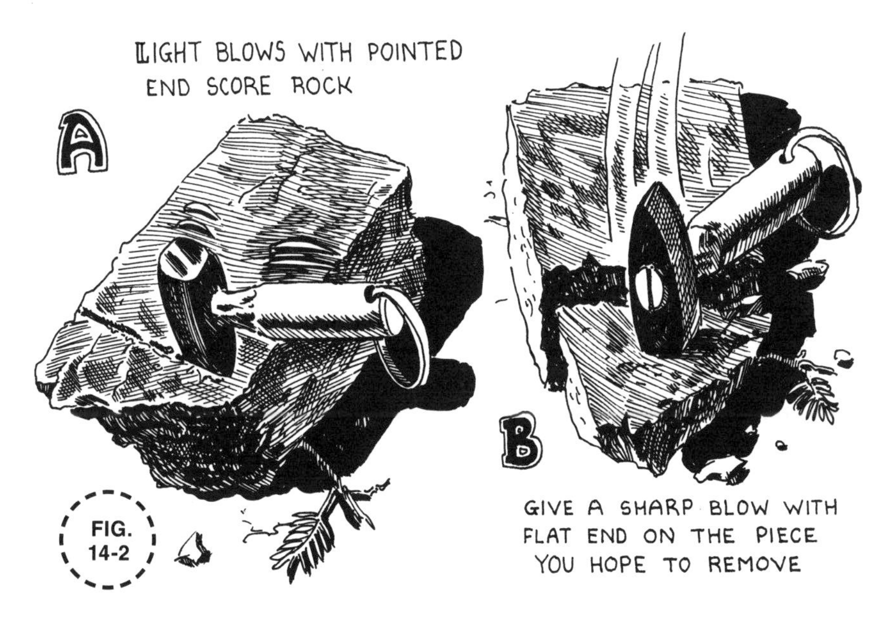

FIG. 14-2

There are other tools that are useful in cutting stone. The stone chisel is excellent for scoring and cracking. With it, one should have a sledgehammer light enough to swing often without getting overly tired.

Needless to say, we became more selective, gathering rocks that we did not have to shape. We searched for large rocks with at least two flat sides, preferably with a long, flat face and a flat seat that was no more than 9" in depth to fit our design specifications. We needed good-size rocks, but none heavier than what we could lift. It was important to do everything with the rocks ourselves, without the interference of machines and tools to lift.

We found many rocks from a nearby construction site where they were blasting in a new road. Most of these rocks were exceptionally angular and proved to be ideal for the foundation walls; in fact, they enabled us to build incredibly smooth, flat-faced walls that we could never have made without such material. We employed precision straightness because we were finding rocks that expressed that style. I'd never suggest this design unless there is optimum material at the builder's disposal, and plenty of time. This project took about 150 six-hour days to complete. Six hours is a full day of masonry work.

There is one basic preconstruction hint to keep in mind: Learn to select rocks as you go on trips to town, outings, or on visits to friends' homesteads. Be constantly on the lookout for possible rocks. Sometimes a simple trip up the road would take us hours, but would reward us with several usable pieces. This is all part of the organic process of development. Nature is never in a hurry—only people are.

Again, in choosing a rock, make sure it has at least two flat, or nearly flat, sides. Try to figure out the face, the broad surface that will be seen after the rock is set in; the seat, or bottom, that will adhere to the mortar below; and the top, the resting place for the seat above. If the top slopes outward, don't choose it, because there is no possible way you can secure a rock above it and expect it to hold the immense weight of the structure it will be supporting. If the top slopes inward, that's okay, the space left can be filled with smaller pieces to provide a flat ledge for the rock above. Of course, the ideal rock to work with would be a square or rectangular right-angled brick, but there aren't many of those around. And if there were, you would become limited in the design. The next best rock would be one whose top slopes slightly back with sides angling into the backing (**FIG. 14-3**).

FIG. 14-3

Gathering took us about 50 percent of the actual working time. We used 50 tons of material that was handled at least three times. Some rocks went through the whole history of the building without being used. So be wise and select the proper rocks in the beginning. It will make your job so much easier.

These are some of our names for several of the usable rock shapes: A "piece of cheese" is a wedge-shaped rock slice with a broad face, flat top and bottom, and sides that slant inward, meeting at the rear. A "wedgeback" has a flat face, top, sides, and bottom, with a back that slopes down radically. "Bulas" are smaller pieces of any usable shape (bula means "hard bread" in Jamaican). "Brutuses" are big rocks. And "bricks"

are the very square, right-angled rocks, which are a dream to work with (**FIG. 14-4**).

When planning a wall or foundation, try to avoid openings and corners as much as possible. Corners take five times as much attention because cornerstones are special bricklike shapes which should alternately "tie back" into both walls, and keep the walls from falling out under a load (**FIG. 14-5**). Corners should be planned so securely that they could stand free to create the total support of the building above.

To the layperson, the hardest part of working with rocks is interlocking the pieces. If you are not good at putting together jigsaw puzzles, don't attempt rock masonry. Here is a small exercise for those interested in checking their skill of visualizing in three dimensions. If you can figure out this problem, you are ready to work with stone. If not, then you should take the time to familiarize yourself more with this material before attempting to build any supporting structure out of it.

You have up to three cuts to shape a 1" round rod of any length so that it will pass through a circle with a 1" diameter, a 1" square, and an isosceles triangle with a 1" base and a height of 1". The rod has to pass through each of these flat geometric designs perfectly, leaving no excess spaces, and without destroying the shapes and designs in any way. You can practice on doweling or other material.

Before beginning to set up the walls, get to know the individual rocks as you would pieces of a jigsaw puzzle. Lay them flat on the ground, attempt to create designs with them—mix larger ones with smaller ones, triangular with square—visualize what they would look like on the wall, see which ones complement others. Once you mortar them in place, they are permanent—there is no changing them around. After you are satisfied with a section, try setting it up as a dry wall to see how it stands. When you are pleased with how each rock sets and balances among the others, take that section down, making certain not to disturb the positioning, and mortar them together. Remember, never hurry—be extremely patient. It can take a long time for the ancient wisdom of these rocks to communicate, for they have been silent since long before any of us have been around. But, in time, they will express their own design, one that most complements them (**FIG. 14-6**).

At this time, before actually setting up much of the first wall, be sure to figure out all the openings, vents, electrical holes, drains, water pipe positions, etc. It would be a nearly impossible task to install these necessities as an afterthought once the walls are set up. If you are not using a small cement mixer, it is wise to make a mixing pad of some sort before you begin setting up the walls. Ours is a 2"- to 3"-thick pad of hardened Portland cement, made from a mixture (1.5 parts sand to 1 part cement). It is 6'

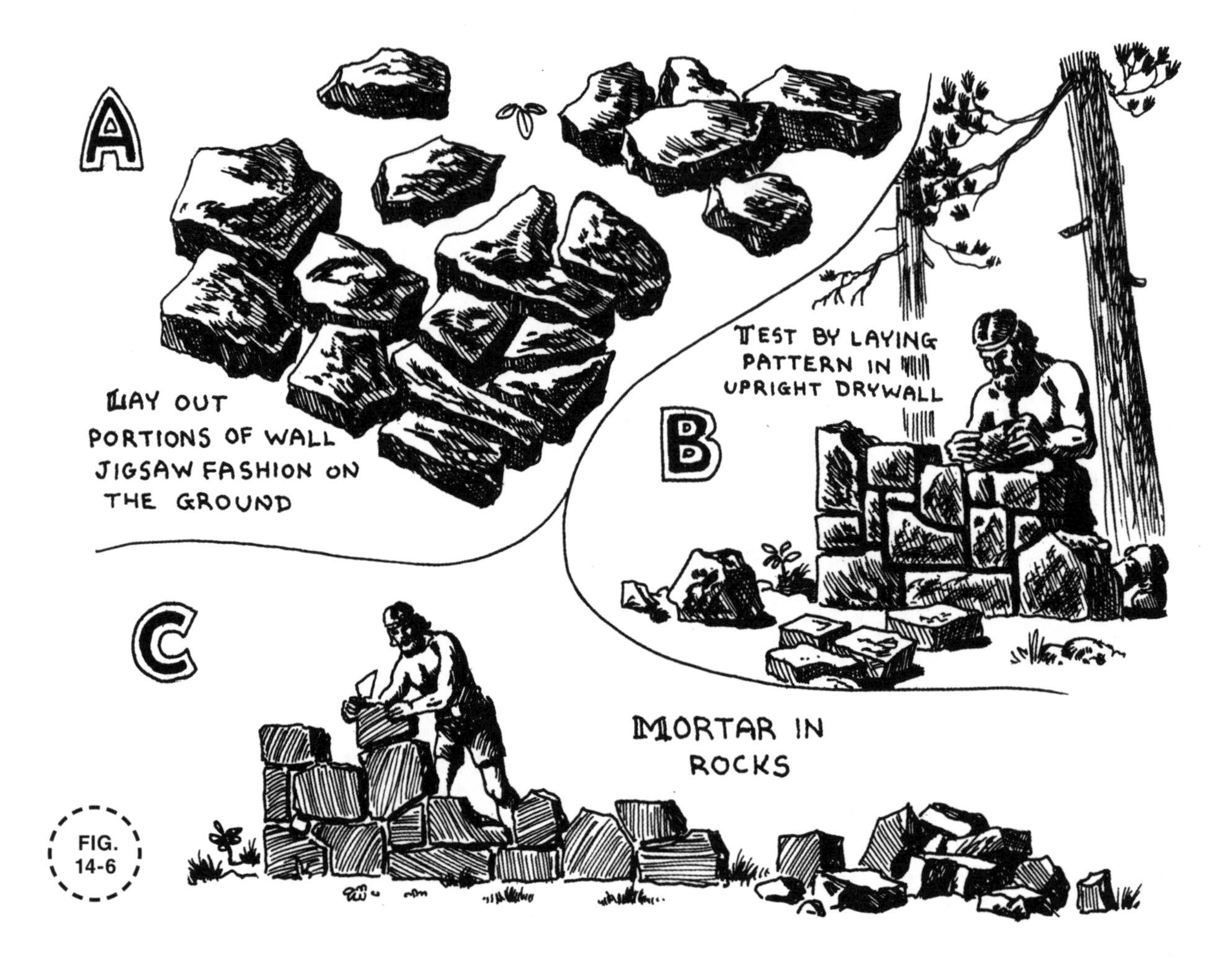

FIG. 14-6

in diameter, poured right on the ground. On this, we mixed the mortar. The mortar is a mixture of 1 part masonry cement to 4 parts sand. The sand should consist of ⅔ fine river sand and ⅓ coarser sand. Since the adhesion of mortar to sand is critical, a test should be made to make sure the sand is clean enough to use. Fill a quart jar ⅔ full with sand, then add water to fill the jar. Shake it up and leave it for a few hours. If more than ⅛ inch of silt settles atop the sand, then it is too dirty to use without first washing—an arduous job requiring a large flow of water and hand turning of the dirty sand.

We first mixed the mortar and sand dry in the center of the pad, making certain that it all had a consistent mortar grey color. Then it was spread out over the pad and drilled with a fine spray of water, and all brought back to the center and worked, gradually adding more water, until it had the desirable sticky, clinging ("cloy") consistency (**FIG. 14-7**). Do not add in any lime to achieve this stickiness. Masonry cement has enough lime already mixed in with it—any more will lessen the strength of the mortar. Also, be sure the mortar is never richer than the 1:4 mixture. The more excessively rich the mixture is, the larger the coefficient of expansion, thus the greater will be the chance that it will crack and shrink.

Starting anywhere along the wall with the first course of stone, we worked toward the corners. This is because the cornerstones are usually the hardest to stand up freely. Having the wall next

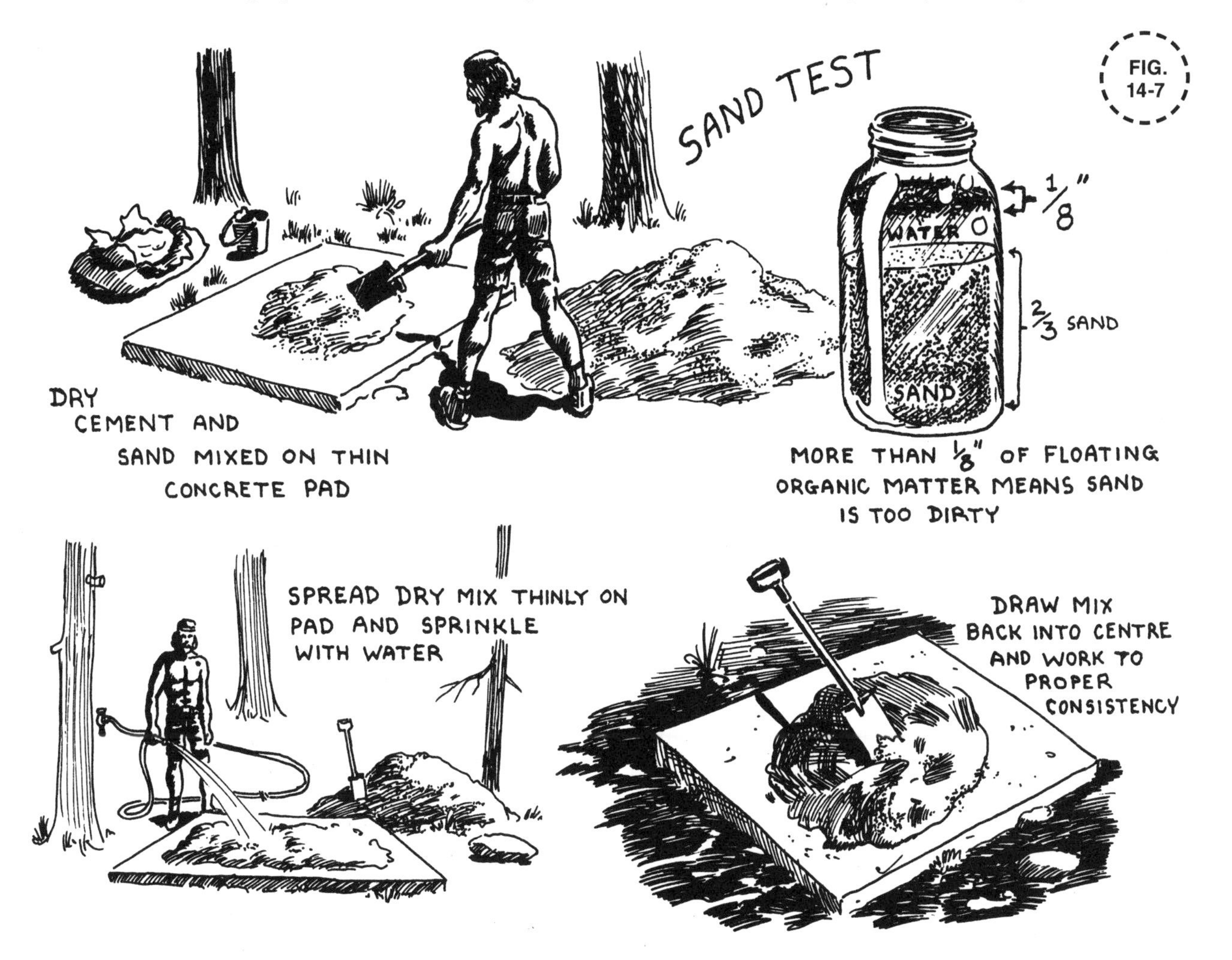

to the corner already complete aids in supporting the cornerstones. However, it is often necessary to maneuver the penultimate rock that fits next to the cornerstone, as the corner is a fixed point and can only go in one position (see **FIG. 14-5**).

The outer dimensions of the foundation are 20 × 26 feet. To figure out the rectangle, we simply made a scale drawing and found the length of the diagonals from opposing corners. That length was converted to actual size. The exact center was marked and the crossing diagonal was also staked. A 20' side was measured and the stakes were adjusted accordingly. A right angling 26' side was then checked and a string was placed around the perimeter. The corners could be confirmed by running a 6' line along one side and an 8' line along the intersecting side. If the diagonal of those two lines is 10', then the corner is a right-angled square corner.

Once we found the perimeter of the rectangle, we set up the vertical templates or upright guides to make certain our walls would be plumb and straight. We used two template frames at each corner, which consisted of two parallel vertical boards spaced the width of the wall (18") apart, and put in a couple of feet directly beyond the end of each wall face to allow us enough space to maneuver the heavy stone without colliding into them. They were rested on the ground and braced securely to stakes, stumps, or trees so that they would remain exactly in place in case we bumped or fell into them. We even left some trees around to brace these posts. After they were secured in place, two horizontal strings, about 2' apart vertically, were extended from template to opposing template. The strings were both plumb above the face of the wall they guided and could be moved up as the courses were laid. The templates and lines were difficult to work with because we were always becoming entangled with them as we struggled with the heavier stones. But never attempt a structural support such as this foundation without such guides, because it can't be done (**FIG. 14-8 AND 14-9**).

The walls are 18" thick all around, consisting of an outer face of no more than 9", a Portland cement and bula backing, a vapor barrier of ½" Styrofoam, an inner wall backing, and an inside

wall face. The two faces were set up first. The Styrofoam was then placed in the trench created by the opposing faces and pieces of heavy-gauge wire were poked through the Styrofoam on an approximately 2' grid. Each bent end was sunk into the fresh cement backing between the inside and outside face, creating a tie for the two walls. The backing mixture was poured on either side of the lapped Styrofoam sheets before the surrounding masonry had a chance to completely set up, allowing it all to monolithically tie together, usually about one hour after doing the faces (**FIG. 14-10**).

Note: On rainy or very humid, damp days, it takes as much as three times as long for mortar to set up, depending on the moisture content in the air. Be extremely careful when attempting to lay stone during a light rain, because the heavy moisture content will cause the mortar to sag and droop under the weight of the stone.

Styrofoam insulation is brittle and cracks quite easily, but has proven to be an excellent thermal and vapor barrier. This vapor barrier prevents the movement of warmth and humidity from inside to outside. If this movement was not prevented, the warmth would meet the snow, ice, or frost on the outer wall, which would then seep back into the wall, possibly refreezing and causing the wall to crack.

If moisture were allowed to penetrate through the walls, splitting and cracking of the mortar would also eventually occur. We installed the

Styrofoam mainly as a thermal insulation because stone has a very low "R factor" (18" of rock provides less insulation than 3½" of fiberglass). But even with the thermal barrier, some cold still manages to come through the walls during the winter. So ½" inch of Styrofoam didn't prove to adequately solve this problem. If we had used 1" of Styrofoam, I think this chill factor would have been greatly reduced. Besides being thicker, that size is much easier to handle and it wouldn't have cracked so readily as we worked with it. We are not advocating its use, though; you can use anything that proves to be a good vapor barrier with insulative value. You can even fill garbage bags with old cardboard or waxed cartons for this purpose.

Before going on to the next course of stone, be sure to clean off the rocks that were just set in place, with a steel brush. If you wait much longer than two or four hours, the mortar will have set up on the faces, thus making it necessary to employ an acid chemical (muriatic acid) to remove the dry mortar. This chemical is a costly, odorous, and unpleasant alternative to taking time as you go to clean off your stone. Besides, the cleaning process will give you a little time between courses to sit back and appreciate your work. Many such breaks are imperative when indulging in any creative endeavor, especially one that will be so permanent. Also, the best time to do the pointing or cutting back is just before the mortar completely sets up. Pointing is the process of trimming the excess mortar that sticks out between the joints of stone ("snots").

There are several styles of pointing. Some masons prefer a deeper look. This provides rich, full-shadowed joints and is best for contrasting the dimensions of the rocks. It also weatherproofs the mortar by cutting it back, sheilding it with overhanging stone (**FIG. 14-11**).

There is a noticeable difference between the lower portion of the north wall where we started and the upper area of that wall. It looks much more raggedy and rough than the others. We got progressively better with experience, concentrating more on the finish as we felt more comfortable with the mortaring (**FIG. 14-12**).

FIG. 14-11

FIG. 14-12

Before taking the corners up again, we reinforced them with twisted strapping bands. These bands were 5' in length, extending at right angles 2½' on either side from the corner, one on each side of the styrofoam. They were laid in at vertical intervals of about 18". The twisting provided them with much more bonding strength, making them immovable within the concrete (**FIG. 14-13**). We searched through several

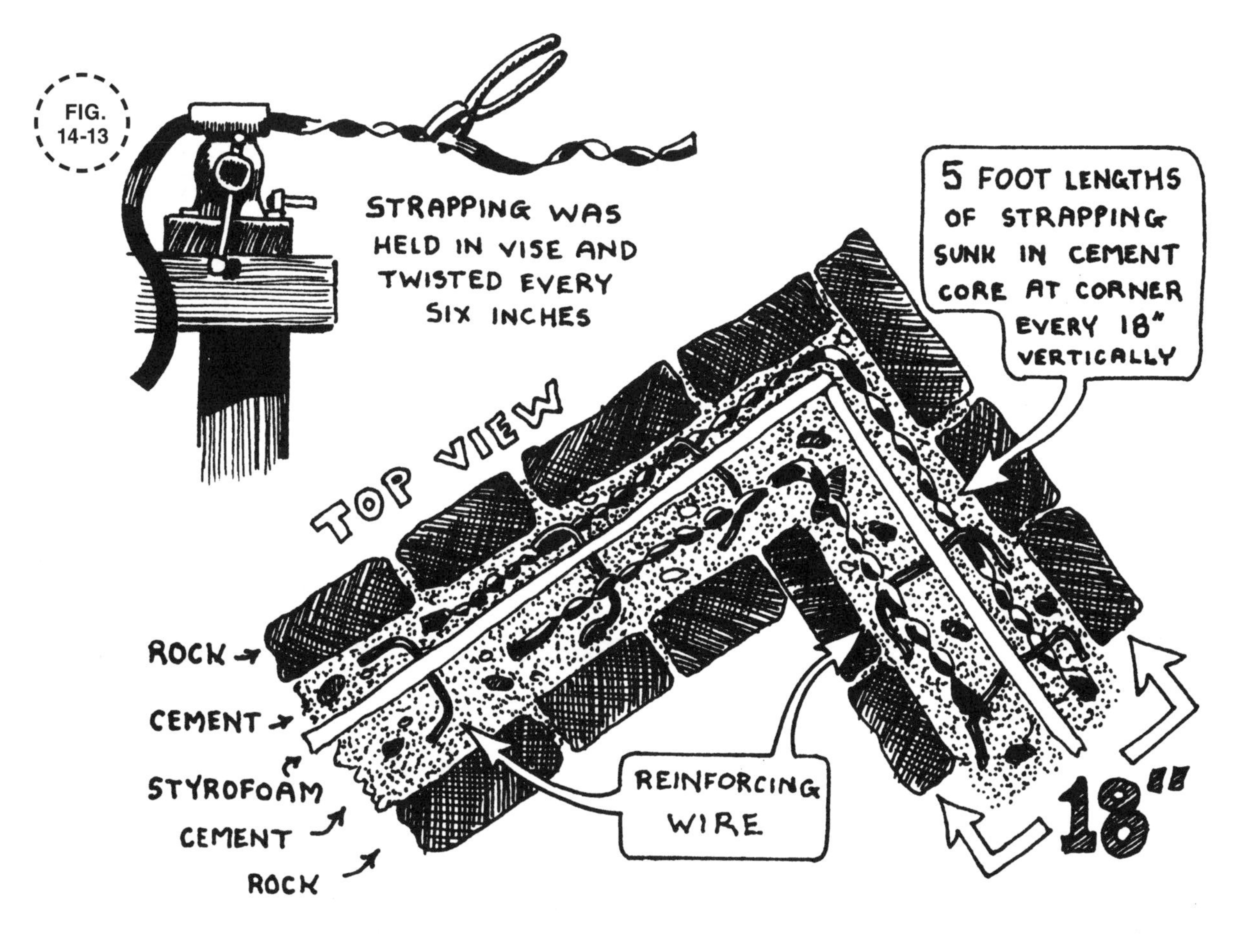

building supply and hardware stores to find a material for this purpose, but everything was so outrageously expensive, it was prohibitive to use. Outside of one store, we found a pile of strap-metal shipping bands. We were told we could have them. We searched for more in other places, including the local mill, and soon brought home all we needed for free.

The walls went up to about the 3½" level where we inserted the floor joists, cellar vent openings, and door frames. This height was chosen because the high point of the bedrock extended 3' higher than the low point (**FIG. 14-14**) and we wanted to allow another 6" to bring the joists above the ground. The joists were spaced at 24" intervals, extending from a short distance out past the front wall to 8' beyond the rear wall as supports for an outside porch area (**FIG. 14-15 AND**

FIG. 14-14

FIG. 14-15

14-16). These longer ends will be sheltered from the direct weather by the porch, but the short stubs in the front east wall have no protection; they were left exposed (FIG. 14-17). If we were to put them in again, we would just extend the front ends to meet the styrofoam; this way, they would have a vapor barrier protection and still have plenty of support from the 9" inner wall section.

There is no creosote or other preservative around the joists where they come in contact with the mortar and concrete. We didn't feel it necessary, since they are of durable cedar and we don't have to worry about getting capillary action from the soil beneath, because we have a floor of bedrock. But even if the joist ends rot over time, which they are bound to do anyway, our children or grandchildren or whoever will be around then can just pound them out and insert replacements, trimming them where they meet the walls for proper fit.

Below the floor level, we put in a 6" vent on the east and west walls to allow the warm air to escape from below the floor. These vents were formed with 6" stovepipe pieces which were left in. Earlier, we put in openings at the bottom of these walls to allow the cold air to come in and circulate through the basement. This air flow provides proper ventilation and prevents mildew and vapor buildup (SEE FIG. 14-21). This will probably be a great root cellar once we complete the house.

The door jambs were also put in at this level. We used uncreosoted railroad ties for this purpose because they were durable and available. They have spikes driven and bolts drilled into their sides to tie into the masonry and join them to the wall logs. These frames were made plumb and they acted as a guide for the new inside corners of the rock work (FIG. 14-18 AND 14-19).

After the frames came the water pipe opening. This was put in above the floor, in the west wall.

FIG. 14-16

FIG. 14-17

FIG. 14-18

FIG. 14-19

For this, we just sank a piece of ½" galvanized pipe, which we could later deal with (**FIG. 14-20**). Be sure to plan all openings before you begin construction, because there is no way to accommodate them once the walls have been set up.

We slowly raised the walls a bit higher, carefully considering the design of each as we went. By now, we had gotten emotionally involved with some of the rocks, we had handled them so much. But many of them just didn't seem to fit anywhere. Up to this point, we were able to use whatever size rock we could lift into place—the larger the better—but now we had to prepare for the sill logs, which meant narrowing down our supply to the special shapes needed, without disturbing our design.

The long sill logs that rested above the east and west walls were set in place on the center of the wall and held there with braced uprights. The shorter side logs were round-notched onto them. This meant that the two long walls had to be raised to one height and the side walls had to extend a few inches higher to accommodate the sills above them (**FIG. 14-21**).

FIG. 14-20

FIG. 14-21

It was a difficult, painstaking operation to fit the last courses under the logs. But at least we had a permanent horizontal line to work up to instead of an estimated line that could not duplicate the exact contour of the log. The styrofoam barriers were set in so they touched the bottom center of the sills, then we experimented with several rocks until we could find the right combination for the first outer wall. We set up the outer walls first, then filled in the backing from inside. This was a tricky manuever, which I'll explain in turn. After we roughly figured out the last course of an outer wall, spikes were driven into the bottom of the sill that rested over it. We were careful to position the spikes so they would not interfere with the last course of rocks, but would instead extend out into the mortar and cement (**FIG. 14-22**). The outer wall rocks were then mortared into place. Now the tricky part: The styrofoam was bent back, which sometimes broke it, and the rubble and concrete for the outer wall facing was shoveled in and tamped hard around the sill log to provide a solid seat for it (**FIG. 14-23 AND 14-24**).

The styrofoam was then bent into vertical position, and the last courses of the inner face were carefully dry fitted. The spikes were driven in and a few of the top course rocks were mortared. The rubble and cement backing was then tamped in behind the face and a few more rocks were set in. This process continued until the last rock was placed in. Then, we moved onto the next wall.

Before finishing the ledges above the inner walls, we sank "fillets," or small wedge-shaped pieces of wood, into the mortar to provide nailing blocks for the wood pieces which would cover them (**FIG. 14-25**). These wooden ledge

covers would be useful caps, providing needed shelf space. They also shadow the very top of the inner stone wall and give it a contrast of timber with stone.

As you can see, our careful meditative approach proved to be not in vain. We have succeeded in constructing not only a useful, strong foundation, but an aesthetically beautiful monument that should still be standing long after we are all dead and gone. I wish you could see the richness and array of colors in these rocks. They are so pleasing to look at that I can sit and meditate on them for hours, and I do just that, sometimes. This is the feeling I want every time I complete a project—how about you?

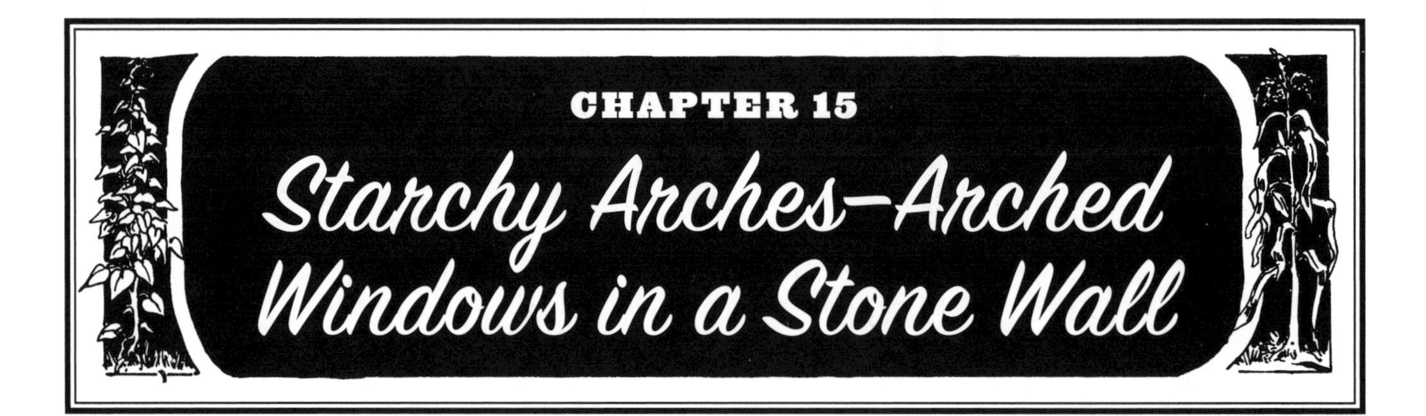

CHAPTER 15

Starchy Arches—Arched Windows in a Stone Wall

Along the tiny unpaved roads that surround our little bakery, there are many dwellings and homesteads, which have been carefully constructed out of the available materials we are so lucky to be endowed with in this area. Some make good use of the forest, some make good use of the stone, some combine the two to put a bit of beauty around them. So I wanted to contribute to this beauty by adding a little of my own heritage into the bakery. And what more Scottish building material is there than stone?

I had a design in mind for a dream house in the round, built out of stone. I made a little model of that round house and put in a few round windows. I still haven't made it a reality and probably won't for a while, so I decided to give the bakery a touch of my dreams by making the kitchen front out of rock, with arched windows (**FIG. 15-1**).

I constructed the building using a post-and-beam style, leaving the front kitchen wall open between the 6' spanning posts. I wanted to give a contrast to the logs and stuccoed exposed beams, which comprise the rest of the face, and I wanted

FIG. 15-1

this contrast to be stone. Though I hadn't really worked much with stone, I knew enough principles to get started. First, I made sure the bottom course was the thickest and wide enough to support the weight that was to be put on it. Second, I made certain that the outside face was very straight and plumb. I knew my upright posts were plumb, so I nailed one-by-six boards across the posts, all the way up the front. This was my outside form. I then collected as many of the straight, flat face rocks as I thought I might need. This was quite easy since the nearby lakeshore is full of such angular and extremely colorful rocks.

I piled the rocks jigsaw-puzzle style on top of each other, making sure the flattest side faced the front and the bigger rocks were at the bottom. I tried the rock that fitted in best, then mortared it in, giving each course a chance to set a bit by going along the whole wall before beginning another course. For this purpose, I made my own mixture of Portland cement and lime. Pre-mixed cement is like cake mix—if you want to buy it pre-mixed, you have to pay for it. I mixed 2 parts lime to 1 part Portland cement and added 9 parts of sand. This made a good, workable mortar. Be careful when you use the lime, because it burns your fingers. I worked with rubber gloves to prevent the lime from burning me.

The stone wall was built in this manner for about 30" in height. At that point, I wanted to put in my windows. To prepare a sill for them, the mortar was formed and flattened above the rocks. This way, bolts to hold the sill boards in place could be sunk into the mortar instead of into the rocks. It is much easier to sink bolts into wet mortar than into granite rocks (**FIG. 15-2**).

The trouble with using forms that cannot be easily removed before the mortar has a chance to set is that mortar, no matter how careful you are, builds up around the rock. This leaves ugly

clumps in the joints that cannot be cleaned off because the mortar has already hardened by the time the form is taken off. If a person were to use removable forms here, this problem could be avoided (FIG. 15-3).

Because the building is post-and-beam, none of the structural weight is on the rock wall, it is all on the outlining posts. This gave me freedom to experiment with the windows. My fondness for round and oval architecture inspired me to devise a way to put arched windows into the front wall. This method actually worked quite well. First, I took a 4 × 8 foot sheet of ½" plywood and laid it on the floor. I found the exact center along the length and drove a nail in at that point. Then I attached one end of a 4' piece of string to that nail and the other end to a pencil. I drew an arch, starting from one corner to the middle of the side opposite the nail, and down to the other corner. I didn't like that shape, because it was too wide at the ends, but the height was just fine. So I moved the nail 1' closer to the apex of the arch, and shortened the string to 3'. I drew that arch and left the bottom foot, beyond the nail on either side, straight to make an elliptical arch instead of a half-round (FIG. 15-4).

FIG. 15-3

After the shape was established, I cut it out with a sabre saw and finished the edges with a hand plane. I then took a thin strip of 3"-wide veneer and tacked it along the edge of the half-oval form, off-centering it to one side to affix the window framing to it. I felt it wise to make the framing now and attach it to the form and casing. This would give the form, casing, and frame additional strength, tying it all together as one interlocking unit to hold it all in place (FIG. 15-5).

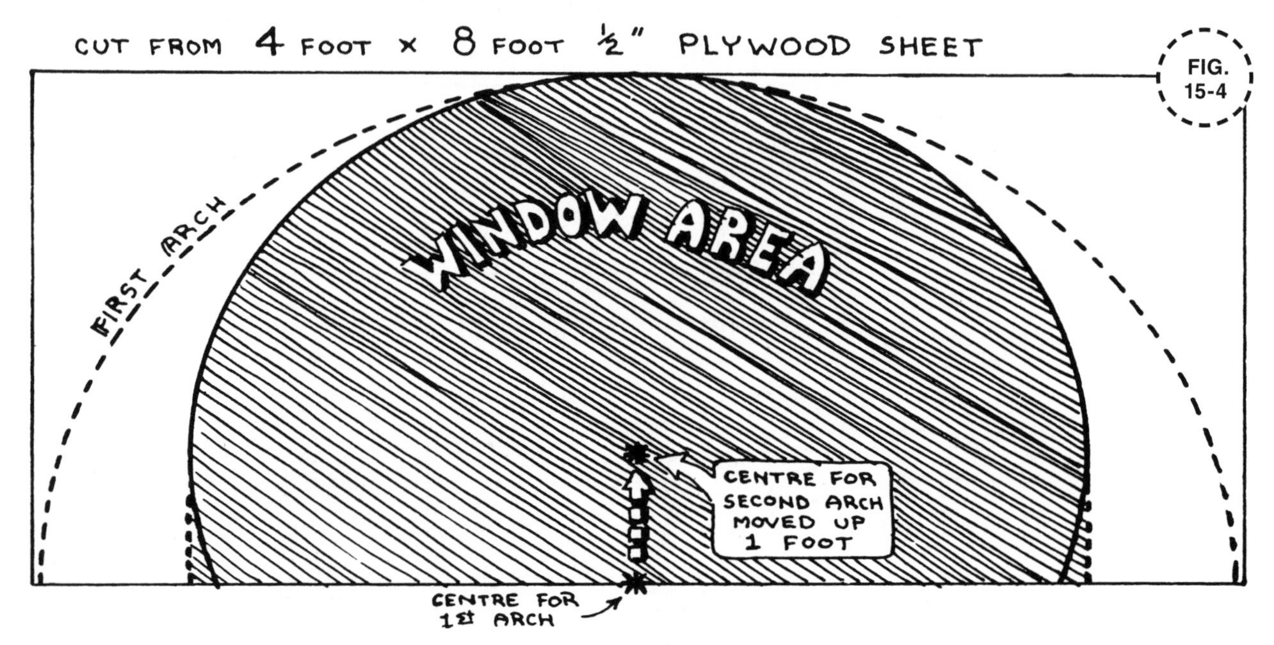

FIG. 15-4

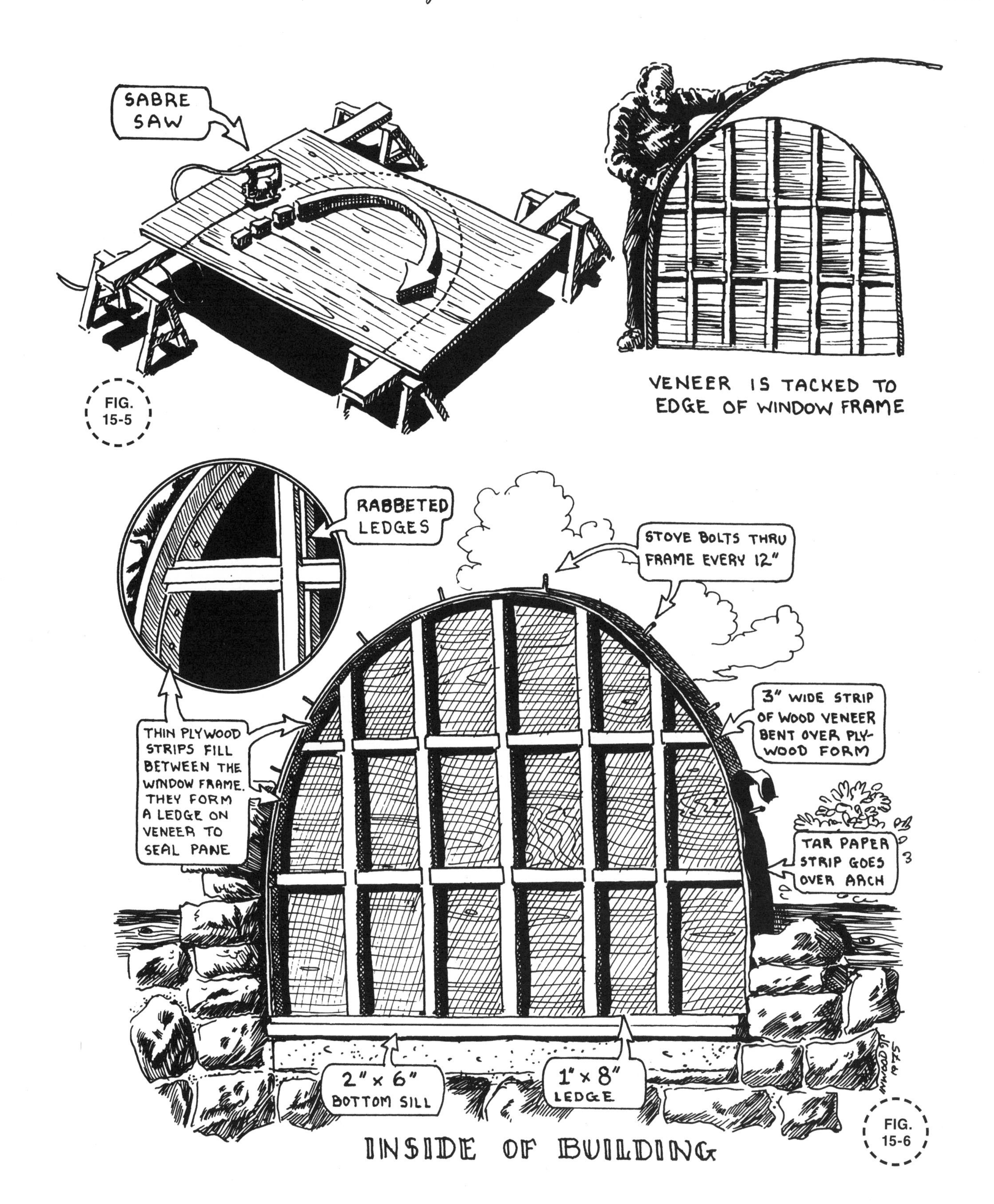

FIG. 15-5

FIG. 15-6

For the window frames, 1 × 1 inch strips were used. This framework is composed of five upright pieces of varying lengths following the arch. The outside edges of these uprights were dadoed ¼" on either side to provide laps for the 12 × 16 inch window panes, leaving about a ½" tongue in the center to hold the panes in place. There were also grooved strips that interlocked with them to frame the panes. These 12" horizontal strips were rabbeted on the outside edge and a sill strip was nailed in place across the bottoms of the upright pieces. Holes were drilled 1' apart around the veneer border for stove bolts, which would be sunk into the setting mortar to hold the strips in place around the window frame. The spring tension of the bent strips would help to keep it in place after the form was removed, because the constant compression against the rocks would prevent it from coming out. Each of these three window forms were made ready before going any further (**FIG. 15-6 AND 15-7**).

The two-by-six sills were bolted over the formed mortar seats below the window opening and the completed window forms were placed above them, with the plywood facing the inside. A few temporary nails were then attached to the outside form to keep them in place, and the rock work was mortared around the arches all the way up to the two-by-six wall plate above. The only tricky part of this whole operation was slanting the rocks above the arches and still making the design blend in with the other stones. It all worked to my satisfaction. After the mortar set for a while, the plywood window forms were removed. Care was taken not to disturb the veneer and one-by-one window frames. The outside forms were also taken off at this point and I got to see the mortar-clumped rocks. I was very careful to trim the mortar between the inside rocks and clean them off, leaving a nice finish on the rocks. After all, I would be looking at them several hours each day as I prepare the baked goods for the upstairs bakery (**FIG. 15-8, 15-9, 15-10**).

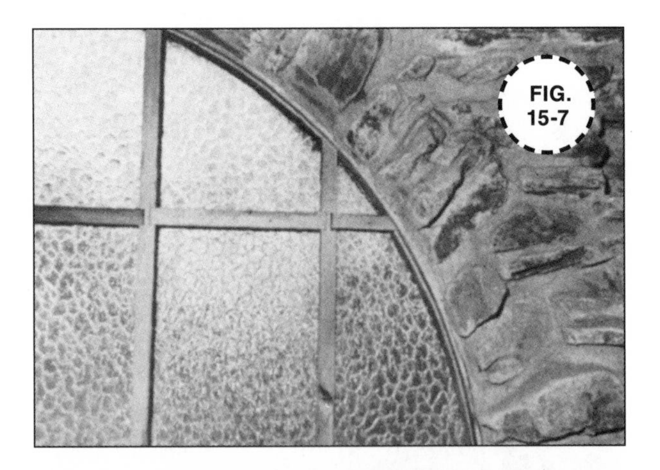
FIG. 15-7

FIG. 15-8

FIG. 15-9

I used opaque, corrugated plastic instead of glass for the windows to cut out the glare, and because I didn't want to go through the trouble of cutting rounded glass. It is hard enough to cut straight glass, let alone get fancy with it. These windows were held on the inside with curved slats that follow the arch. They were nailed along the veneer strip, which covered the bolts and holes (see **FIG. 15-7**). The windows were puttied on the outside. Putty did not work so well because it did not properly adhere to the plastic.

A question might come up about the insulative value of this rock wall in a cold area such as this. I realize that stone is not a good insulator, but I haven't been uncomfortably chilly at all. Once the ovens are going, the kitchen remains quite warm and what little heat that might escape through the stone is welcome to do so.

I am quite pleased with the way the rock wall came out. It has even temporarily satisfied my desire for a round stone house with round windows.

FIG. 15-10

CHAPTER 16

Slipform Cellar and Sauna

Since the surrounding hillsides, lakes, and streams are so richly endowed with beautiful multicolored stones with stripes and other designs, I thought it only natural to use this available material in the foundation walls of our house. In searching around the lakeshore, I found hundreds of flat-faced stones that were smoothed to a polish and washed clean by the flowing water. I gathered several loads and trucked them to the building site. Then I spent some time figuring out just how I was going to create walls that had inside surfaces of the clean, flat-faced rocks. (**FIG. 16-1**). I knew that I had to either attempt to build a wall without a form, or figure out some type of sectioned form, which I could periodically remove to clean the concrete off the rock faces before it dried and covered the rocks with permanent ugly stains. Because of my lack of experience with stone masonry, building a foundation without forms was out of the question. I soon figured out a slipform system that solved the dilemma. It is a simple design, consisting of a long continuous one-by-four form, 22" high, braced by upright two-by-fours every 24". This is about the maximum height the forms should be for each pour, to allow easy tamping of the mortar around the rocks.

FIG. 16-1

The 16"- wide by 4"-high footings were poured first, and, after they dried, we set up the first course of forms, one on either side of the proposed 8" wall. We were attempting to have only an inside stone face. The outside face would be of the rubble-concrete filler that backs the stone face and cements it in place. Straps of banding steel were cut long enough to extend beyond the sides of the form uprights, holding both sides in place around the 8" long, one-by-four spacer blocks. These blocks kept the forms the right distance apart and would later be knocked out. Two holes, 1" apart were drilled through an end of the straps—one for the nail that will be stretched by the crowbar and one for the nail that will secure the forms together. Two holes were also drilled in the other end of the strap, and that end was nailed to the outside form. The strap was then brought underneath to the far side of the inside form. A nail was put in the first hole for the crowbar to grip against and stretch the strap tight while a second nail was pounded

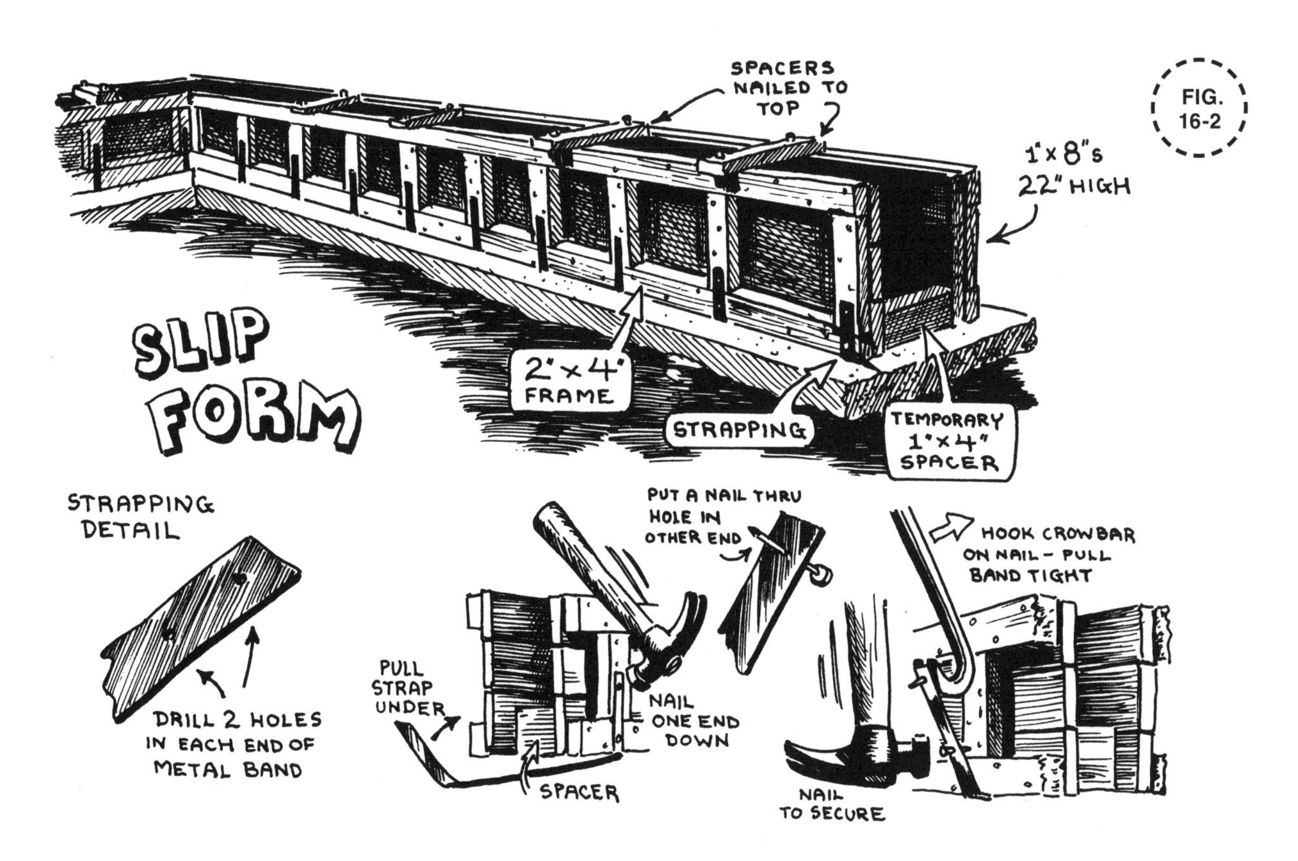

into the upright. This was repeated along the 16'-long form.

It is a lot easier to drill the holes prior to putting in the nails rather than struggling with the crowbar while you are awkwardly attempting to start a nail hole. These straps are removed while mortar is setting. The top of the form was then held at an 8" width with another set of spacers that were nailed in above. Be sure that the forms are absolutely plumbed before the pour because you don't have to be out much in 22" to throw the whole wall off. When you are satisfied that the forms are plumb, secure them with diagonal braces so they do not move during the tamping. It would be wise to set up some kind of plumb vertical brace the length of the wall to provide a guide for the stacked forms. I wish I had taken precautions; it would have saved a lot of time, rather than having to plumb each set of forms individually (**FIG. 16-2**).

When the bottom forms are set up securely on your footings, put about 2" of pretty soupy concrete into the bottom of the formed space and begin setting in a line of wall stones. Be sure to jamb them tight against the inside form with pieces of rubble rock to prevent the soupy concrete from getting in front of the face of the wall stone. This will create a flat stone face wall and make your cleaning task much easier. Then just set in some more wall stones and pour in the rubble-concrete mixture until the forms are filled.

To reinforce these wall sections, I backed the face stone with two runs of wire rebar per pour. This rebar was homemade, utilizing a roll of 8-gauge spooled wire. The wire came off the spool six strands at a time. I unwound 100 yards, trying to keep all the strands together as I tied the ends to two vehicles. One vehicle remained stationary while the other moved forward, putting tension on the strands. Then I stuck a piece of wood through four of the strands and twisted it all together. It stayed tightly entwined and didn't unravel even when I cut it. The rebar went completely around each pour and the ends were spliced together with smaller pieces so they all remained securely in place.

Be sure to leave a few wall stones sticking up along the length of the top of the forms so the next course of forms have something to grip onto. And be very careful that they don't go out past the face or they will throw the wall out of plumb.

Tamp rigorously throughout each pour so the soupy mixture can work its way between all the face stones, otherwise there will be a lot of "beehive" holes in the face. I just keep poking the mortar with a stick, pushing it into every crevice. Wherever necessary, I also tuck it into narrow openings with my fingers or with a pointing trowel, making sure no spaces are left unfilled. I recommend wearing a pair of rubber gloves for this operation because the wet mortar is very abrasive (**FIG. 16-3 AND 16-4**).

The mixture I used was 2 parts Portland cement to 5 parts sand. Even without the lime the mortar adhered well to the clean rock, and was less expensive. The strength of concrete is not in the amount of cement but in the amount of water used. The soupier it is, the better it all mixes. But you don't want it too soupy or it will leak out of the forms. Coarse sand and gravel is ideal for the rubble-concrete behind the face, but for a smooth face wall finish you need a finer mixture. I put in at least one shovel-full of blue clay with every mix for a finer consistency.

After the walls were left to dry for about 12 hours, the inside form was carefully removed while the mortar was still green but firm enough to not slump. This is the best time to clean the cement stains off the rocks. I scrubbed them thoroughly with a wire brush, then hosed them off really well. This method is a lot easier than having to later treat them with muriatic acid, a highly toxic and dangerous chemical that eats through the cement. (Just think what it could do to your hands.)

We waited 48 hours between pours to give the lower section a chance to set up a bit. Don't wait much longer than this or you will not get a good

FIG. 16-3

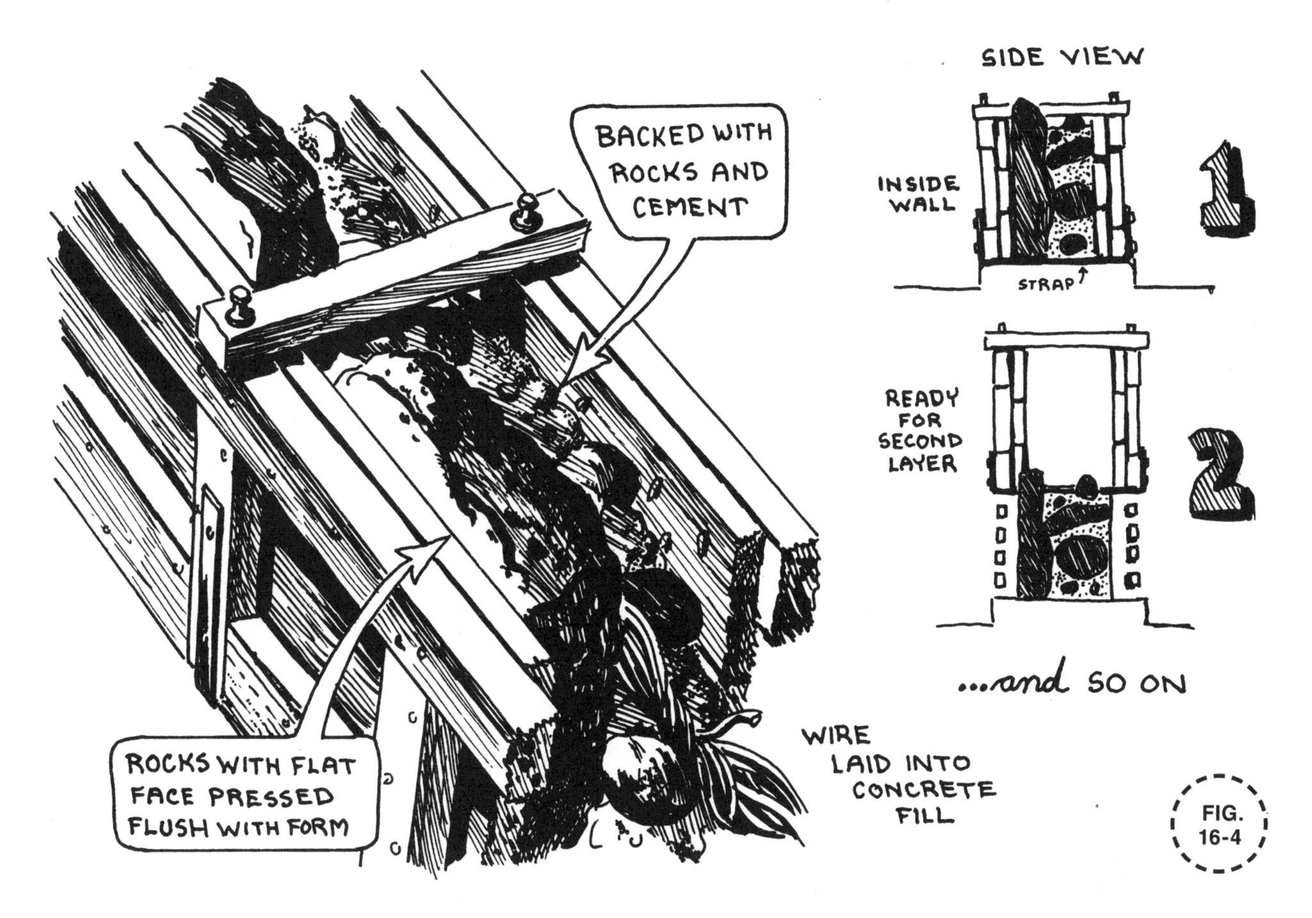

FIG. 16-4

bond between the courses. The following courses of forms were held together in the same manner as the first, using the protruding wall rocks from the section below to give them something to grip onto. The courses went to the top of the side and rear walls and to the window openings in the front wall. Four 18 × 28 inch windows were placed along that wall to bring in the morning light. To ensure a good bond between the 8" window frames and the mortar, 4" ardox nails were pounded into the frames. The frames were then set into the forms, and the surrounding stones and masonry was set in (FIG. 16-5).

Though a wall like this is very beautiful, it is very difficult to make it entirely watertight. I used a thick coat of block foundation coating, but it still leaks periodically. (Actually, the only time it does leak is when someone leaves the hose running alongside the wall.) This problem could possibly have been remedied if drain tile had been used just outside of the footing to catch the excess runoff. Instead, I used only loose gravel, hoping that would be enough.

FIG. 16-5

FIG. 16-6

FIG. 16-7

Though 8" of stone and mortar or 8" of concrete have an insulation factor equal to only ¾" of wood, the basement remains very toasty throughout the winter, because the furnace is located in that area. The only place it loses heat is through the windows and the upper 2' of the front wall that sticks up above gound level.

The stones below ground level actually help to keep the basement warm for several hours after the furnace fire goes out since they retain the heat so well. Because of this factor, I decided to construct a sauna at the southeast corner of the basement. The sauna area gets lots of light from two front wall windows and remains warm for a long time after the cookstove fire dies

FIG. 16-8

out. The only problem is it takes a long time for the fire to get the rocks warm. But when it does, you can just throw some water on them and really get the place steaming. Part of the hassle of getting the sauna warm enough is the flat-surfaced, small fire box cookstove we use (**FIG. 16-6**). This can be replaced—sacrificing its flat surface which we used to put rocks on for additional steam—with a more efficient, airtight heater that would throw off a lot more heat.

The sauna compartment is about 6 × 8 feet, consisting of two one-by-four tongue-and-grove cedar slat bathing benches. The lower one is 18" above the concrete floor and the other is 18" above that. Both are held up by two-by-four horizontal braces at the inner cedar wall and two-by-four uprights at the outer wall. The ceiling is of one-by-four cedar with a raised section above the windows to provide indirect lighting (**FIG. 16-7 AND 16-8**).

The sauna works well, except that it takes so long to heat. It is really convenient to have such a relaxing, meditative healing and bathing place right in the house, with a nearby shower area to cool off in. And the entire basement, though it is an enclosed secondary cellar space with minimal direct lighting, is rather comfortable and private. It is an excellent space for a guest or for just getting away from the mainstream of the house activity. The colorful exposed stones tone down that dingy, cold feeling of most cellar spaces (**FIG. 16-9**).

FIG. 16-9

CHAPTER 17

Putting Metal Through Changes–Making a Forge

The basic function of a forge is to heat up iron so it can be formed and shaped. It should be able to contain a fire that is hot enough to get steel or iron to the workable temperature, using coal, coke, or some such fuel. It is best to use a soft or bituminous coal that is relatively free from impurities. The impurities, when heated up to malleable form, will combine together to make clinkers. Clinkers are globs of crystallized matter that weld together as the coal burns. They gather at the tuyère or air nozzle and block the flow of air coming in from the hand blower. They should be cleaned out often so the air can properly be fed to the fire. A commercial blacksmith coal is preferred to regular hard coal.

Coal alone is not enough to heat the metal to the proper workable temperature. You need an oxygen source to supplement the fuel. I use a hand blower to regulate the air. The hand blower is attached to the right of the top edge of the 20" split rim wheel drum, which I use for the bowl or hearth of the forge. The air from the blower is directed down and then over to the center of the forge, then below the bowl, then it blows up through a tuyère, into the drum. The tuyère is the perforated steel plate at the bottom center of the drum, which allows a controlled amount of air into the bowl or drum.

The drum is raised to 32" or table height and is supported by an 8" core pipe that extends up from the base. This height is comfortable to work with and is great for tool placement. The rock work was mortared together with a mixture of gravel screened through a ¼" mesh screen, Portland cement, and a small amount of fire clay. The fire clay makes the mix muddier and stickier, enabling it to adhere well to the rocks. I soaked the rocks for at least ten minutes so they would take on moisture. The moisture also helps the mortar stick better. I cleaned off the rocks with a trowel and sponge as I finished each mix. This rock work extended behind the drum, high enough to support the hood above (FIG. 17-1).

A 3" air intake pipe made from an old driveshaft comes down from the bottom of the hand blower.* It supplies plenty of air through its wide diameter. It is 12" in length, long enough to extend to below the bottom of the drum. At this point, it is welded to an elbow pipe that brings the air to the center of the forge. Another through pipe comes down to meet the air intake and to drop the clinkers and slag down to the cleanout at the bottom. I built a little catch shovel below this intersection to direct the air into the drum when the blower is working. This tool prevents air from being lost down the cleanout. When the forge is not in use, I just pull this shovel back

*EDITOR'S NOTE: A foot pump blower leaves both hands free.

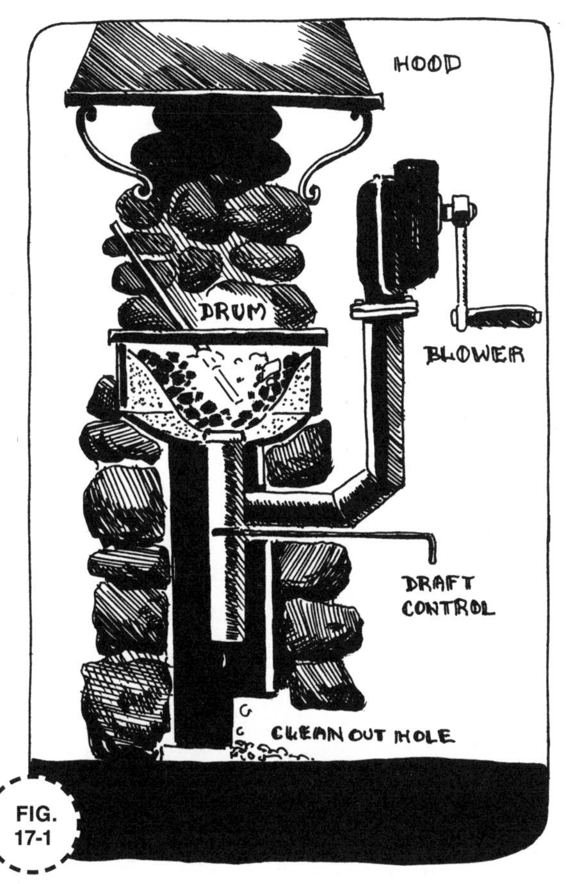

FIG. 17-1

a couple of inches and let the slag and other detritus that went through the tuyère drop down into the cleanout hole (see **FIG. 17–1**).

The drum bowl is wider and deeper than the hearths of most portable forges. To concentrate the fire more within an 8" diameter and protect the metal drum, I lined it with a river clay refractor. The clay comes in flat for 2" from the edge of the drum and slopes down toward the bottom of the drum for another 3 or 4 inches acting as a reservoir for the coal. This provides the fire with more depth than the usual setup. A short pipe extension comes up from the bottom of the drum to the top edge and is easy to clean out (**FIG. 17-2**). Once the clay gets hot, it vitrifies and becomes a great refractor to protect the drum. This inner barrier doesn't have to be clay. It could be any material that serves as a

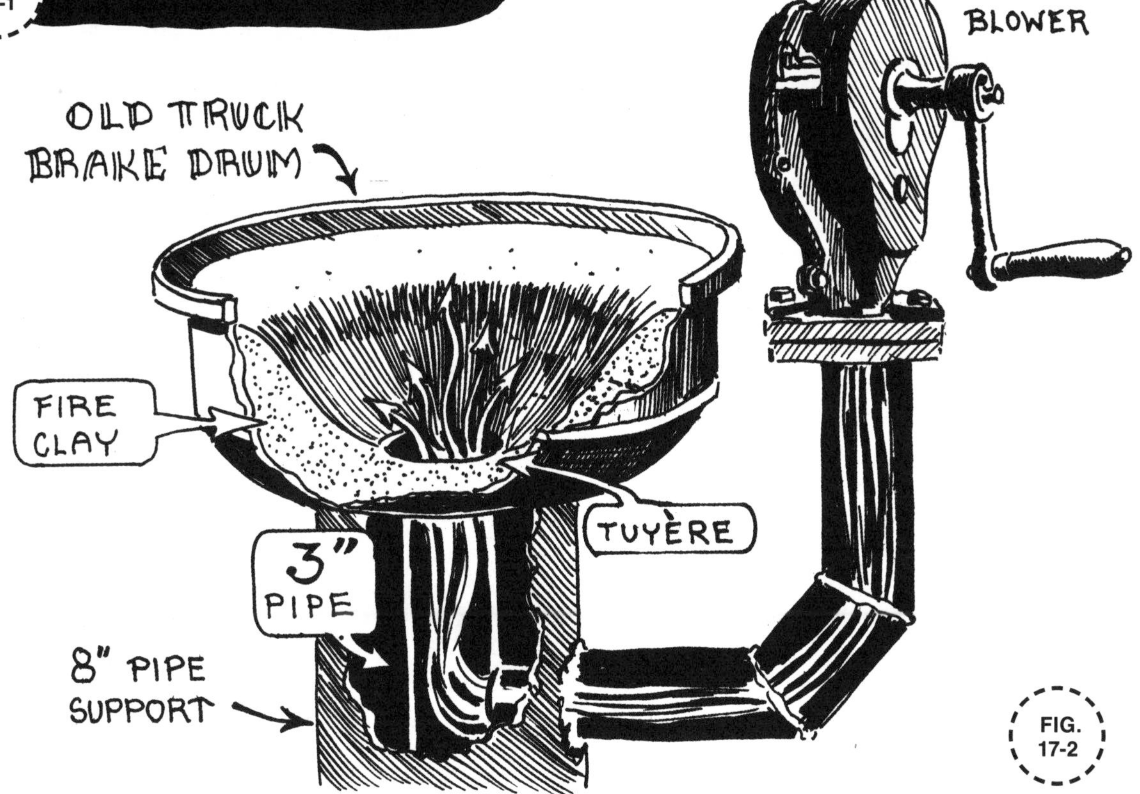

FIG. 17-2

refractor, is resistant to heat, and provides insulation for the drum.

The hood is made of plate steel and is 3' long by 2' wide at the bottom, tapering in 20" to a narrow 8 × 10 inch opening at the top to accommodate the stack. It is 30" above the drum to provide proper draw without being too much in the way. This hood also picks up the fumes if I'm welding nearby. The flue stack should be at least this 8 × 10 inch size for sufficient draft. Anything smaller wouldn't effectively do the job. The hood is attached to two bolts that stick out from the rear concrete support. It should also be braced at the stack. This stack will be welded to the top of the hood and will extend straight up through the peak of the roof (**FIG. 17-3**).

I use the forge for making hand tools and manipulating and forming steel. It is cheaper and easier than welding in certain operations. With welding, you can change the shape of a piece of metal but you cannot alter the cross section. If you have a 1" square stock, you can cut it. If you heat that 1" bar stock in a forge, you can actually make it fatter by dropping it (beating it) on the anvil. You can also draw it out and make it thinner.

There just aren't that many fine blacksmiths around anymore. So much has been lost that books on the subject haven't been able to recapture. But it's actual experience that counts—and I just love to see metal go through changes.

Here is a simple project involving a few of the basic forging techniques. This will give you an idea of how easy it is to make your own useful tools.

Start out by purchasing some blacksmith's coal from a local hardware store. This product is best for the job because it contains fewer impurities than regular hard coal, so it leaves fewer deposits or clinkers.

To activate the forge, make a well in the center of the bowl and clear the tuyère of any foreign particles. Wad up paper in the well and build up the coal pieces around and over it. When the fire is lit, push the coals in from the sides to supply the well with new fuel. Turn the crank of the blower and supply as much air as needed to really get the coals hot. When the fire spreads throughout the bowl, dampen the periphery of the well to confine the fire to the center. At first you'll get a lot of harsh smoke, but it will soon clear as the residue burns off and the coals reach the hotter temperatures. Watch where you set up. Be careful not to set up too close to a neighbor's clothesline, lest the soot from the forge gets all over the neighbor's laundry. It is also advisable

FIG. 17-3

to protect your eyes with goggles to prevent sparks from burning them.

Once the coals are red and the actual fire has subsided, you are ready to put in the metal. For this project, a length of ½" round stock was used. A 12" bolt or most other salvageable iron of this thickness could be used. The bar was heated up to a very hot white heat so it could be easily cut to length on the anvil. When heating metal, create an oven with the coals by enclosing the metal with them. Enclosing the metal in this manner also slows down oxidation. Oxidation causes pits, which show up on the surfaces. These pits are unsightly and are hard to grind out. They leave an uneven finish, especially in finer projects like knives. Leave a little corridor so you can watch the metal turn from red to yellow and finally to white. Work the coals with a stoker, bringing the new coals into the well. Take out clinkers as you go and clean out the trap below the forge to maintain a proper flow of air to the fire. Keep the fire steady. Do not put too much of a blast of air on the metal, but slowly crank the blower to provide a continuous supply of air. Be careful not to get the metal so hot that it begins to melt. Check it when you feel it is hot enough. If tiny sparks come off of it when it is lifted out of the well, this means it is beginning to melt.

Since the round stock was too long for the hook, it had to be cut on the anvil. The anvil should be solidly mounted on a heavy stump or round of hardwood. It should be raised to the height of the blacksmith's knuckles so he or she can work continuously and comfortably for several hours in a straight, standing position and not have to bend over while working. Cutting is usually done on the hardy attachment, which fits into the hardy hole at the butt of the anvil surface. If this attachment is not readily available, cutting can be done on the shoulder behind the horn. Cutting should never be done on the surface plate face because that section of the anvil is of harder steel than the shoulder. This hard surface will dull a tool if it hits it. The shoulder is of a softer temper and will give when hit; even so, when you are cutting, try not to let your tool go entirely through the metal you are working on. Cut almost through it, then bend it back and forth at the cut until it breaks.

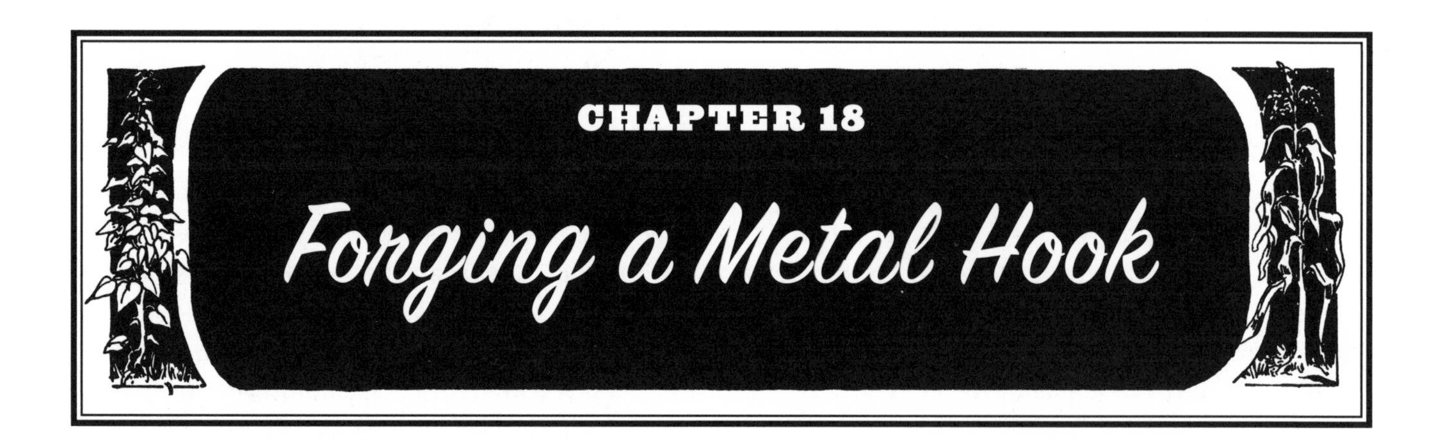

CHAPTER 18

Forging a Metal Hook

The first step of the actual hook is punching the ring or eye. This employs the technique of upsetting. Upsetting is bringing more metal to a given area by beating down on it. In this case, we want to gain material for flattening. Upsetting requires a very hot heat, close to the melting point. To upset, bring the metal to the anvil with hot end down. Beat on top until the opposite end begins to flare (**FIG. 18-1, 18-2, 18-3**). Don't let it bend (**FIG. 18-4**). Once it begins to bend, keep turning the bar and beat on the bottom. Bevel the ends as you work to constrict the flow of metal. Bring the flow into the center instead of making the sides mushroom out. Use several lighter taps with the ball peen

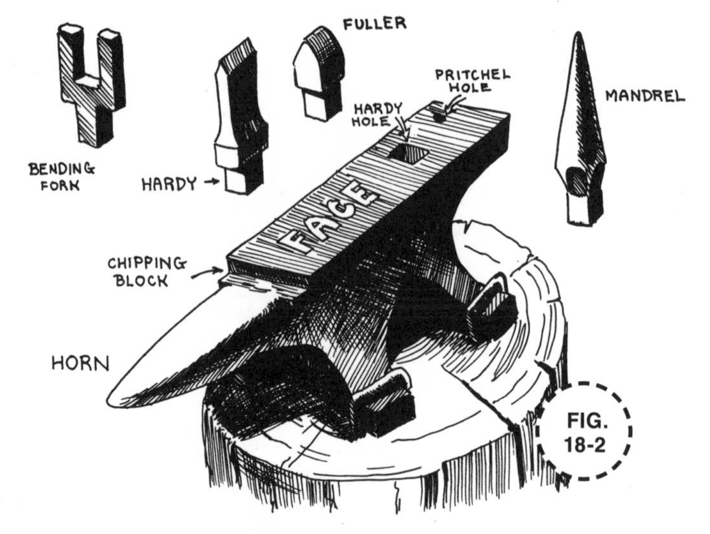

FIG. 18-2

FIG. 18-1

FIG. 18-3

instead of smashing it hard. The pounding effect rarely reaches the core of the bar, anyway, so save your energy.

Some blacksmiths use the ledge of the shoulder as a guide while they beat on the rod, twisting in a circular motion as they hit (**FIG. 18-5**). This gives the rod a shelf to be pounded against.

If the tip starts spreading quicker than the area where the ring will be, quench the tip immediately to cool it off so the area behind it will spread instead of the tip. When you quench, move the metal around instead of keeping it still, to make it cool off more slowly and evenly. The faster metal cools, the quicker its molecules freeze and become brittle. This might cause your project to crack at the most brittle part.

Other causes of cracking are letting the piece burn when heating, making it lose its temper and become too brittle; working it until it gets too thin; or working the piece too cold.

Once the tip of the rod is upset enough, it is ready to be flattened with a large ball peen hammer. The flattened tip of the rod is then placed over the hardy hole of the anvil and a hot punch is centered on the peened area. The punch is hit several times with a hammer until it makes a dent into the peened area (**FIG. 18-6**). Be careful

not to drive the punch through the rod or else flaring may result. Instead, turn the flat side over and find the nub from the punch, then go through the piece from this opposite side. Again, be cautious not to work the metal too cold, especially when it becomes this thin.

Now widen the hole, using the front tip of the horn. Put the horn tip through the eye while holding the unworked end of the hook with tongs. Hammer toward the anvil face first with downward pounding motions, then work around the horn, turning the eye as you hammer around the tip and sides (**FIG. 18-7 AND 18-8**). The tip should have the greatest concentration of heat.

FIG. 18-8

Since the horn tip makes a tapered hole because of its shape instead of an even one, the eye should be worked around a drift to make it the same size on both sides (**FIG. 18-9**). Be sure to work the sides flat as you widen the hole, to prevent them from getting too thin. This can be done by intermittently flattening the sides on the anvil face with a heavy hammer. After this process, the outer rim of the sides should also be beveled in for additional strength, with the drift pin reinserted in the eye so it is kept round. Repeat until even and proper sized. After the eye has been worked to the desired diameter, the sides and tip can be shaped in the same manner. Then the project should be hammered over the horn to true up the base of the ring. To finish the eye and sides, round off any points with a small hammer (**FIG. 18-10**).

FIG. 18-7

The next process involved is drawing out. This process makes the cross section of a given area of material thinner while it lengthens the whole piece. Drawing out should be accomplished over the horn, pulling the rod toward you with tongs while pounding the top of it with a heavy hammer. This enables the metal rod to be worked over the convexed high point of the horn, stretching it out as it passes over the contour. The project should be turned as you work it, to get an even stretch on all sides (**FIG. 18-11**).

FIG. 18-9

FIG. 18-10

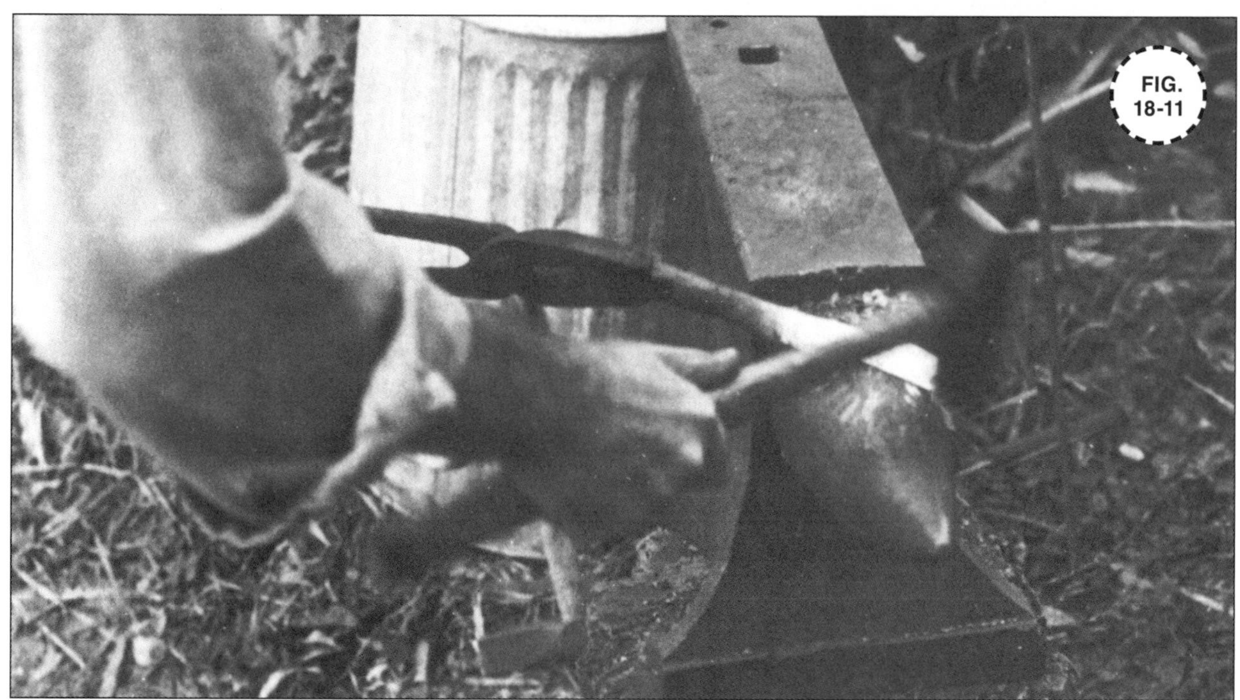
FIG. 18-11

After it has been drawn out a bit, take the rod to the anvil plate and true it up to keep it straight. Repeat the drawing out process until the rod is the desired length. Then bring the tip to a point and taper it from the shank by beating around it on the anvil plate until it becomes the proper shape.

Now comes the bending: Bending doesn't require as much heat. Heat up the tapered tip until it is just beyond cherry red. Position it on the anvil horn and beat the tip until it bends slightly with the contour of the horn (FIG. 18-12). Heat the whole shank to the same cherry red color. (Be careful not to heat and weaken the eye.) Then turn the rod over and shape the shank around the horn in the same manner (FIG. 18-13).

From this point, the finished design is up to the individual. If you want a hook that hangs straight, you have to create a curve in the hook shank that will evenly distribute gravity. Gravity seeks the lowest and highest points of a hanging object, so you want the gravity line to bisect the eye and the curve of the hook shank. After the project has been worked to the desired shape, flatten it out and bevel it for strength. If necessary, tap it again to realign it and always be careful to maintain the balance. Then dress it up with light taps with a ball-peen hammer for a rustic finish. Finally, heat it up one last time all over and quench it completely to provide an even temper (FIG. 18-14).

FIG. 18-12

FIG. 18-13

FIG. 18-14

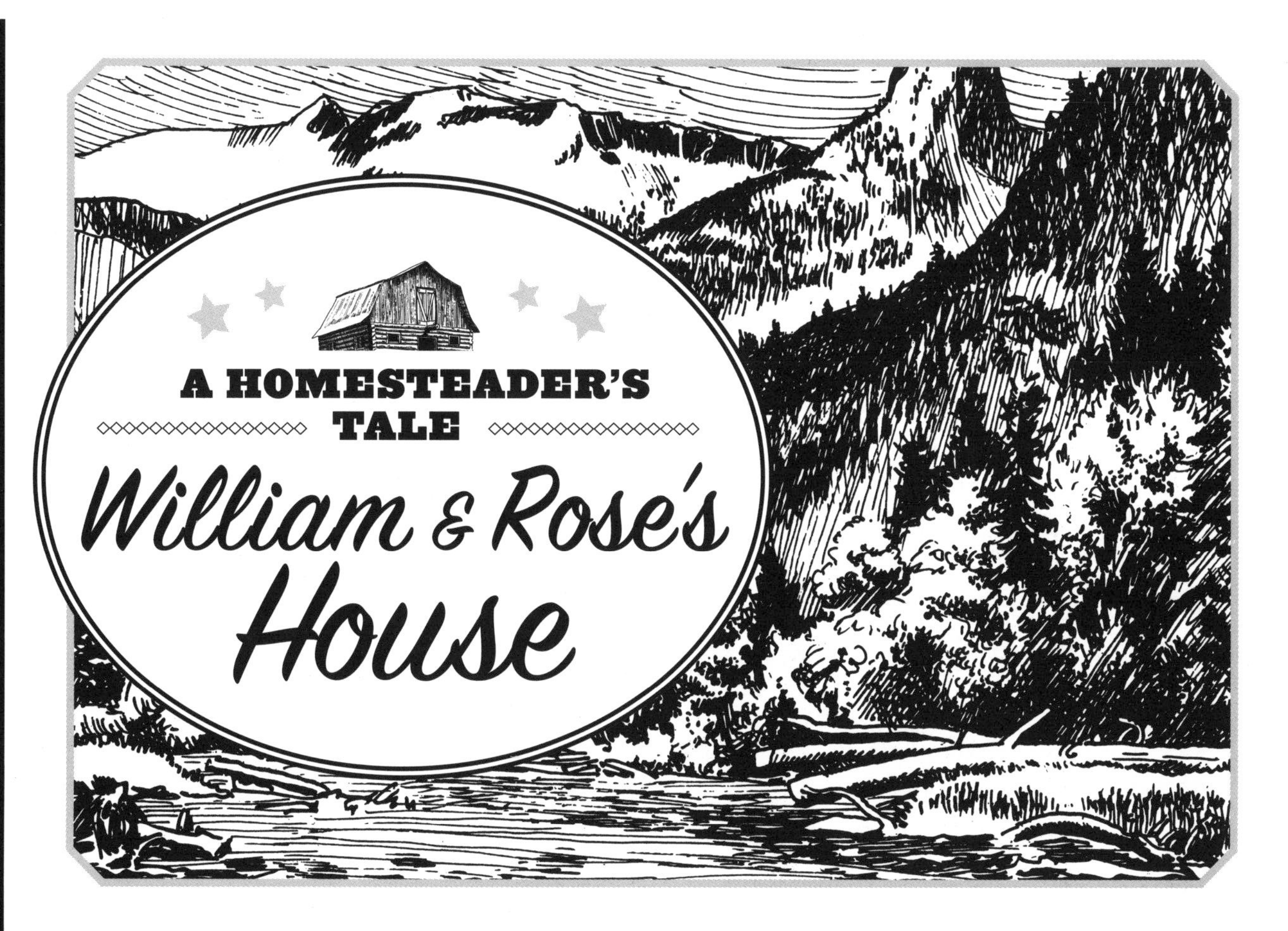

Instead of constructing their house themselves, William and Rose hired an architect to design it and inexperienced carpenters to build the structural portion of it. Yet they wanted their home to be special, something uniquely their own, as do most people. Being practicing artists, they desired their house to be aesthetically interesting in itself. To achieve that, they put in the finishing touches themselves, using the mediums they know best. They enhanced their home's structural beauty and made it a visible example of their own creative personalities.

Most architectural offices refuse to design small, individual houses because it is too much trouble for not enough profit. But there is usually a young architect around, a few years out of school, apprenticing in an office, who would love a chance to design a house (traditional moonlighting). The particular fellow they found was working as a night-shift janitor in a local pub when they hired him to design the house. In fact, he did most of the designing in between his janitorial duties. He had "dropped out," as they used to say, leaving a large architectural corporation because he decided "it wasn't good for his health."

Rose and William wanted a multi-bedroom house to fit their meager budget, so a design was

Evolution of floor plan

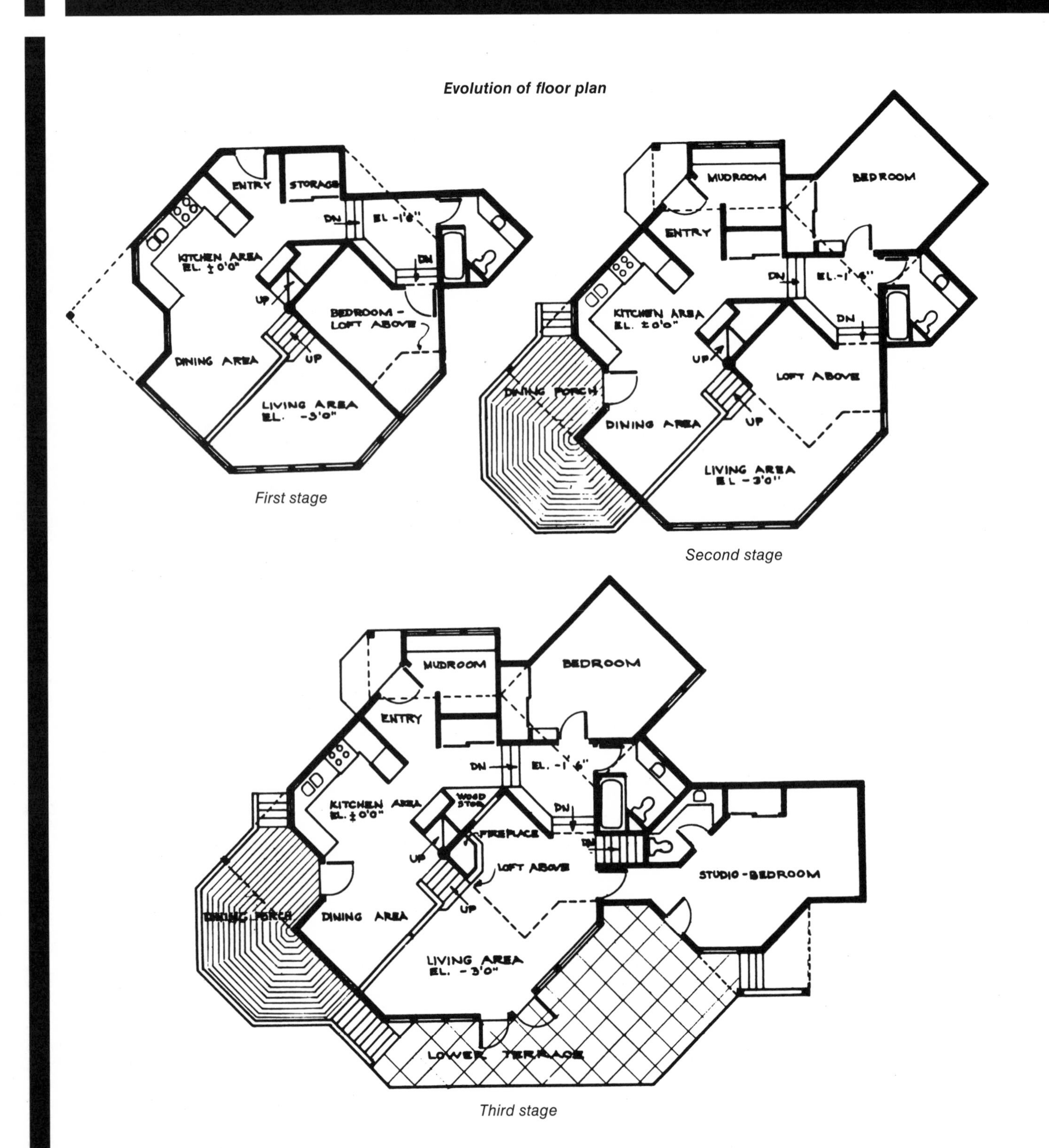

First stage

Second stage

Third stage

developed that could be built in stages. The first stage was an octagon, 12' on a side and about 30' across. It would contain a completed kitchen and dining area, bathroom, sleeping loft, and living area. The living area would be temporarily partitioned off to provide a closed-in bedroom under the loft. The octagonal core would have one high, open ceiling space, and the areas would be divided by changes in floor levels, each level spiraling around the center pole. The first stage would be well within their budget.

The second stage would add a mudroom and a pantry at the entry, another bedroom, and a small, partially covered terrace deck off the kitchen. The partition in the living area would then be removed, and that space would double in size when the bedroom was added. They chose to go this far with the structure at the beginning.

The third stage allowed for a large terrace off the living area and another bedroom which would extend from the east side of the octagonal core. This section could be built onto, almost indefinitely, as funds allowed.

The owners chose a site on the slope which provides complete southern exposure. It includes a view of the surrounding mountains and overlooks the nearby river (the river view has since become the site for a huge canal project that wiped out its serene beauty). The southern exposure also provides the greatest amount of year-round sun.

Though the slope has a 6' in 30' grade, this did not cause a problem. In fact, it inspired the multileveled floor layout which follows the slope and sets the house into the landscape. A single-level floor design would have forced the house to be lifted out on cantilevers and look like an object that did not belong. For the structure to properly

Southern slope

Northern face of building

fit in with its surroundings, it had to be built as close to the ground as possible, with each level staying with the contour of the land.

Since it was difficult for Rose and William to visualize the architect's ideas from the drafting plans, as it is for anyone without proper training, the designer built a scale model of the house and explained its spaces and relationships. The house and its inner areas were positioned in relation to their exposure to the sun and the airflow. The southeast bedroom would get the cool morning sun; the living area, because of its southeastern, southern, and southwestern exposures, would get the sun at different angles throughout the day; the dining area and kitchen, to the west and northwest, would get only the afternoon sun as it began setting behind the sparse forest.

Eaves, 3' long, were designed to extend over the southern and western sections to shade them from the direct sun but allow sufficient light to enter through the many windows. The low-opening windows at the south would allow cool air to enter the house, while higher vents would let the warmer air escape.

The design and the actual building practices were changed to suit the owners' style. Since William is 6'4", the ceilings of each enclosed section were lifted 4". High shelf spaces were allowed above closets and partitions for open storage of his pottery. The loft was designed as a semiprivate place in which Rose could do her writing. Many shelves were built into the loft railing for her books.

The whole central core has a spiraling effect around its center-pole axis. Each space moves into the next by natural progression, yet each is distinctly individual. The spaces are broken up for quietness and privacy by varying levels and short partitions. There are no dead ends within these areas.

William and Rose were pleased with the designer's concepts, except for the lack of basement space. The site was wet in the springtime with runoff, which flowed down from the background mountains. This made such a space impractical.

Construction started just after the summer solstice. The crew consisted of three relatively inexperienced builders, and a master carpenter who visited the site every two weeks to aid in whatever problems arose and to give advice where needed. The crew was headed by a man named Joel, who, though not a journeyman carpenter himself, had apprenticed under the master carpenter in the past. The owners' son was the second crew member. The young architect was hired as the third member, to assist the others in carrying out his design. The trio and a few of their friends shared a bunkhouse, tents, and an old house on the property. They each received the same wage. Rose became head cook and vegetable gardener, and though it was suggested that she too get paid, she refused. Everyone was involved on equal terms. The sense of building a truly creative structure as a communal effort made morale high.

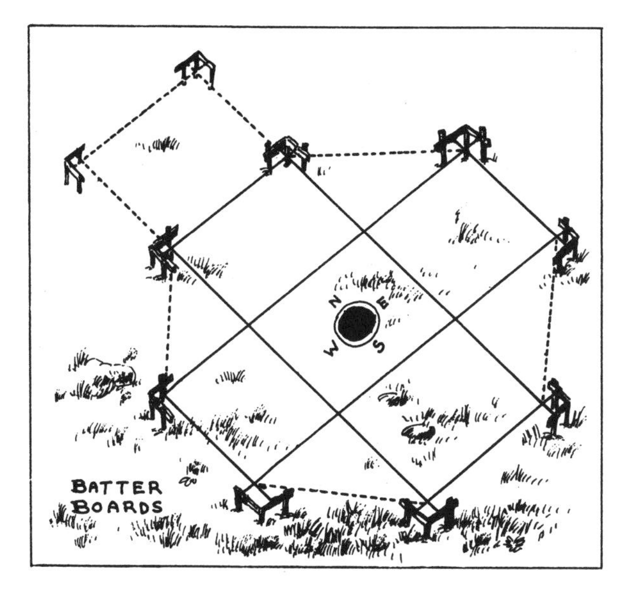

Batter-board layout for octagon

The construction began with a traditional batter-board and string layout, but, because of the many angles of the octagonal core, it proved to be more difficult and complex than usual. The builders roughed out an area approximately 12 × 45 feet and set up the batter boards at the four corners. Another rectangle, about 12 × 30 feet, crossed the first one 8' from the southern end, making the basic octagonal shape. These rectangles were then checked to make sure they were true.

After the area was laid out, an 18'-long trench, 3' deep, was dug for a concrete block stabilizing wall. It ran southwest to northwest in the middle of the octagon, where the central level change occurred. The stabilizing wall proved to be structurally unnecessary. The fireplace pad, which was incorporated into it, could have been made separately as an additional footing.

Footing holes were dug at each of the twelve intersecting points of the octagon and at each angling point in the remaining sections—twenty-two in all. Their holes were 3' deep to insure their being safely below the frostline. The forms only surrounded the 12" squares that protruded above the surface, because the sandy ground below acted as a natural form for the rest.

The cement for the footings was poured. As the concrete began hardening, spikes, which would later be driven into the six-by-six upright posts that supported the floor beams, were sunk into the tops of the pads. The varying heights of the upright posts were determined by the surrounding terrain. If a footing was lower because the ground dipped, its post would be longer to

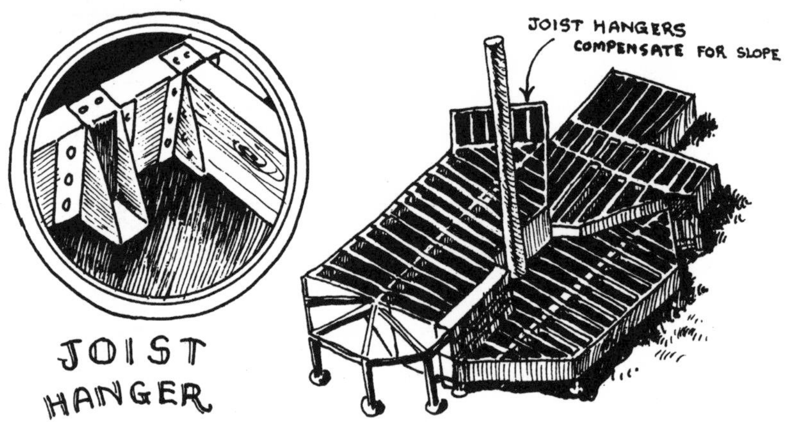

compensate for it and to make it level. The posts have asphalt vapor barriers below them to prevent the sweating concrete from prematurely rotting them.

The design of this house incorporated three separate levels to keep the structure as close to the sloping ground as possible and to act as a space divider for the various areas. The living area and fire-nook floor is the lowest. The hall, bathroom, and bedroom level is 18" above it. The kitchen, dining area, terrace, and mudroom are on the upper deck, which is 18" higher than the second one, and 3' above the living area. To insure that each level would be as close to the contour of the slope as possible, joist hangers were used as needed. In areas such as the main bedroom, which is very near to the ground on the north end, the hangers made it possible to keep the joists at the same height as the floor beams, instead of lifting them 8" above these beams. Joist hangers are widely used and are known to be structurally sound.

The double two-by-eight floor joists were spaced 16" on center, and 4" of fiberglass insulation was tacked between them. A one-by-four subfloor was then nailed over them.

As the carpentry got more complex, Rose and William felt as if they were spectators to something that was running away from them, rather than being the controllers of what was happening. William was teaching pottery at a nearby college and Rose was working on her doctorate, growing vegetables, and cooking for the crew. The house seemed to be the private adventure of those who were working on it; Rose and William did not feel it was theirs. In fact, they found their own living quarters and privacy overrun by the crew and their occasional visitors, who freely used the old house's facilities whenever they felt the need. At times, William even found it necessary to lock himself in the bathroom to escape the omnipresent crowd. What was happening on their own land was indeed a cultural shock to them. They were not used to living as communally as they were being forced to. But they could not complain; the work was going well. After all, how else would they get a custom job like this done for such a small monetary investment?

Next came the framing of the walls. The crew decided that the easiest and most accurate method of assembling each wall was to first lay out its pieces horizontally on the level deck, nail them together, and then raise the finished frame and fix it into position. This method made tedious toenailing unnecessary, and provided the carpenters with a straight surface on which they could line up and measure the two-by-fours.

The studs were placed at 16" centers and appropriate spaces were allowed for windows and doors. Where a window was to be put in, the cripples were cut to meet the horizontal sills. A cripple is any vertical stud which is interrupted

Wall framing

by a horizontal piece before it spans the top and bottom plates. Single two-by-fours were used to frame the windows and doorways. Any additional framing and spacers to absorb the settling of the completed structure were put in later.

The raised walls were plumbed with levels and the bottom plates were nailed through the subfloor to the floor beams. The wall sections were then joined together at the ends with flattened tin oil cans and they were capped with a second top plate of doubled two-by-eights on end. The two-by-eights lapped the first plates on the wall ends to provide an additional tie to keep the walls in place.

Because most of the inside partitions were not bearing any structural load, they were framed with single studs, placed at 24" centers, and topped with single plates. When the framing was finished and the walls and partitions were in place, the once simple, visually appealing, multilevel decks were transformed into a maze, which looked like a forest with trees going every which way. There did not seem to be any continuity within the structure. But after the plywood sheathing was nailed onto the outside of the walls, the maze developed meaning: Aha, there's the kitchen and the mudroom; this mess is really becoming a house.

Loft Area

After the decks, walls, and partitions were in place, it was time to provide the growing organism with a center-pole axis to support its roof, and convert it into a spatial masterpiece. The raising of that 16' cedar pole brought a welcome celebration. As it was put up, cheers filled the air. Rose prepared the festive table. By then she was ready for anything. There was plenty on hand to make the celebration last all night, and it did. The bunkhouse and old house were filled with merrymakers. Soon, exhausted by the event, Rose found her way to a quiet hillside and slept in the bracken. (William was in the hospital at the time, having an operation.)

The next morning, the weary crew began assembling the loft area. They cut spaces for the huge, custom-made beam hangers into the center pole. These hangers hold up two of the double two-by-ten beams which frame the loft. Four of these beams surround the loft and span from the center pole to the southern and southeastern walls. A ridge of two-by-fours was nailed to the inside foot of the beams to support the heavy, four-by-six double tongue-and-groove deck. A low railing was later placed around the loft area. The crew then set a bookcase into the front railing for Rose's books.

Truss supports for roof

Trusses and center pole

View from entryway

Before the actual construction began, Joel and the architect spent many hours trying to figure out a suitable roof support. One night, Joel began thinking about airplane hangars and how those roofs are held up. After studying that type of truss system, he and the master carpenter designed a similar system, which would intersect the original octagon and not detract from its open, spatial beauty by cluttering it with a network of ceiling joists.

The four trusses were hung from a four-way iron hanger located at the top of the center pole. Each truss spans to opposite walls. Their main function is to hold up the collar beams, which form a 12' square around the center pole to brace the roof. Two of the trusses also act as gigantic hip rafters, which support the hip roofs. The other pair are hip rafters until they meet the collar beams, then they continue as ridge boards for the gables over the front door and loft areas. Each has two double two-by-six frames with webbing bolted between each side.

The roof in this dwelling is multilayered. The interior roof is made up of two-by-six nailers spaced at 36" centers. These nailers are visible from the interior and span from the collar beams outward. Not only do they enhance the spiral effect of the ceiling, but they also act as nailing strips for the materials above. This gives the carpenters the advantage of not having to nail upward from below; instead they work comfortably from above. That feature alone justifies the extra use of materials. The nailing boards also provide a deceptive sense of structure, which confuses even skilled carpenters. From the interior, they look like rafters placed too far apart, holding up an uninsulated cedar ceiling.

After the interior nailing boards were in place, the ceiling of one-by-eights was put on. Economy-grade cedar was bought for this job. The usable, unrotten, and unsplit sections of each board were sawed out and nailed up. The rest was used above the outer rafters to nail the shakes onto. Nothing was wasted. The cedar one-by-fours were spaced ⅜" apart to allow for shrinkage. Since shrinkage is inevitable, this spacing is a means to develop uniformity in the size of the distance between the boards.

When the ceiling was finished, the architect lay on the loft floor and studied the fissures of light which shone in through the spaces between the boards. He watched as they seemed to spin around the center pole, giving the roof the desired spherical appearance. To finish the ceiling, the shining light had to be shut out. A layer of black building paper was stapled to the boards to cover them and to act as a temporary vapor seal until the finish roof was put on.

The outer two-by-six hip rafters and ridge boards were cut and placed. The roof rafters were then spaced at 24" centers, nailed over the ceiling, and supported at the outer walls, collar beams, and hip rafters. The angle to which each rafter was cut depended on the pitch of the roof and the angles of intersection with the ridge rafters. The rafters and ridge boards extended 3' beyond the walls as eaves to shade the interior and protect the exterior lap siding from direct precipitation.

Blocking was put in between each rafter directly above the collar beams to keep the long spanning rafters from spreading. Six inches of fiberglass insulation was then tacked between the rafters. Though some may think this thickness is unnecessary and prevents air circulation, the roof is excellently insulated and the snow does not melt on it, preventing damage from ice collection.

The one-by-four cedar scraps left over from the ceiling material were nailed across the rafters to support the finish roof of cedar shakes. The 24" shakes used on the roof were store-bought, but they could easily be made from cedar bolts. The shakes were lined up on the cedar nailing strips, which were spaced 6" apart.

Normal preparation for plumbing and electricity was made before the interior one-by-six cedar ship-lap paneling was nailed on. It was then time to begin the interior finishing. Very little money was left, and the crew decided it was time to depart. The coming months of severe winter would make it impossible to work. The following spring, Rose, William, and their daughter became more involved in the house and finally felt it was theirs. They learned to use hammers and saws. They put up most of the plywood and became skillful in paneling with cedar.

For a long while, they could not decide on the material they were going to use for the finish floors. William was creative, and employed the art form he knew best. It took 500 pounds of clay to make the unglazed tiles that he used, but he would have needed more if he had not decided to leave the plywood and subflooring under the carpets untiled.

He also made copper-red glazed-tile counters for the kitchen and bathroom. Rose finished off the bathroom shower area in a tile mosaic made mostly of unsatisfactory pots, which, in the potter's tradition, had been hammered into pieces. Not everyone is a potter, but most people have some special skill. Many other materials can be used for the finishing touches.

The architect is pleased with the way the owners incorporated their own talents and imagination into the house. It would never have occurred to him to use handmade tiles for the finish flooring or make cupboards out of the same

Homemade tile floor

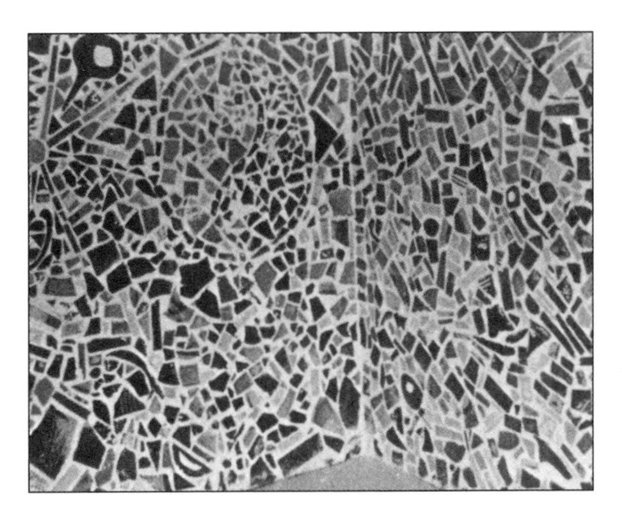

Tile-shard mosaic in bath

paneling as that of the rest of the interior.

This wood house, built into the slope, succeeds in becoming a masterpiece of spaces and relationships and blends well with the background forest and mountains. Its gabled roof entranceway protects visitors from the weather even before they enter into the backlit mudroom area where they may remove their overcoats and kick off their shoes. The spaces beyond the entrance foyer are shielded by a partition. From above this shield, the center post, trusses, and living-room lights can be seen; voices can be heard from those areas. One has time to become oriented and choose one's route. To the right, past the open cupboard featuring many of William's useful ceramics, is the kitchen. To the left, down to a lower level, are the bathroom and bedroom.

The kitchen is an open area, with windows that frame the hillside apple orchard and expose a narrow corridor into the woods. From this area, you can see into the dining area, which is only

Schematic layout of floor levels

Living room

divided from the kitchen by its own rectangular shape and its different floor. From the kitchen, you can also see out the high windows of the lower living area.

Between the dining and kitchen spaces, there is a Dutch door, which leads out to the roofed section of the terrace, past the west wall. The terrace is partially covered by the end of a hip roof. Its open deck extends from the west wall out to a railing, which would soon be sheathed-in with cedar siding to protect it from low breezes. The terrace overlooks the large mountain range and a nearby stand of aspens.

The afternoon sunlight streams in through the many windows of the dining area and makes the huge carved-wood table glow within this space. This table is where Rose chooses to do most of her writing. This space is divided by a short partition and a few steps leading to the lower living area. One can stand at the partition and view the whole lower area or sit at the table and feel completely closed off from it. From the

Loft

View from loft

living area, the spiraling ceiling and the large, spanning trusses seem to be much higher up than they really are. This effect is created by the many windows which surround that area.

Beside the living area is another space about the same size. Yet, because of its low ceiling, it gives the feeling of being much smaller. This area is a fire nook, which was soon to be enhanced with a beautiful brick-and-clay fireplace. The upper loft towers over the other levels and allows a view down from its sides of the spaces of the basic octagonal core. When standing, one can even gaze out of the high living-area windows. When sitting, one is in a private space created by the encompassing railings.

Two other separate areas with closing doors are additions to the original octagonal core. The bathroom is a tiny, five-cornered room, fully equipped with bath, shower, sink, and toilet. The wall behind the bath and shower is decorated with a tile mosaic pieced together by Rose. The sink has a copper glazed-tile counter. For the purpose of viewing the upper meadow, the toilet is located in the center of the room, facing the window.

The adjoining bedroom has a shed roof which slants down away from the octagonal core. Its Gyproc walls break the cedar-paneling style and help to make this room different from the rest of the house.

Rose and William are now living in this house. They are delighted with its views and are very comfortable within its spaces. They feel it is truly a house tailor-made for themselves and they regret that they did not participate as much as they could have in the actual building of it. They discovered that it is more worthwhile to make something you want than to earn money with your time so you can pay someone else to make it for you.

A HOMESTEADER'S TALE
Bob & Lynelle's Sauna

Since we needed a place where we could cleanse and rejuvenate ourselves as we planned and developed our new homestead, one of the first and most important buildings we constructed was the sauna. I personally do not like tub baths because bathing in a tub of still water doesn't really get you clean. Bacteria remain in still water and coat your skin with an oily film, so you are never really rid of the dirt you started with. A sauna cleanses your skin of its surface dirt as well as ridding your body of its inner impurities. It also helps you unwind after a hard day of working and provides you with a warm, peaceful chamber in which to meditate.

A sauna cleanses you from the inside out by flushing your blood with its heat. This stimulates your sweat glands, and your sweat washes through your pores as it flows outward to cool your skin. Then, after your body is saturated with sweat and you cannot stand any more heat, you scrub down with soap and plunge into a pool of cold water or stand under a cold shower and rinse off. The cooling effect of the water momentarily stimulates you and closes your pores. Afterward, when you reenter the sauna or go to another place to relax, you find that the previous combination of stifling heat and cold shock induces restfulness and loosens muscular and nervous tension.

Saunas have other healthful advantages that are beneficial to country living. They help heal surface wounds by bringing blood to afflicted areas. They act as vaporizers to clear out stuffy sinuses for sufferers of head colds. (Note: if you have a cold, icy plunges or showers are not recommended after this treatment.) They relieve hangovers because they work out all the alcoholic toxins from your body by stimulating circulation. Saunas are also perfect for massages and are great places to get to know people.

I built my sauna into a hillside, close to where I planned to build my house. It was set into the hill to make full use of the insulative property of the earth. My design included precautions to make sure the building would be well protected from runoff. I dug out a hole, 9½ feet wide, 11 inch long, and 5 feet deep, for the structure—all by hand. This, needless to say, was a tedious task that took several weeks to complete. But I did not want heavy machinery leaving its destructive scars on my land. I needed a hole this size because I wanted the sauna to be large enough to accommodate six to eight people, but small enough to retain heat well. Many of the people I know have saunas which are 8 × 8 feet. That space seemed too small to me because it allows only a few bathers to be far enough away from the source of heat to not be scorched by it.

I used the dirt I shoveled out of the hole to create the walls of the pool, located on the downhill side of the sauna. Four upright posts with crossing boards prevented the earth from sliding any farther down the hillside. The ends of the posts were buried into the ground and the boards were nailed across them to hold back the dirt. On the uphill side, the ground was leveled so the runoff would have a tendency to flow down the sides of the hill instead of draining in back of the sauna.

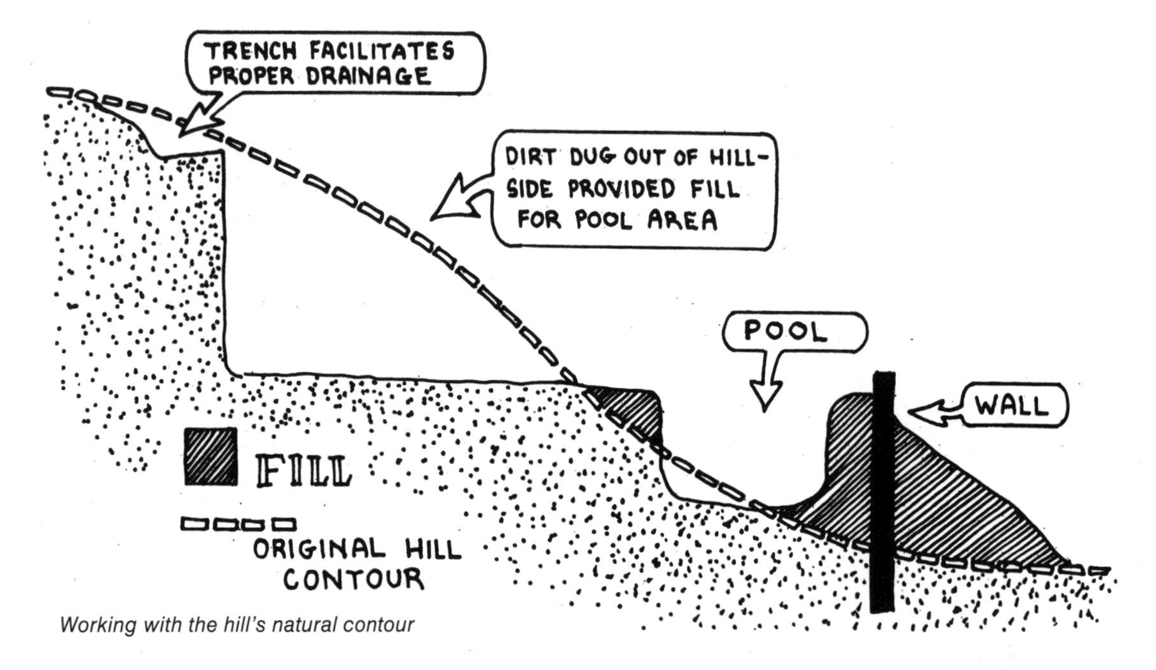

Working with the hill's natural contour

Since this sauna was to be my practice building, I wanted to experiment with many different techniques. After becoming familiar with them through practice, I would be able to choose which mediums I preferred working with and incorporate them into our house.

The walls were framed with the post-and-beam method, and were filled with dimension lumber, cement, and a stucco mixture. To start them, I dug four corner holes in the 9½ × 11 foot trench. Each was 3 feet deep and wide enough to accommodate an end of a creosoted railroad tie. These 8-foot-long railroad ties were discards I bought for very little. They are excellent to use when in contact with the earth; they were thoroughly soaked in creosote and will not rot out for as long as I'll be around to use the sauna.

After burying the bottom ends of the upright ties 3' into the ground so their top ends would be flush with the uphill side of the hole, I ran two-by-fours above each of them and made sure, with a carpenter's level, they were all the same height. I then packed sand and gravel around them and plumbed them on two sides with the level.

The front uprights were temporarily held together with two-by-fours so they would stay in

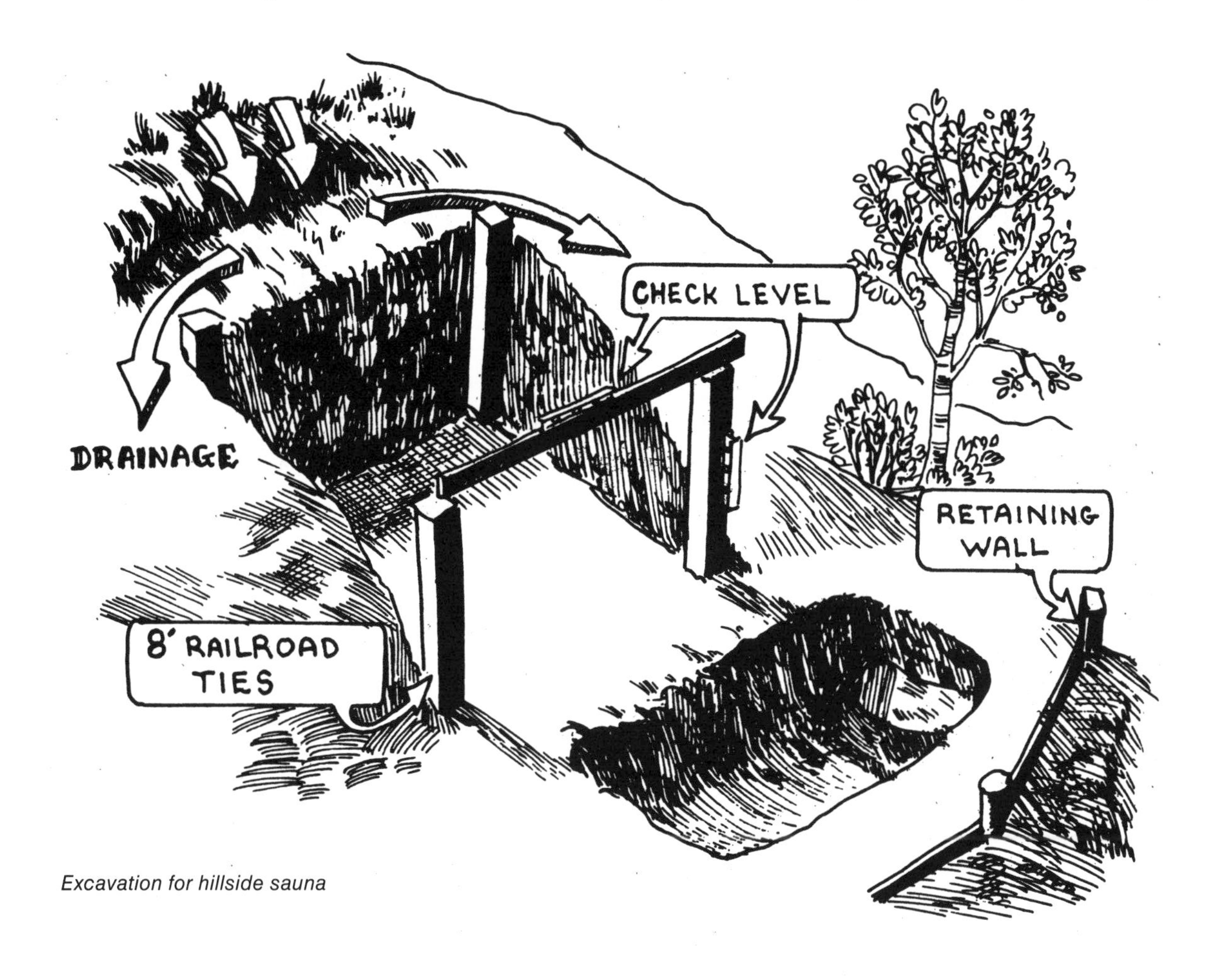

Excavation for hillside sauna

place while I dug an 18-inch-deep trench between them. This trench was for a short retaining wall. The wall would extend from the bottom of the trench to 12" above ground. I drove spikes into the sides of the uprights, and then nailed one-by-four forms between the ties. They were braced every 3' with narrow slats. I drilled holes into these slats and ran lengths of baling wire between them on either side. The wire ends were then wrapped around nails to hold them in place. These braces effectively prevented the forms from bulging as we poured in the cement. The mixture we used was 5 parts sand, 1 part Portland cement, and as many flat rocks as we could put into the mortar without weakening it.

The two side retaining walls were dealt with next. We built these walls to a height of 2½', using square rocks that we trucked in from wherever they were available. The walls were mortared in without forms. Whenever you build such a wall, it is important that you use flat, wide rocks that stack up well. We dug 10" channels between the side uprights, and nailed spikes along them to tie the walls together. A 4" layer of mortar—3 parts sand and 1 part masonry cement—was poured into the channels for the bottom rocks to set into. The remaining wall rocks were mortared in above them until the walls were built to height, then they were checked for levelness with crossing two-by-fours.

Foundation forms

Above the upright posts, I put in a double-layer top plate of interlocking horizontal logs. These logs were 10" cedars that were peeled and ready. They brace the long, overhanging roof and raise the walls another 18". The first layer of side logs was notched over the uprights. The crossing logs were then round-notched onto the side logs. To make my notches, I eyeballed the contour of the lower logs, figured the approximate depth and width of the notch, and marked that on the log to be notched. Then, with a swede saw, I cut deep scores into the notch area to the marked outline, and knocked the pieces out with a chisel. The upper layer was notched over the first layer in the same manner. The top side logs were 17 feet long, extending 2' beyond the rear wall and 4' beyond the front wall to support the roof's overhangs.

Front and side foundation walls in place with uprights

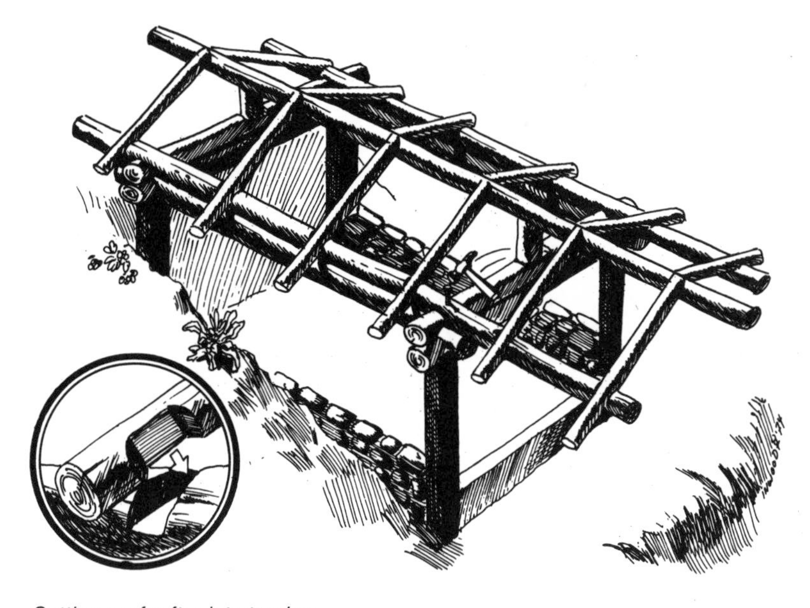
Setting roof rafter into top log

At this point, because the weather was getting cold and rainy, we decided to put on the sheltering roof instead of filling in the walls. I figured out a comfortable pitch for the roof and spiked in two 18" upright ridgepole supports above the front and rear top wall plates. These uprights allowed the roof a slight pitch of less than 15°. I didn't want the pitch any higher than that, because heat would be more apt to escape through a higher roof. Also, I planned to finish the roof with sod, and if the pitch was any steeper than 15°, the loose sod would have more of a tendency to erode off.

Built-up roof

A 17-foot-long ridgepole was notched over the short uprights, and the rafters were seated deep into the top side plates and the ridgepole. These six rafter sets were each spaced at 36" on center. I lap-notched them deeply to make them level with the tops of the side logs and the ridgepole. This eliminated having to use spacers between each rafter to fill in the usual gap between the subroof and top wall plates.

The lap joints were easy to make. I simply sawed and chiseled out the square laps at 36" intervals on the side plates and ridgepole. I then figured the angle for the rafter top by resting the rafter bottom above the side plate so the top met with the ridgepole. I eyeballed the angle and sawed into the rafter top. After spiking it into the side of the ridgepole, I eyeballed the angles and depths of the bottom cuts in the same manner. If I were to seat the rafters again, I would hew the sides of the ridgepole just slightly instead of notching them. The deep notches weakened the structural soundness of that primary beam.

Because of the heavy sod roof that would be above it, I made sure to construct a strong, vapor-proof subroof. The first layer was of two-by-eight cedar boards, which were nailed lengthwise across the rafters. Over those boards was a layer of heavy building paper and another layer of two-by-eights, which crossed the first. I stuffed thin strips of fiberglass insulation between both sides of the peak and over the side wall plates to seal off these areas. Then a ridge cap of aluminum flashing was put over the boards to prevent the

Front entry of sauna with knee braces supporting overhang

rain from entering through the peak gap.

To seal this subroof, I spread melted tar over the top boards using an old broom. The tar came in 100-pound lugs which I melted down in an old bucket. I needed 200 pounds to initially cover my roof. Another thin coating would go on it when I spread the coarse gravel. This gravel will embed into the soft tar to keep the sod on the roof. Also to hold the sod, 4" poles were nailed around the outside edge of the roof eaves. They cross at and extend over the peak for an ornamental effect.

Filling in the walls

Even without the final sod covering, the sauna roof did not leak at all. When the weather is hot, uncovered tar has a tendency to get soft and sticky, but it doesn't drip or run. The only problem that did occur with the roof was that the 4' front overhang, which shields the dressing area, sagged under the heavy winter snow load. I corrected that structural problem by notching in two diagonal braces to support the overhang. These braces spanned from the side plate extensions to the center of the corner upright posts and successfully corrected the sag, while adding a nice aesthetic appeal to the front section.

I planned to put 2" to 3" of good soil on the roof for the sod covering. It will be seeded with white clover so the roots will help hold the sod together. The flowering clover will make the roof flow into the landscape of the hillside.

When the roof was on and the sauna was protected, we returned to the walls, handling each in a different manner. Above the short rock walls on either side, we laid an 8" strip of fiberglass. A horizontal larch log was placed over each strip and was toenailed to the corner upright posts. A shallow notch was cut into the center of each of these logs and into the center of the top plate logs above. Two 8 × 8 inch upright beams were then spiked into these notches. This created four smaller rectangular spaces, two at each side wall. I wanted these smaller spaces because the wood I was using to fill the walls was short scrap stuff that was lying around. None of it would go the entire distance between the corner posts.

The outside edge of each of these rectangular areas was framed with 2 × 2 inch nailing strips and three of the four sections were filled

in with horizontal four-by-four planks. The fourth rectangular section was divided again with two 4 × 4 inch uprights. These uprights support the small, hewn poles which framed an 8 × 10 inch window.

To brighten the outer rectangular sections of the side walls and give them a rammed-earth effect, the spaces created by the 2 × 2 inch nailing strips were filled with a stucco-like mixture. To prepare the sections, I stapled a double thickness of chicken wire into each of the areas framed by the 2 × 2 inch nailers. The thickness closest to the inner wall was ¼" away from the paneling; the outer one was about 1" away from it. The strips were also lined with several bent, rusty nails to hold the stucco mortar in place. The mixture I used was the basic 3 part sand, 1 part masonry cement mixture. The sand was a fine river-bottom type. It was mixed dry so there would be less chance of shrinkage. I added water until the mortar was a pastelike and malleable consistency. I worked it into the chicken wire until the mortar covered it and came out to the edge of the uprights. Using this method, I was able to build up the thickness over only one day. Applying the mixture in several coats, as is usually done with stucco, takes much longer.

Burlap was hung over the stuccoed wall and the drying mortar was sprayed with water for several days to cure it and prevent it from setting

Textures on sauna walls

Insulating and sheathing the front wall

Completed sauna structure

up too early. It set up well, without obvious shrinking or cracking, and the wall spaces were sealed in efficiently.

Many artistic touches can be added to the stucco as it is drying. You can draw or paint on it, or you can embed pieces of colorful glass into it.

To fill in the front section, I first laid burlap on top of the short retaining wall. The burlap acts as a gasket to prevent air from whizzing between the retaining wall and the six-by-eight horizontal beam toenailed in over it. This beam provided a bottom wood surface for the eight-by-eights which were to frame the doorway. The uprights divided the front wall area into three rectangles, the middle one being the entrance to the sauna. The other two areas were framed with two-by-fours that were nailed to the centers of the surrounding burlap-lined beams. Rough-cut one-by-four sheathing was then nailed to the outsides of the two-by-fours, closing off these two rectangular sections.

A layer of building paper was tacked onto the inside of the two-by-fours and an inner sheathing of two-by-six was nailed over it. I left a small area open near the top of the inside wall, to allow sawdust insulation to be shoved in between the walls. This sawdust was packed down as tight as possible so it wouldn't have a tendency to settle. After the sawdust was packed in, the top sheathing boards were nailed in place.

The center 32 × 48 inch entranceway was then framed in and a door of two-by-six tongue-and-groove cedar was hinged in place. These cedar boards were held together with two-by-fours, which were doweled into them at top and bottom.

The back wall had to be structurally sound and able to repel any drainage that would come in from the hillside behind it. To create an adequate vapor barrier, I nailed a big sheet of 6-mil plastic between the corner uprights. This vapor shield covered the entire wall section and acted as an outside form to protect the cement wall from the earthen hillside in back of it. The cement-and-rock portion of the wall was built high enough to shield the blazing heat stove, which would be directly in front of it. An eight-by-eight beam was set above the retaining wall, and the remaining upper section was framed and filled with dimensional lumber.

Gable-end closure

The front and rear gable sections created by the roof's pitch were the last areas to be filled in. It took me awhile to figure out how I was going to deal with them. I had many ideas but few materials. My decision came at a friend's summer solstice party. There before me, lying all over the field in front of the musicians' platform, was the material I needed: I collected several cases of discarded beer bottles and took them home for the sauna gables. The gabled areas were prepared with chicken-wire strips, which I stapled from the top plate log to the underside of the subroof boards in a zigzag pattern. Bent nails were then added to also grip the mortar.

The chicken wire not only held the concrete, but it also stabilized the large number of bottles as they were being cemented in. The job was simple and went very well. The setting mortar was sprayed with water for about a week to prevent it from drying too fast or cracking. The end result was—and still is—a beautiful amber light shining in through the many tightly cemented bottle bottoms.

I began the interior by laying a floor of shale. The shale pieces that made up this floor were 1' to 2' slabs, ½" to 1½" thick, which were outcroppings from a hillside in an area where this kind of rock occurs. Before I laid the shale mosaic pattern, I prepared the floor area with a 3 inch thickness of sand into which the slabs would set. The

Shale floor inside sauna

Inside sauna

Rafter supports and beer-bottle masonry from interior

spaces between them were then mortared over so the sand would not get on people's feet.

Something I had noticed about other saunas that annoyed me was that when people ran in and out, to plunge and return, they let so much heat out that it would lower the temperature of the room uncomfortably, and even cause drafts. I prevented this problem in my sauna by building a framed-in entry chamber to absorb the temperature change. This area is only 2' long and as wide as the doorway, but it works well. To frame this area, I set up two upright poles for the inside corners. I figured that as long as interior uprights were being set up, they might as well have some structural significance. I ran a horizontal rafter brace between two rafters on either side of the area and notched the uprights under them to hold them in place. This produced an extra support for a large section of the heavy sod roof. Horizontal poles were then nailed across the uprights, just above the top of the doorway. These poles created a ledge above the entrance chamber. A horizontal pole, spanning the side walls, was put in to continue the ledge to these walls. Slats 1¾" thick were notched into this horizontal pole on either side of the chamber ceiling. This raised platform is for people who like superhot saunas.

The outer chamber itself was then walled and roofed with one-by-four boards, which spanned from the inside posts to the narrow nailing strips that framed the doorway. A floor was put in this area so there would be two gradual 6" steps into the sauna instead of one giant step.

Two vertical bench ledges were nailed to the back wall just above its 2½ feet-high cement section. An end of each of the inside horizontal bench supports was nailed above them. These bench supports span from the rear wall to the front wall and were notched into the entrance chamber uprights so they wouldn't have to span that entire distance unbraced. A two-by-four was then run along each side wall just above the cement retaining walls. They were approximately level with the inside bench poles and provided a shelf for the outside end of the 1¾" bench slats.

Wide-angle view of sauna interior

The inside ends were notched into the horizontal support and their tops were trimmed with a saw so there were no rough edges sticking out.

These slats were spaced 1" apart to allow heat to come up from underneath and circulate around the whole body of the bather. These spaces also provide an area where the excess

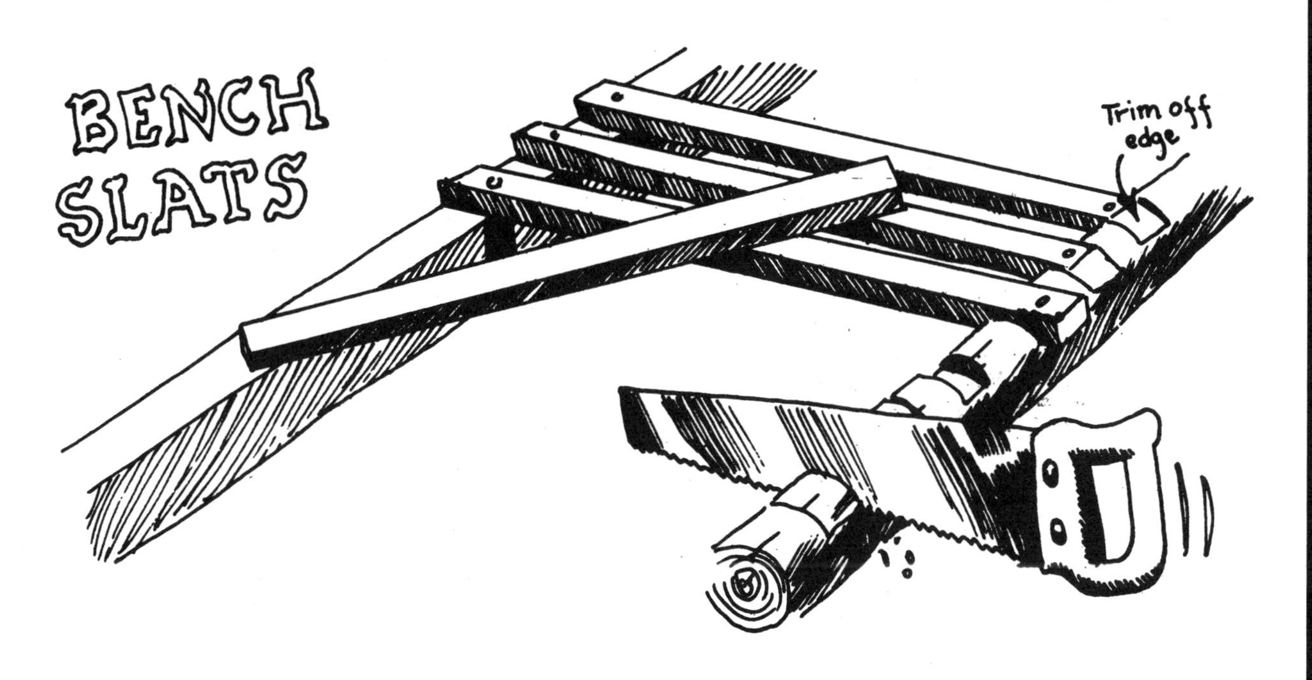

moisture can drip off a person's sweat-soaked body so he or she doesn't have to sit or lie in the wetness. The disadvantage of this slat design for the upper platform is that bathers on the lower benches sometimes complain about such drippings. Diagonal boards can be fitted between the slats so the bathers can lean against them at a comfortable angle.

This sauna is heated by a super-duper, homemade 15-gallon-barrel wood heater, designed and built especially for the structure by a close friend. It sits on top of a cradle of 4" poles, which is surrounded and covered with rocks to help it reflect heat out into the sauna. The stove's chimney goes straight up through the roof, rather than making a bend into the wall. I think this is a safer method because there is no chance of creosote collecting in the bent areas and creating chimney fires. When creosote builds up in a straight pipe, this pipe can be cleaned out from above without having to be taken apart. The sooty deposits will fall back into the stove and burn again. To keep creosote buildup down, burn dry wood, and let a lot of air circulate through the fire box.

Where the chimney goes through the ceiling, it is reinforced with a 36" length of Yukon-type, double-walled, galvanized pipe. This pipe incorporates a 4" dead-air space between its walls for protection. I filled that dead-air space with clay to give it more of an insulatory value and packed additional clay around the stack where it comes out of the roof, to seal that opening.

We throw water on the rocks to create steam for a hotter wet-vapor bath. The sauna is large enough for the bathers to be a comfortable distance away from the stove so they won't get roasted by its radiant heat. I personally like to start off dry and, after a while, throw a lot of water on the stove to fill the room with steam. This steam raises the temperature level and induces me to sweat until all my pores are saturated. Soon my body can't stand any more of the stifling heat. Then I run outside, jump in the plunge, and stay in the cold water until that too becomes uncomfortable, and I repeat the process till I feel totally clean and relaxed.

We used a temporary cooling device until the plunge was completed—a showerhead connected to an upright pole, which was fed by a plastic water pipe running down from a nearby mountain creek.

Saunas can be used for other purposes beside bathing, massaging, and getting to know people. We use ours for a fruit-drying shed during harvest time. Since harvest time in this area comes after the short, hot summer, we cannot leave the apples, apricots, and plums out to dry. Instead, we put the fruit on screen racks and keep the wood heater going for a few days until all the fruit is dried. A sauna is also a handy place to dry clothes when the weather is foul and there are no facilities around for that purpose.

We enjoy our sauna and have had many compliments on its design. It is comfortable and spacious, and heat seems to distribute very well throughout it. Even with six or eight bathers, it is not too crowded and is easy to keep hot.

When the sauna was first completed, we used to bathe in it as many as three or more times a week. We soon noticed that we were getting dehydrated and seemed to always be thirsty. It is unhealthy to take so many saunas in such a short time, because the heat pulls out too much moisture from the body. We found that one a week is sufficient and is best for health.

While on the subject of precautions, I want to mention a couple more hints: Do not eat immediately before a sauna, because the high temperature speeds up your heart just as heavy exercise does. This overworks your digestive system and makes you nauseated. It is also unhealthy to take a sauna at the onset of a cold because the shock of the chilly plunge will be too much for such a distressed system to take. The sauna itself, followed by a warm rinse instead of a plunge, is beneficial in the later stages of a cold because it clears out nasal stuffiness and cleans out pores.

We are very pleased with the sauna and have been regularly enjoying its health-giving benefits. We consider it a necessary and important structure that should be an integral part of every natural homestead.

Section 5

HEATING STRUCTURES & ROOT CELLARS

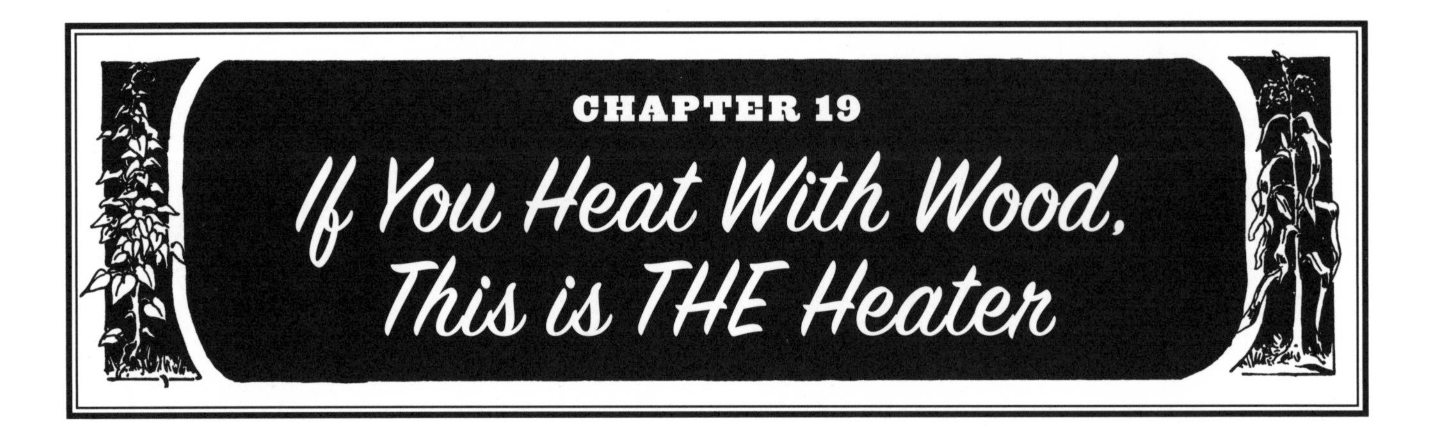

CHAPTER 19
If You Heat With Wood, This is THE Heater

Where I live, wood for heating is abundant and costs only the labor of hauling it. There are many roads being built nearby and much of the removed timber is pushed into piles to be later burned. There are also several second-growth forest areas surrounding us which are filled with dry, easy burning deadfall and standing snags—ours for the taking. It would indeed be wise to use the cheapest, most available resource to heat my house, but I didn't want to deal with the hardship of cutting and splitting small rounds for a conventional wood heater. Also, I can appreciate that my wife has no desire to be constantly sweeping up wood debris and washing down soot-covered walls. She has enough thankless tasks.

When visiting friends who heat with wood, I usually notice a premature drowsiness that comes over the group until someone takes the initiative to open a window or two. The fatigue is caused by the lack of oxygen in the room. As wood burns, it uses up the same oxygen we breathe. But opening the windows, of course, allows cold air in. The stove then has to be adjusted to put out more heat, using up a greater amount of fuel.

After much thought, solutions to each of these problems came to me. To prevent pollution of debris and soot, the ideal wood burning stove should have an outdoor loading chute and must be airtight, having no leaks to the interior. To prevent oxygen depletion, this stove must draw its oxygen from other areas instead of the area being heated. To save physical energy, the stove should be able to take large pieces of wood so one does not have to split short rounds. I soon developed a workable design for the ideal wood burning heater.

Since there is a basement in my house, I decided to locate the damper there instead of exposing it

to the variables of the sometimes severe weather. The oxygen comes up from this area to feed the fire. For convenience, I knew that I would have to devise some method of working the damper from the living area; that way, whoever was enjoying the comfort of the warm living area did not have to leave it to adjust the air intake (**FIG. 19-1**).

I began figuring out the workable dimensions for the wood heater. Since the only real constant was the size of the open 45-gallon drum heat exchanger unit, I had to first do the necessary welding on it. This consisted of cutting through from end to end and opening it up to a half-oval shape. After achieving that shape, I set the open drum on the floor and estimated the outer dimensions of the brick enclosure that would house it. The size of the hole would be the area of the drum plus 2" of air space around it plus the width of the surrounding brick enclosure. The 39 × 39 inch opening was then cut out (**FIG. 19-2**). This facilitated the following operation of constructing the concrete pad, which is located less than 2' below the floor. Cutting the hole first allowed me plenty of room to stand as I built the forms and poured the slab. Its thickness is irrelevant as long as it can support the weight above it—6" to 8" is adequate. The pad is made up of an inexpensive mixture of sand, Portland cement, and a large quantity of rock for volume. The perimeter of the pad is the same as the perimeter of the hole in the floor because the pad will be supporting the base and the brick enclosure.

FIG. 19-2

After the pad set up, a few courses of brick were mortared inside its perimeter. An inner lining of firebrick was then stacked along the inside surface of the red brick and was cemented together with a commercial fire clay. Mortared rock could substitute for the outer brick, but because rocks have a greater tendency then red brick to crack by expansion and contraction, they should not be used in direct contact with fire.

Below the floor, low enough so the entire firebox will be in the basement area, I installed a four-holed, manifold-type damper system. I tried to locate an old truck manifold for the job but could not find one the proper size, so I had one made up. This manifold had to be mortared into the brick base in such a way that instead of any structural weight being on it, all the weight was carried on the bricks between the four tubes.

Bolted onto the damper and sealed with an asbestos gasket is a commercial air intake vent. This vent has a tongue that protrudes above a pivot pin. When the tongue is depressed by the vent rod, the vent opens and allows oxygen in. When pressure is removed, gravity automatically shuts the vent. I prefer utilizing gravity in closing the damper to prevent it from being left open by accident.

FIG. 19-3

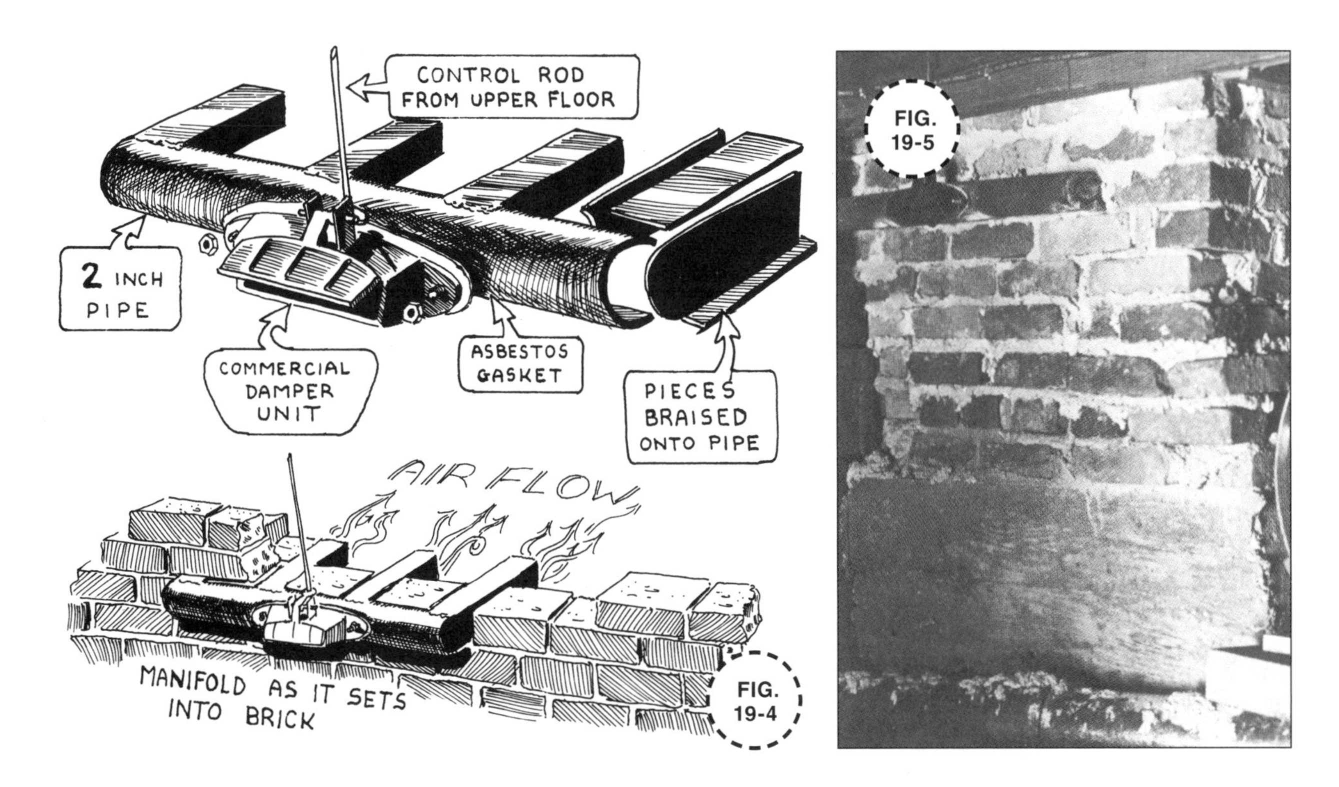

On the outside of the basement, below the loading door space, I installed a cleanout slot to collect the accumulating ashes. This slot can be any size as long as it can be easily covered. It should be located just under the level of the damper to prevent clogging. This cleanout hole has a removable cover, which is held on by two wing nuts and sealed with an asbestos gasket (**FIG. 19-3, 19-4, AND 19-5**).

After the manifold damper was installed, the red bricks and firebricks were set up to floor level. At this point, the firebrick was completed and two ¾" angle iron ledges were welded on the bottom ends of the drum so it would fit directly over the firebox. Care was taken to make certain the inside diameter of the firebrick agreed with the inside diameter of the drum ledges. The outer red bricks were then extended up into the living room area. Having the firebox below the floor gives the house a warm basement and a warm floor area.

The heat exchanger drum was at this point made ready for installation. After the ledges were welded on, an 8" hole was taken out of the top front of the drum. A thick-walled length of pipe, 8" in diameter by 10" long, was welded above the hole. A commercial 8" damper was then set into the flue support pipe, and a square base of 1" angle iron was welded to the top of it. This angle-iron base allowed the flue liner to temporarily be supported by the heat exchanger until the bricks were mortared around it.

A heavy ¼" iron frame was then welded onto the front of the drum to support the salvaged boiler loading door, and a 14 × 16 inch hole was cut out for it. A lip was added around the 14 × 16 inch opening to enclose the door and two rod supports were attached above the opening for it to hang from (**FIG. 19-6**).Because the door is held tightly shut by its own weight working with gravity, it keeps the wood burner airtight and free from drafts (**FIG. 19-7 AND 19-8**).

FIG. 19-6

FIG. 19-7

The gravity closing feature is also a safety factor. When loading dry branches into the firebox, the existing fire sometimes explodes and leaps out. If I had to take time to stop and think about shutting the door when such an explosion occurs, there'd be a good chance I might be singed before being able to do this. With gravity, the door shuts when I let go of it.

My original thought was to make the heat exchanger out of brick instead of using a 45-gallon drum, but I later realized that brick takes much longer to heat up than the drum. When you come in from the cold, you want to be able to open the damper and get immediate warmth; you don't want to have to wait a half hour before the bricks get hot. True, bricks would last almost indefinitely, but so will the drum, if it is not allowed to get red hot. Since the drum is only a heat circulator, it does not come in

FIG. 19-8

contact with the actual charcoals, so it will not burn out from contact with fire.

The ornamental brick enclosure around the heat exchanger and the chimney were then continued up from the base. I wanted adequate air circulation around the heat exchanger, so I allowed 2" of air space between the drum and the bricks. This air circulation factor allows for maximum heat circulation throughout the living area. Before raising the brick enclosure, I made sure to allow several openings between the bottom bricks to draw in cold air from the floor, which would prevent draft problems. Without this precaution, the cold air would creep up to the top and get heated without properly circulating.

The brick enclosure and chimney were then raised and a tapered (for aesthetic purposes) chimney front was built above it. This front rests on a heavy 3" angle-iron support that extends from outside enclosure brick on the opposite side. When completed, the enclosure was high enough to easily sit on. Directly above the drum, a heavy-gauge mesh screen was fitted in. It is a safety screen and sometimes seconds as a drying rack for clothing and other items; but by no means should the air circulation space around the heat

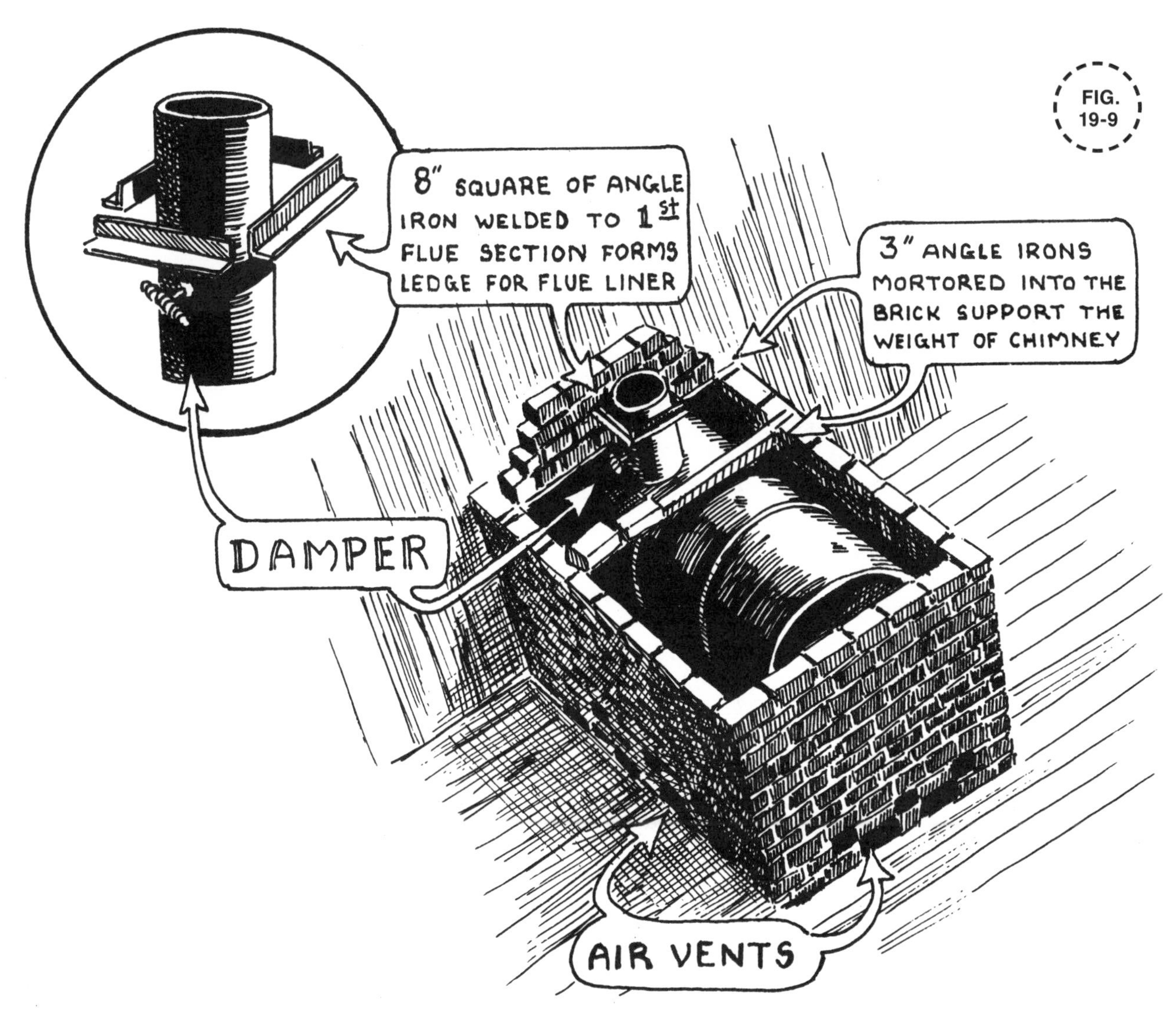

exchanger be interfered, with for obvious fire-hazard reasons.

This brick enclosure is the favorite spot in winter for my friends. First thing they do is come in and warm themselves, then, after everybody thaws out a bit, the conversations begin (**FIG. 19-9 AND 19-10**).

The flue lining was then stacked above the support and the chimney bricks were mortared in the usual alternating tier method. The chimney was raised to a height of 10" above the peak of the roof for safety (**FIG. 19-11 AND 19-12**).

A word of caution to those who do not want to foot the price of a commercial flue liner. If you build your own flue, make sure it is smooth inside and out, and it is of an equal diameter top and bottom for proper draw and easy cleanout. Creosote sticks in crevices and can be very difficult to remove from a rough surface. Also, if you don't use a ceramic or stacked oil-can flue liner, there is danger of the liquefied creosote seeping

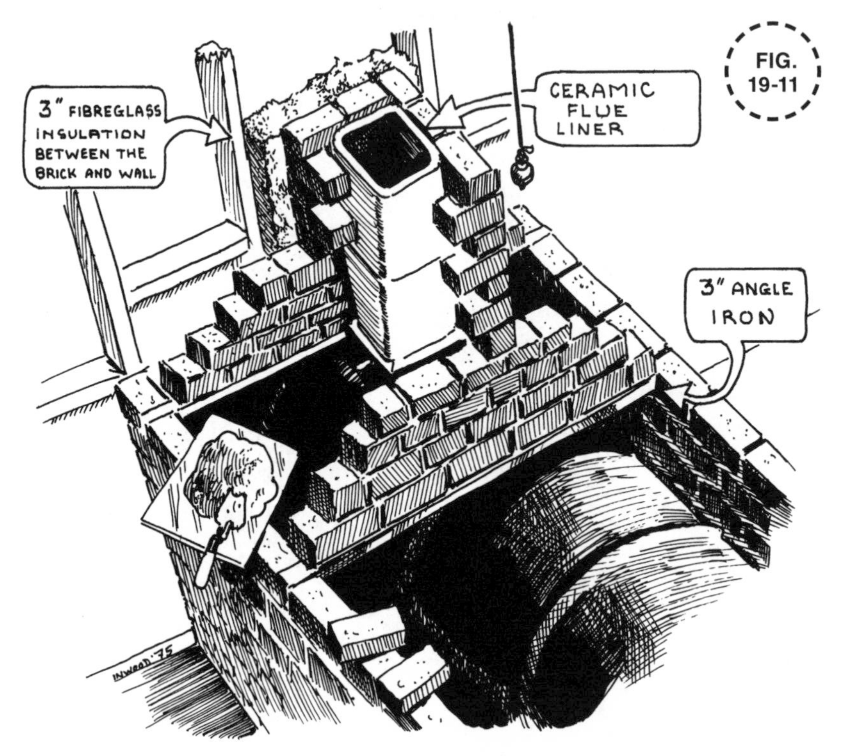

into the mortar and discoloring your bricks or rock work. Creosote is made up of very small molecules and can seep into the tiniest cracks. Because of this characteristic, it is an excellent wood preservative. This is one place to definitely not chance a shortcut.

In my opinion, the safest and easiest chimney to deal with is one that is straight up and down. A straight chimney allows creosote to drop into the firebox instead of collecting in angled areas. To check for creosote buildup, I simply open the door, stick my head into the furnace, and look up. If I can see the sky, I don't clean the chimney.

I have had little success with burning creosote out of chimneys. It is a waste of fuel and it is bad for the heat exchanger to get the furnace hot enough to liquefy the residue. Even if you do get the furnace hot enough, you never get all of the creosote. The residue that remains crystallizes and becomes part of the flue. This buildup eventually disturbs the draw and must be removed. For cleanout, I use a 20' long 1 × 1 inch stick with a flat scraper on the end (**FIG. 19-13**).

Even though this heater is designed to burn wood slowly, I have to clean out the flue only every six weeks because the chimney is well insulated. There is 3" of insulation between it and the outside wall to slow down the condensation of smoke. Creosote is condensed smoke. If the chimney is kept warm, your creosote problem is cut in half.

The width and length of a chimney are not critical with this heater design as long as the opening is uniform at top and bottom for draw. A friend of mine has a similar setup with a 3 × 6 foot chimney in which he built a smoke chute. He put in a large airtight door enabling him to walk into the stack, and he installed grates for hanging bacon and sausages for smoking. This proved to be an excellent smokehouse because creosote smoke is cool and does not cook the meat. The draw is not interfered with either. Smoke curls out of the heat exchanger drum and curls through the huge opening.

As a final touch to protect the loading door from the weather and to ensure that it is airtight, an outer door was added. This painted plywood door encloses the loading door and its inside is lined with aluminum foil to reflect any escaping heat back into the heat exchanger (**FIG. 19-14 AND 19-15**).

The heater was then ready for the big test. Before lighting the first fire, I protected the

FIG. 19-14

bottom cement pad with a 4" layer of sand for the charcoals to rest on. With the sand, there is no need for a grate. I used any type of wood that was available—cedar, fir, birch, larch. Cedar burns fastest but not hottest and birch is excellent when you prefer a warmer house, but I usually do not have time to pick through the pile, so I use the wood as it comes.

I began the fire with very dry wood and once a solid bed of coals was established, I used slower-burning green wood. The initial test proved successful. The unit heated the entire house and absolutely no oxygen was taken out of the living areas. The fire stayed in the firebox and the heat exchanger then circulated heat out into the house. There was no pollution in the house and I could put in logs up to 30" in length, which saved tremendously on wood cutting energy. In fact, I only put in one full wheelbarrow's worth every evening and the fire kept slowly burning. Where it gets its oxygen I do not know because I keep the damper shut most of the time. I only open it to get a sudden burst of heat after coming in from the cold.

The stove's efficiency amazes me. When I put the logs in the burner in the evening, they turn into charcoal during the night and that charcoal

FIG. 19-15

gives off heat practically throughout the entire next day without giving off much smoke. When I open the damper, cold air gets in and takes all the heat up the chimney, so I make sure to keep it closed down as much as possible to save fuel.

It takes approximately two months of constant everyday use before the firebox needs to be cleaned out. When the ashes reach the cleanout slot, I move the hot embers to the rear and take out the cold ash. These ashes are then taken to the garden, because they are an excellent source of lime and should never be wasted. Because the cleanout slot is located a bit lower than the manifold damper, the ashes never interfere with the damper. I also found that creosote collects in the heat exchanger and never interferes with the damper.

Another amazing discovery I made was that even after I left the fire for a couple of days, it did not go out. Very little heat rises because the damper is shut, so the heater is constantly giving off combustion heat.

CHAPTER 20
A Dominant Attraction—The Fireplace

My wife and I were already familiar with the warm, earthy presence of an aesthetically beautiful fireplace and its natural attraction to people like ourselves who enjoy intimate get-togethers with friends. Such a fireplace is a must in our home and became the major consideration in its design. We wanted a fireplace with which we could participate while eating, while conversing with friends, and also while working in the kitchen. It is important for us to have as much exposure to its warmth and nourishment as possible.

Upon designing the house, we figured out the size and location of the fireplace, but had no idea of what the finished product would look like until after gathering the materials. I began working with the materials—and let it flow. Soon it started to communicate and began to determine its own form.

My accumulation of stone consisted of several shallow, flat-faced pieces and a whole lot of rough, square ones. I knew a plain straight fireplace would not be the natural outcome, so I experimented with the stones. I laid them out on a flat surface and played with them as one would a jigsaw puzzle. The flat stones seemed to want to gather together in the center section, leaving the square stones at the ends. I stacked the center stones flatways so only their edges were revealed. This produced a horizontal feeling in the center, which protruded out beyond the receding sides. It emphasized contrast and a third dimension. Then I put in a few niches here and a little shelf there where items like plants would go. I did what kind of felt right as I worked with my materials (**FIG. 20-1**).

There is a popular material that can be purchased for the fireplace. It is a rough, rustic-looking rock that is quite sharp around the edges and is precut to produce a clean effect without exposing much mortar or joints. But this material is pretty expensive while the type of stone I used is just sitting around waiting to be picked up. The purchased product robs you of the chance to be totally creative with your design because it is uniform in thickness and lacks dimension.

FIG. 20-1

I began the actual construction of the fireplace by pouring a very adequate pad for the heavy structure. This pad should not be less than 10" thick and should be reinforced with ½" reinforcing rod, crisscrossed throughout it, especially on the two sides where most of the structural weight is going to be supported. A vertical anchor of ⅜" rebar, coming out to meet the corner concrete blocks framing the firebox, should also be included. This pad must extend at least 12" beyond the perimeter of the fireplace in all directions and could also be the base for the hearth, depending on design.

You want a strong pad, so you need to make a mixture 1 part Portland cement to 5 parts sand. The adhesive strength should not be broken up with too many plumbs (large rocks) or else the whole pad will be weakened. Large rocks do not provide enough tiny surfaces for the concrete to adhere to and cement together (**FIG. 20-2**). Remember, if you are going to go through the trouble of building anything with masonry—masonry being permanent, expensive, and very difficult to alter—build it with the proper materials. Do not skimp—especially do not skimp on the pad. It is the foundation for a very heavy, permanent structure. If the pad cracks or crumbles, so does the fireplace.

Once the pad has been poured, you can begin laying out the fireplace in terms of firebox preference. The two most popular alternatives are the double-jacketed metal Heatilators or heatforms, and the traditional handmade firebrick box.

Heatilators or heatforms are supposedly the more efficient route. They allow quick emission

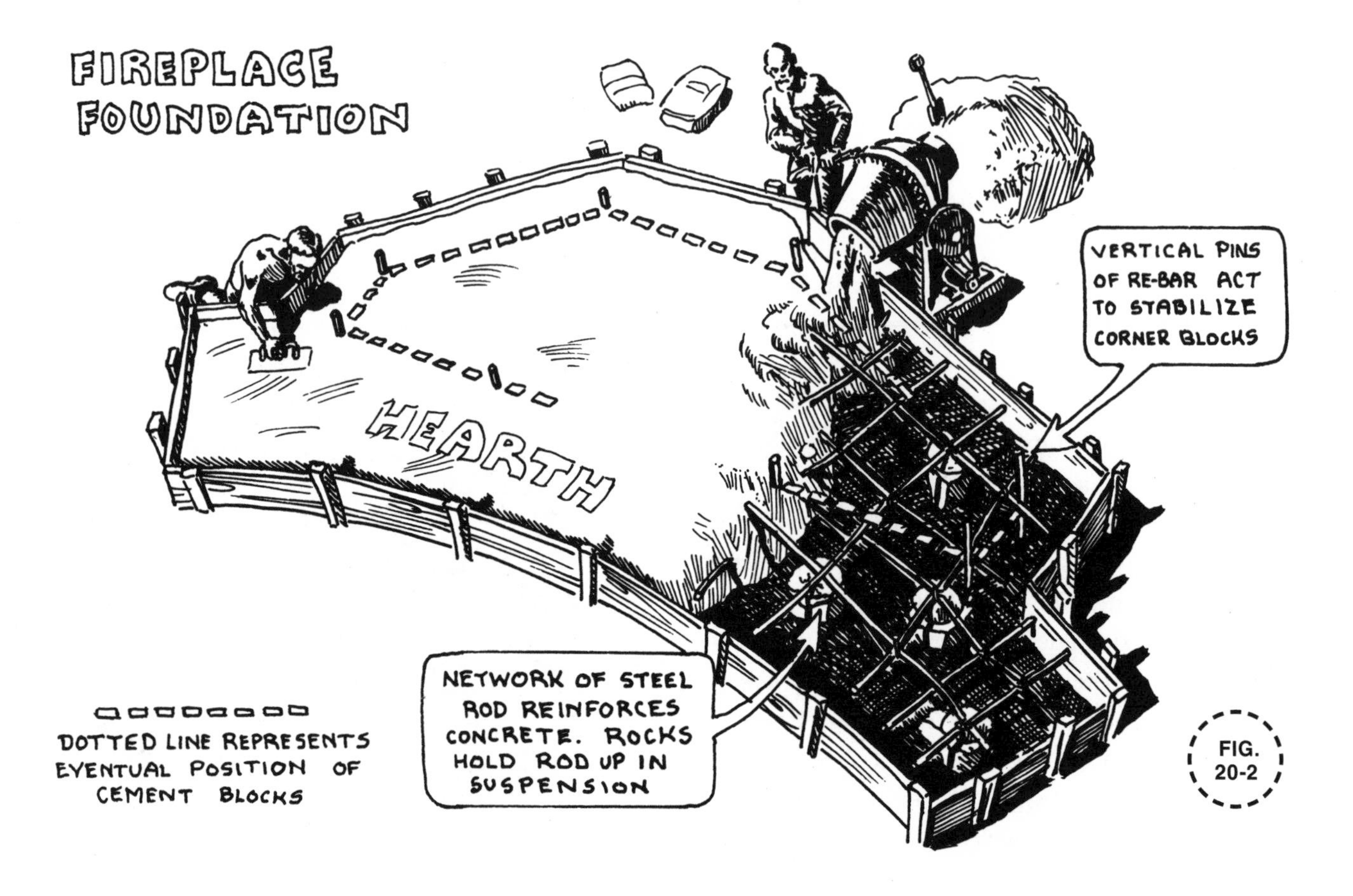

FIG. 20-2

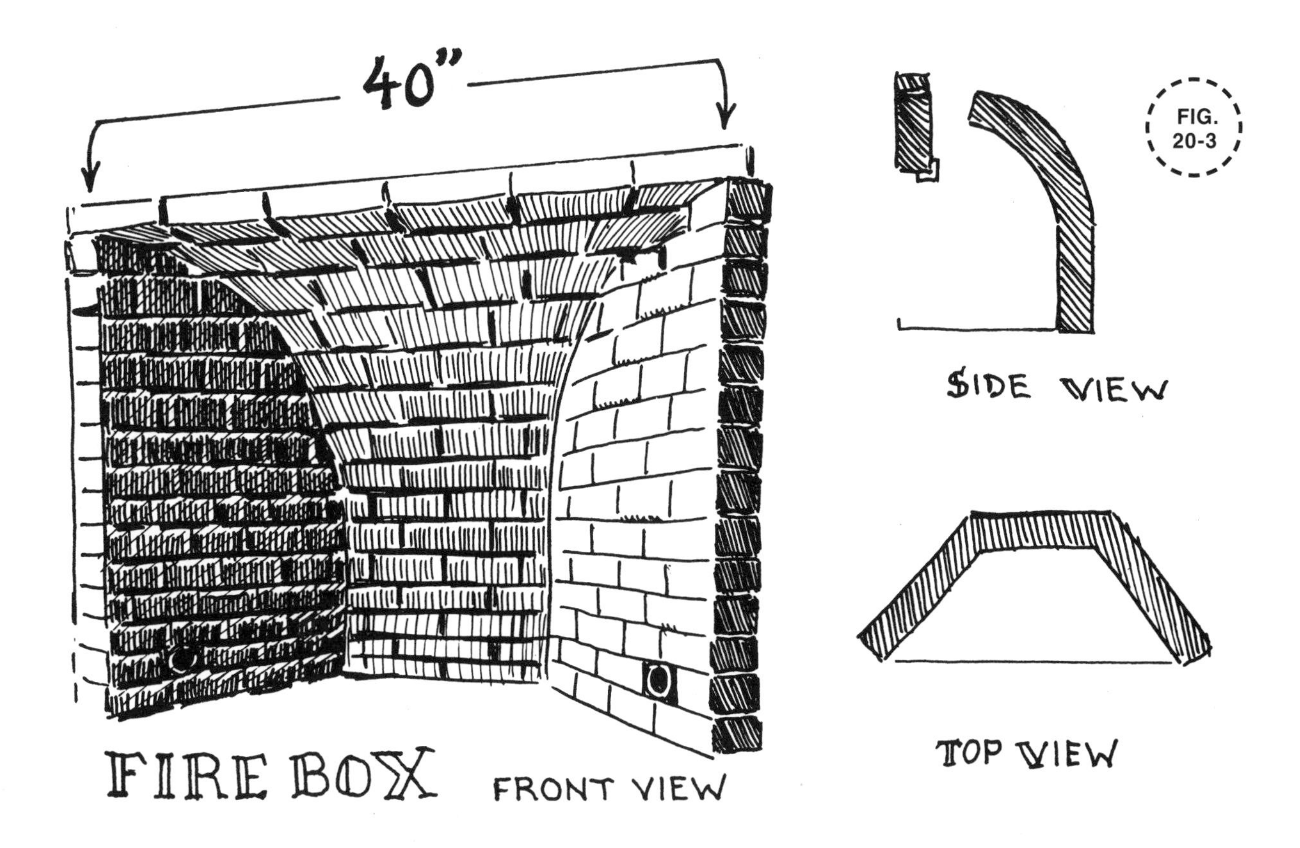

of hot air into the living area, whereas it takes firebrick longer to heat up before it radiates. The Heatilator is easy to install and comes as a whole unit. Firebricks have to be set up, shaped, and adhered together. This usually becomes a long, involved process, but is a very satisfying and creative one.

I personally prefer to build my firebox out of firebrick. It is cheaper than buying the metal unit, is more aesthetically pleasing, and can be shaped to optimum efficiency. Though firebrick takes longer to heat up than the metal form does, it retains heat much longer. The bricks stay warm several hours after the fire has gone completely out, and so they keep out the chill factor much longer than the metal form does. Firebrick will also last up to twenty or thirty years.

My main argument for firebrick is that one can form it to a desired shape. At the rear, I angle my firebox side walls deliberately toward the center. Though there is a 40" opening in the width of the front, the backside is only about 16" to 17" wide (FIG. 20-3). There are several reasons for this design. The reflected heat from this firebox is far greater than from a deeper, conventional one. Thus a small fire in this box will work just as well as a larger one. This is a major fuel-conserving factor, if nothing else. It takes a larger fire to fit a conventional box or a heatform box because their backsides are deeper. There is far less reflective heat coming into the room from the deeper boxes, consequently there is less heat economy. The deeper the firebox, the more the reflective heat rises into the throat and dissipates up the chimney instead of coming into the living area. With a shallow firebox, the reflective area is closer to the living area, so more rising, reflective heat enters it.

The rate of smoke emission increases proportionately with the depth of the firebox, especially in early states of the fire, because there is more room for air to circulate around a deep box.

I tend to make my lintel higher than normal. This provides more of a view of the fire. From the hearth up to the lintel, my firebox is 32" high.

Before actual construction of the firebox, I designed a vent system, enabling oxygen to enter in from the exterior, flow under the firebox, and come out the sides to supply the fire. A standard fireplace will utilize a huge amount of room oxygen, consequently depleting the living area of air. This causes its inhabitants to feel fatigued. Also, without the exterior air supply, chilling foot drafts occur as the fire draws oxygen. A well-known formula explains that a standard fireplace will cause the displacement of over twice the amount of room air required for optimum ventilation; therefore, at least half the amount of oxygen for the fire should be drawn directly from the outside and not be permitted to pass through the room to the fireplace.

The air is brought in from the outside through a 3" can–formed opening that goes from the exterior bottom and comes in from directly underneath the firebox, between it and the slab. I allowed an opening for it there, the width of the firebox and 2½" in height. The air intake is located at the lower right of the stonework masonry exterior (**FIG. 20-4**). One outlet comes through near the right front side of the firebox and the other travels beneath it, coming out on the left side opposite the first (**FIG. 20-5**).

The construction of this air intake system was tricky. The first course of brick around the sides and back of the firebox were mortared directly to the slab, leaving a 3" space near the front of either side as air channels. A piece of sheet metal was placed over this bottom course. It acted as a form for the 2" secondary pad on which the firebrick bottom of the firebox rests. Rebar was distributed throughout the area above the sheet metal and a second course of brick was mortared in place around it. This second course also acted as the side form for the 2" of concrete, which was then poured over the sheet metal. After the concrete dried, the bottom firebricks were put in over it.

This secondary pad construction created a 2½" plenum over the pad for an air chamber, which would allow the passage of air from the exterior air intake through the bottom plenum, out to either side of the firebox. Two 3" cans were used as forms on the lower sides and another course of firebrick was adhered around the firebox. An archway of brick was then mortared

FIG. 20-4

FIG. 20-5

around the vent openings to protect them from the later rubble and concrete fill, which would be put in between the firebox and the surrounding concrete block (**FIG. 20-6**).

If I were able to redo this vent system, I would bring both of the outlet vent 2" forward, more to the front of the box. As it comes in now, the air hits fire from the sides, blowing it a bit toward the hearth. If the vents were nearer to the front, the incoming air would blow the fire to the back where it could more efficiently warm the bricks and radiate its heat.

The sides of the firebox were then laid. The overall dimensions of this area are very crucial, but one is free to choose from several designs depending on one's own needs. I used a bell-shaped curve at the back. The mason from whom I learned prefers this design, and so do I. It provides more of a gradual flow in the back of the firebox, allowing the fire to follow a gentle curve. To establish this curve, you put a little bit more fire clay between the backs of the firebricks than between the fronts. This tilts the top of the brick face slightly toward the front of the fireplace.

When setting the firebrick, a commercial fire clay should be used. You want only a small amount between each brick because it is the firebrick, not the clay, that radiates the heat. In volume, the clay will eventually crumble. It takes only about 25 pounds to do a normal firebox.

Fire clay is mixed with water to a consistency of heavy cream. The firebricks must first be submerged in water for a minute or two until they almost quit bubbling. This enables the clay to stick on the brick like glue. If the bricks are left in the water too long, they have to be dried out a bit before the fire clay will stick.

The bricks were then raised straight on the back and sides of the fire box for 9". Next, I made a template form, shaping the end of a piece of ¾" fiberboard to the bell-shaped curve. The bricks were then laid to follow the gradual curve. I periodically used this pattern as one would a level to make sure the curve was true. This curve

must be a gradual one to prevent any humps from occurring as the brick courses are raised (**FIG. 20-7**).

To narrow the surface of the back of the firebox as it curved forward, the sides were also splined inward at a good angle. Then the back was soon angled outward again above the center so that the back plane would widen almost to the traditional angle at the throat to prevent smoking. This design improved the radiant efficiency.

In other words, when you decrease the width of the side walls by splining or corbelling them, the back wall has to come forward. And since the side walls are angled outward, the back also gradually becomes wider. Corbelling or splining is the process of creating a taper or angle by overlapping each course above it, forming an angled or herringbone-shaped intersection. There was an added complication in my firebox design of the splined sides also meeting the back at various angles depending on the bell-shaped curve, causing the crossing bricks to overlap. This made it necessary to trim the exposed face of every other brick with a forked-head-type chisel, blending it with the others. The backs did not have to be dealt with because they would be covered with the rubble-concrete backing behind the firebox.

To chisel these faces, you first mark off the portion you want eliminated, and softly hit the chisel repeatedly on that line until it gouges a groove all around the brick. Keep going around until you make a clean break. Practice this on a few discards before trying it on the actual bricks you will be using. Remember, it's a gradual process and patience must be exercised to make proper breaks.

To begin the flare, I start setting the brick back a bit just before the halfway point of the backside. This is a simple process. I just use less

FIG. 20-7

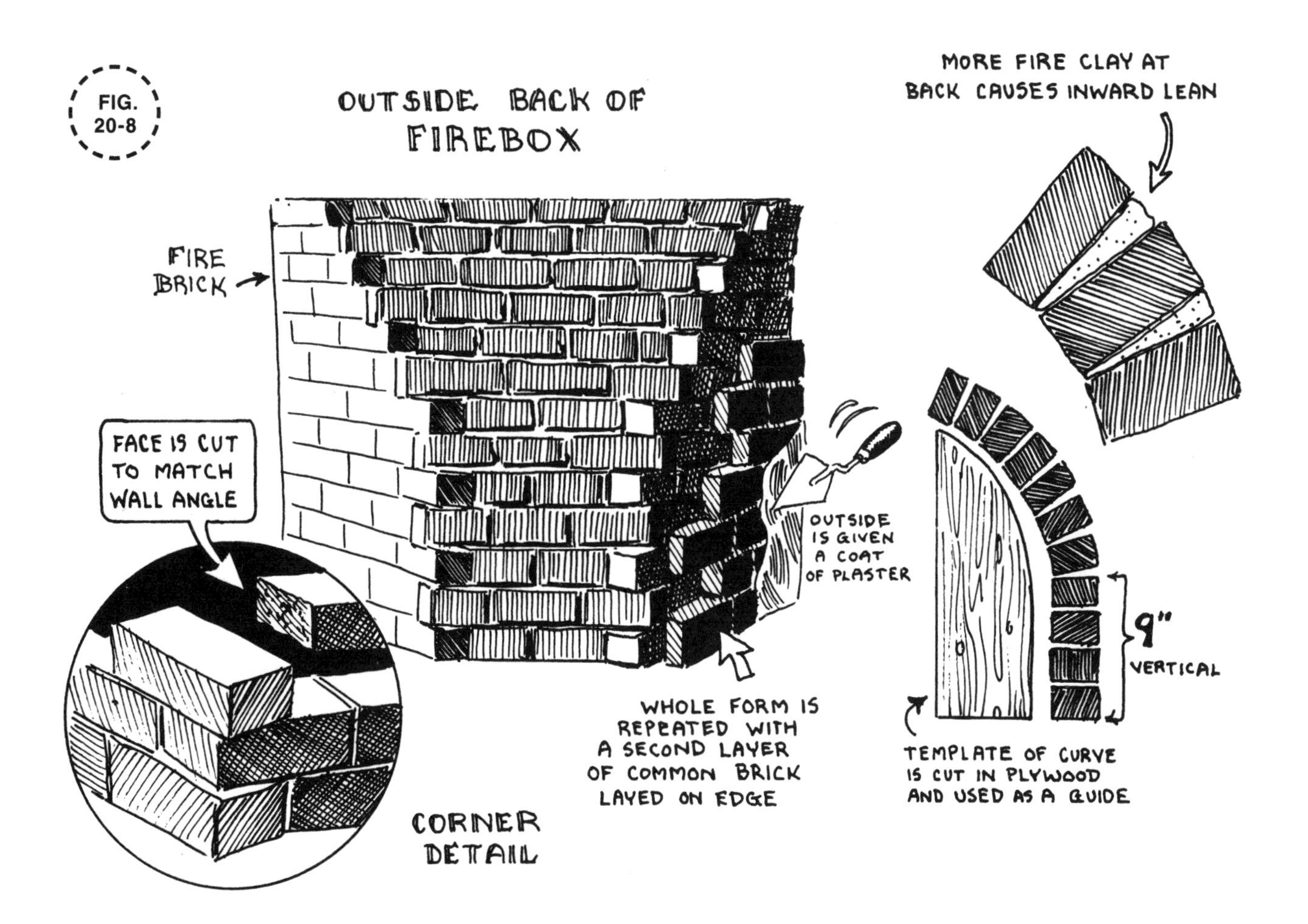

clay on the backside of the bricks until they gradually curve outward. This widens out the back of the firebox directly above the firebox, where the maximum radiation occurs (**FIG. 20-8**).

I took great pains to make certain the opening at the throat was the proper width in proportion with the smoke chamber and the flue area. If it were too narrow, there would be tight spots at either end of the lintel, causing the smoke to come back into the room instead of going up the throat. If it were too wide, there would be an unnecessary loss of heat.

The firebrick ends at the throat of the fireplace. This is the area where the smoke escapes past the damper and flows into the smoke chamber. Here, I put in a reinforced shelf of 3½" x 4½" thick angle iron. This shelf extends 4', supported on either side by the concrete block and rockwork face. It absorbs the weight of the lintel, mantel, and chimney face, taking any structural weight off the firebox. When iron gets hot, it expands far more than the rocks; this expansion must be allowed for. For this purpose, a small amount of fiberglass insulation was put in at either end of the iron shelf. This provides an air space at the ends, into which the expansion of the iron could take place. This shelf is located toward the backside of the fireplace at the beginning of the throat (**FIG. 20-9**).

The throat is about 8" to 10" above the lintel. This is where the damper is located. You have a choice of building your own damper, having

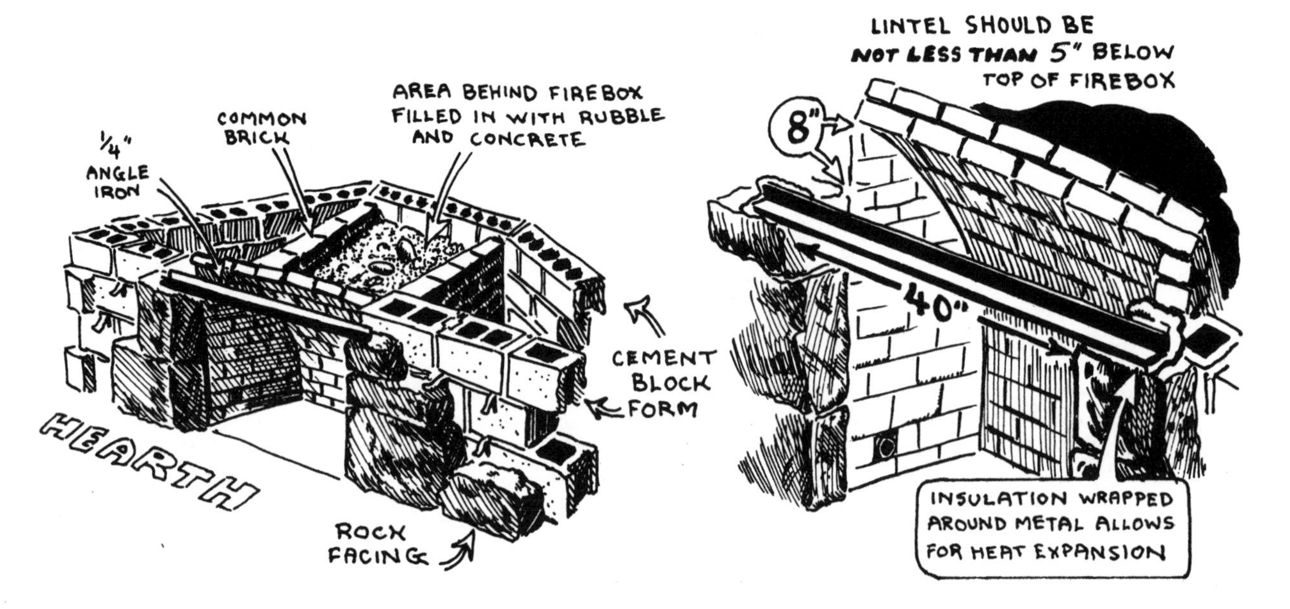

one made to fit the throat area, or buying a commercial damper arrangement. Though the commercial ones are foolproof, they are far deeper than they need to be and need a wider throat to fit into than the one on my fireplace. Our throat is 4" deep from the back of the lintel to the top of the firebox, and is 40" in width. I wanted a wide firebox area for viewing purposes. Because this opening is so wide, the throat could be as shallow as 4" and still be efficient. The throat opening must have at least the same area as the flue opening and it would be if it were even a bit larger. In other words, if a flue were 12 × 20 inches, its opening would be 144 square inches; so the throat area would have to be at least that. The throat on this fireplace has an opening of 40 × 4 inches, which equals 160 square inches; this is more than sufficient.

I chose to make my own damper. It consists of a long rod with a flat 4" piece of iron welded to one side of it and a control rod attachment which is hidden in the stone face of the fireplace. At the end of this control rod, there is a kinked or bent handle that maneuvers the damper flap. When the damper is flopped down, it is closed; when it is flopped up, it is open. The damper should be of a size that when it expands from heat it still opens and closes without touching the throat. During construction, be sure to work the damper daily to free it of "mud" (mortar) that may have fallen from above. I advise installing a damper in, because when you are not using the fireplace, you can close it off in cold weather and the room heat will not go up the chimney (**FIG. 20-10**).

Directly above the throat is the smoke chamber. It extends between the long, narrow throat to the 12 × 12 inch flue opening and is the area in which the size of openings are converted. Because of this conversion, a smoke shelf is created. This smoke shelf stops downdrafts.

FIG. 20-10

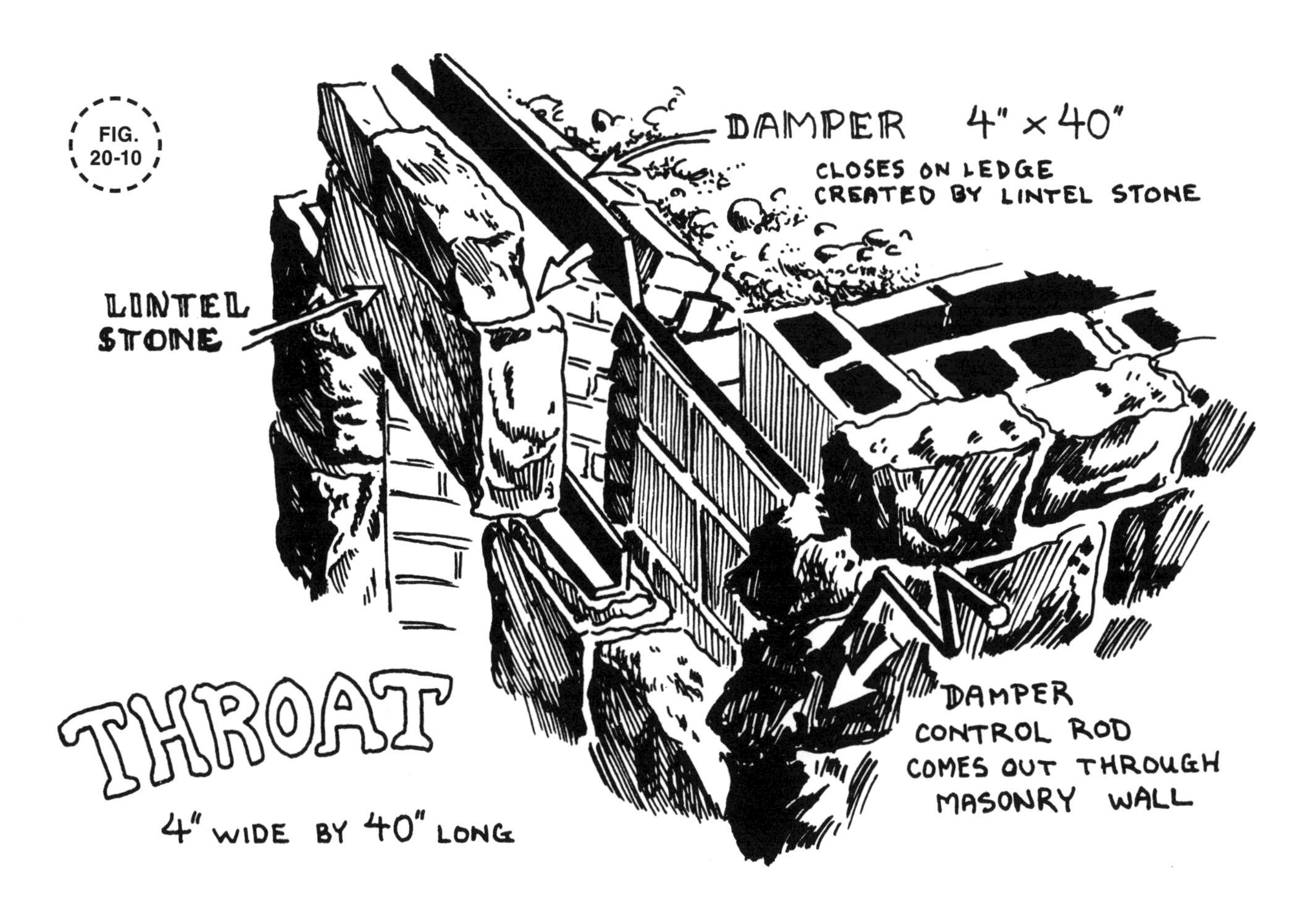

When the downdraft hits it, it curls back up the chimney. It is located directly above the sloping firebox back, which also helps form it.

The smoke chamber and smoke shelf are built with normal brick because these areas do not get hot enough to warrant the use of firebrick. The sides and front of the smoke chamber are gradually corbelled at about a 60° angle until they outline the 12 × 12 inch flue opening. The backside remains pretty much straight up and down.

Corbelling is a popular method of obtaining a tapered slant. The bricks lay flat instead of gradually sloping to the front as in the bell-shaped curve. Each tier of bricks is stepped up half a brick's width until the 60° angle reaches the desired height. The courses of brick are alternated; one course consists of a double thickness of brick with the side of the bricks as face, the next with the end of the front brick as face. This chamber should be constructed with this double thickness of brick to prevent smoke seepage, in case one layer of brick begins to crack. Building codes usually require this precaution. We're dealing with something that can never be repaired, so it must be done right in the beginning.

I plastered both sides of the smoke chamber to prevent any creosote seepage through the mortar. The inside being corbelled resulted in a series of rough, stepped-up edges. It is a difficult job, but this surface has to be made smooth so that the smoke can travel up without any hindrance. The plastering should be done as the smoke chamber is being built. I used a flat trowel and made the plaster good and fat (sticky) so it would adhere well to the joints. Then I plastered the outside, making certain it was sealed thoroughly. You can't be too cautious when it comes to preventing unsightly creosote seepage (**FIG. 20-11**).

The size of the flue is dependent upon the size of the opening of the firebox proper, and the height of the chimney. If the chimney height is 12' or less, you must make the cross section of the flue at least ⅛ of the total cross section of the area of the firebox opening. If the chimney is 15' or thereabouts, the flue opening should be ⅒ of that area; if the chimney is 25' or over, it can be 1⁄12 that area. In other words, the higher the chimney, the smaller the flue liner needs to be because the length supplies the necessary area for proper draft pull.

FIG. 20-11

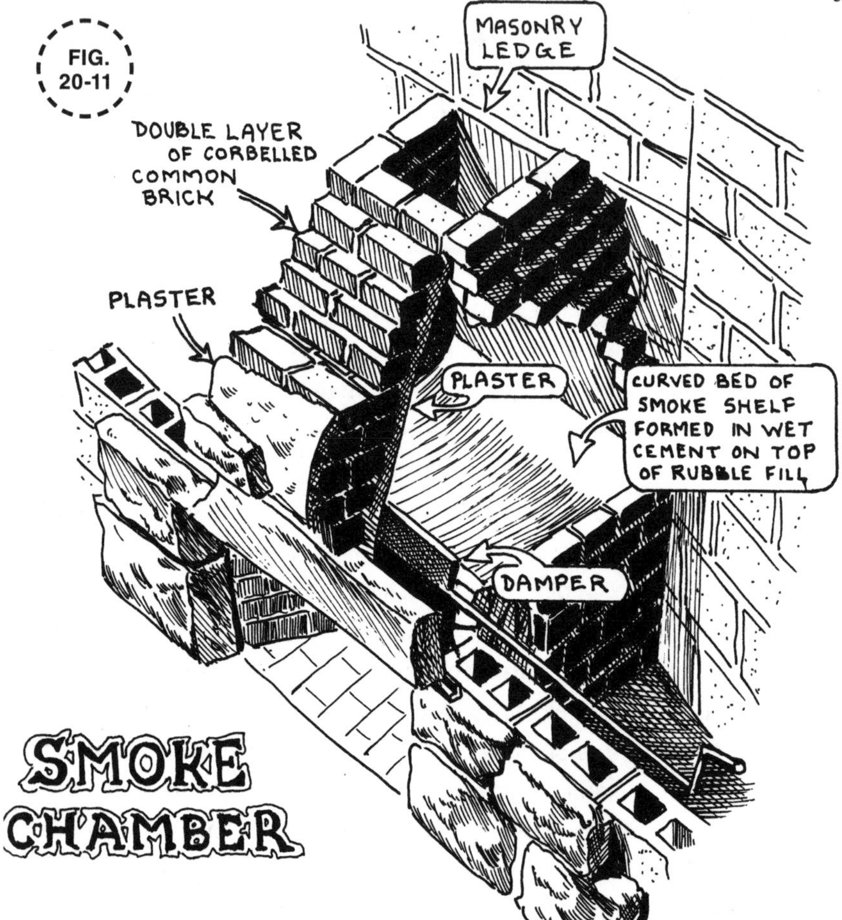

SMOKE CHAMBER

The flue liner can be set on a reinforced shelf of angle iron. This shelf is supported between the stone or block work that surrounds the flue liner. It is located in front, directly above the smoke chamber. It carries a lot of the structural weight of the masonry and brick front of the fireplace.

The flue lining should be raised with the concrete blocks and the outer stonework. To minimize confusion, I'll explain each process separately.

Note: A fireplace should not be built without a flue liner. No building inspector will pass one without it.

I put mortar between each flue lining joint as I stacked them. I was taught to do this and I prefer the results. An architect looked at one of the flues I happened to be constructing and he wanted to know why I was mortaring the joints. He said that within two years after I built it, the mortar would crack anyway. I told him I didn't agree, it all depended on how you did it.

Normally your flue is going to have stonework, concrete blocks, or brick around it. There should be a space allowed for expansion between the flue liner and this masonry. I always insulate this space with no less than 2" of fiberglass insulation—sometimes even more. This insulation also cuts down the creosote problem.

I bring the insulation up to within 1" of the top of the flue liner I am working with. Then I put the mortar on top of the insulation and on top of the

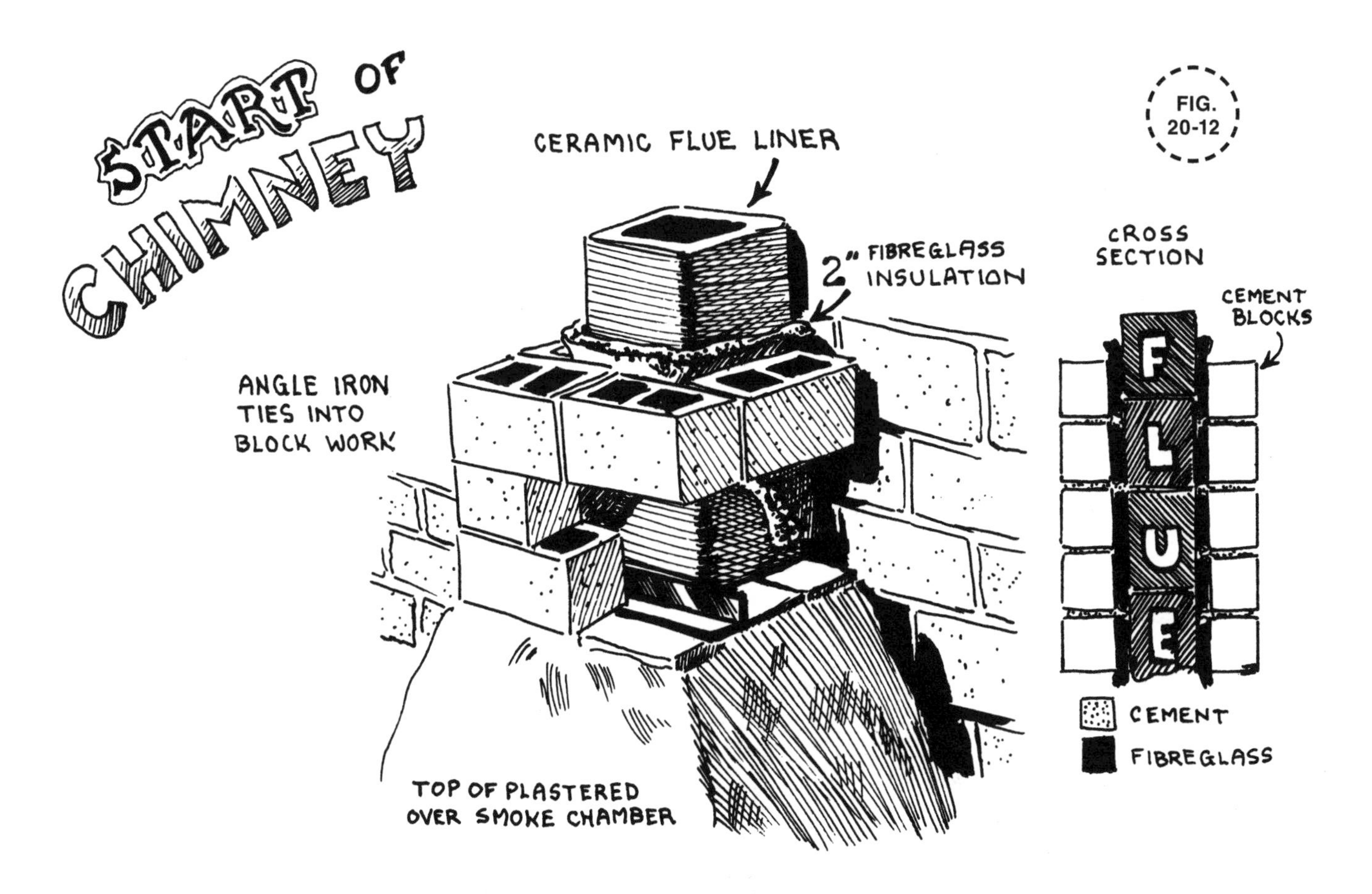

joint. The "mud" is well packed all the way around it, sealing it in. Above all that, I put in more insulation and climb right up with the stack. What I end up with is a well-insulated and mortared flue stack held in place partially by the insulation. It is only tied to the exterior very tenuously by a bit of mud around each joint.

With this method, I supply an area for the flue to expand and contract without cracking anything. If it was made solid, for sure it would expand and crack (**FIG. 20-12**). If you want to bypass the insulation process, you should at least put a couple of layers of paper or something equally as porous around the flue liner to create an air space to absorb the expansion.

The flue liner must have at least 12" of stone, concrete block, or brick around it, or a combination of any of these three materials. For economy, aesthetic creativity, and structural soundness, I chose to use a combination of multicolored, multitextured stone for the face and an inner structural base of concrete block. The concrete block frames the flue and the firebox. It provides the facing stone with a flat surface for stacking and it is a structural necessity when constructing large, wide fireplaces. Concrete block is far easier and faster to construct than mortared stone, and it is an excellent base for the textured stone.

Since the concrete blocks are uniform in size, they can be laid very quickly around the flue and firebox. I suspend a vertical string from an overhanging board at each corner of the area to be covered with the block. A plumb bob hangs from the strings to maintain the true vertical line, then that line is held taut with a bottom brace.

When constructing around a narrow area such as a fireplace or a flue, it is not necessary to have a network of horizontal guides. The blocks can be eyeballed into place horizontally. Care must be taken, though, to make sure there is a uniform thickness of ½" of mud between each layer so the blocks will stack properly. The vertical string guides have to be constantly checked to ensure that the blocks are true. Also, it is a good idea to tie in every third course of block with a binding wire grid to secure the tall narrow stack in place.

Here are a few hints on using mortar: Make sure the mortar (mud) is of a consistency that when placed on a trowel and flipped a bit, it spreads across the tool and the excess moisture

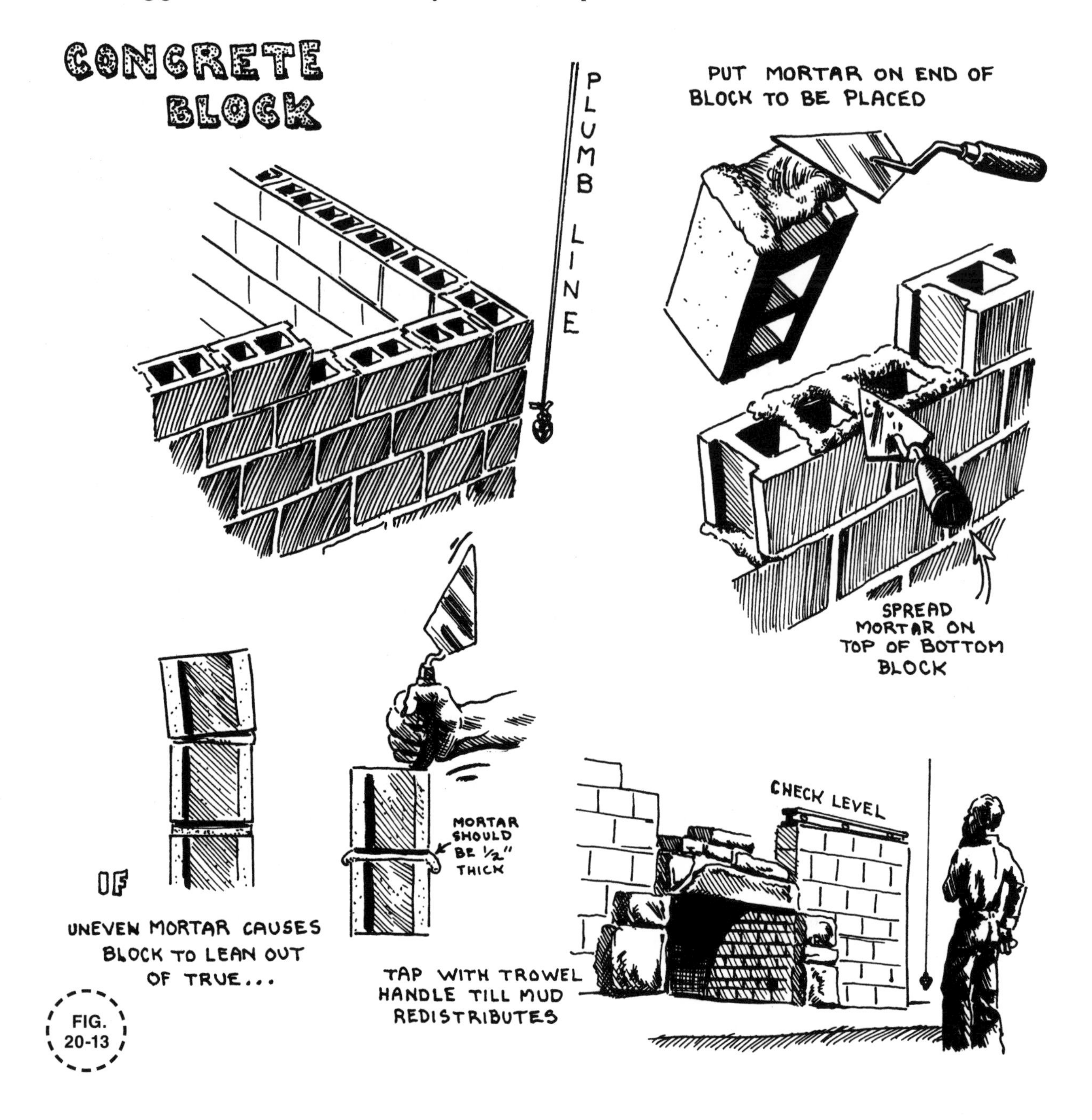

FIG. 20-13

comes to the surface. Don't have it so wet that it just slops over the edge or so dry that moisture doesn't come to the surface when the mud is worked. If it is of a nice sticky consistency, it will hang onto your trowel as you turn it over onto the block. You then spread a large amount on the top edges of the block already in place and the sides of the block you are setting. Be generous with the mud. You can always clean off the excess later.

If the block you have set is lower than the others on that course, lift it and slap on more mud. If it is higher, tap it with the trowel handle until it is the same height as the others. If the back is low, tap the front down to compensate, and vice versa. There are no real secrets to successful block laying; just be patient and remember to use as much mud as possible. It is easier to wipe off the excess mud than to lift the heavy blocks several times (**FIG. 20-13**).

The concrete blocks were raised along with the firebrick in the firebox. Ties must be set into the mortar between the blocks to secure the facing material to them. A commercial flat metal tie can be purchased for this job or strips of rough-surfaced, perforated sheet metal can be used. Also, I filled in the blocks with zonite insulation for added thermal protection of the flue (**FIG. 20-14**).

When setting up the stone facing, you can eyeball it if you want to, but I found that after raising the stones 4' or 5', eyeballing becomes too inaccurate, so I used string guides.

The proper way to build up stone is to begin at one corner and go up 2' or, say, 30", depending on the rocks you are using. If you have two rocks and they fit in real easy—great. Then you move around the whole structure, and by the time you are back at the corner you started at, it is all set up. Sure, it's green and you have to be careful not to bang into it, but you shouldn't have any difficulty. I've gone up as much as 4' at a time. I just started going on a corner and kept right on going. Everything kept falling into place just like in a chess game. Every move I made was just beautiful. I always had backup for it. When you get into that kind of mode, just flow with it. You have to be extremely careful, though, if you are attempting to go 4' at a time. Your rocks can't be too big or too heavy. It's better to stick within 2' (**FIG. 20-15**).

FIG. 20-14

FIG. 20-15

Before striking the joints—that is, before digging the mortar out—make sure it has set a good four to eight hours, depending on the weather (longer if damper). The striking process gets rid of the excess mortar, which squeezes out between the stones as they are set in place. Striking helps reveal the rocks and gives them a more rustic, three-dimensional feeling. I used a simple screwdriver for this operation (**FIG. 20-16**).

FIG. 20-16

The chimney should be capped off with 4" of concrete. It is best to cap it with concrete instead of mortar because concrete doesn't have as great a tendency to crack in mass as mortar does. (Concrete can crack in mass—that's why expansion joints are used in car ports, but are not necessaryin a small place such as above the chimney.)

If the chimney was made of concrete blocks, a form for the cap could be built quite simply. You take two-by-fours and put them around the top of the chimney. Nail one-by-sixes around the exterior of the top of the two-by-fours, letting them extend up 4", then just let the frame hang on nails over the chimney top. This way, the cap overlaps the exterior side of the concrete blocks by 1½", the width of the two-by-fours. Be sure to leave 4" of the flue liner exposed as an inside form.

I bring the insulation right up to within a couple of inches of the top and lay reinforcing rod around the poured cap.

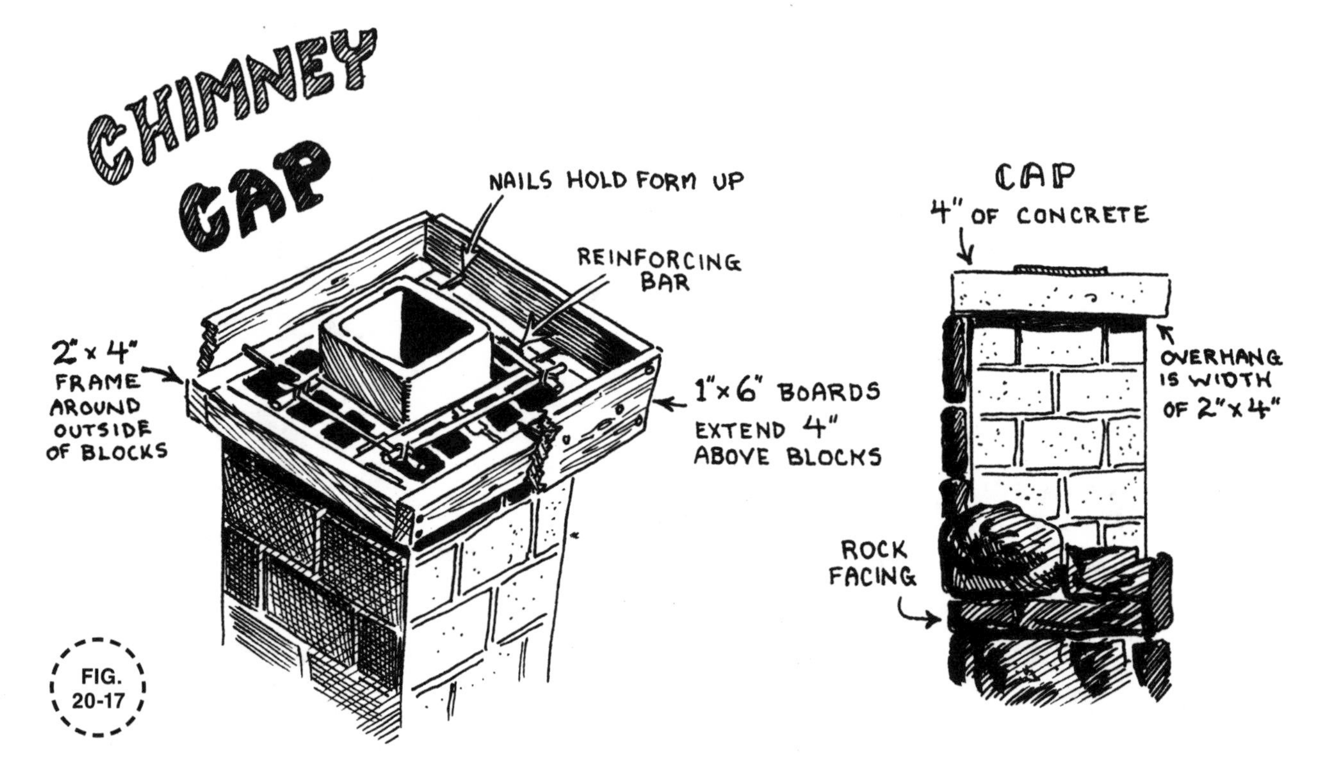

FIG. 20-17

A similar type of form can be used above stone, but care must be taken to shape the top of the chimney stack so it will support such a form (**FIG. 20-17**).

The face and firebox were built simultaneously and everything was taken up at once except the exterior wall. That was constructed later. It took a helper and me seven or eight days to do the entire interior area, except for the hearth, which was also added later (**FIG. 20-18**).

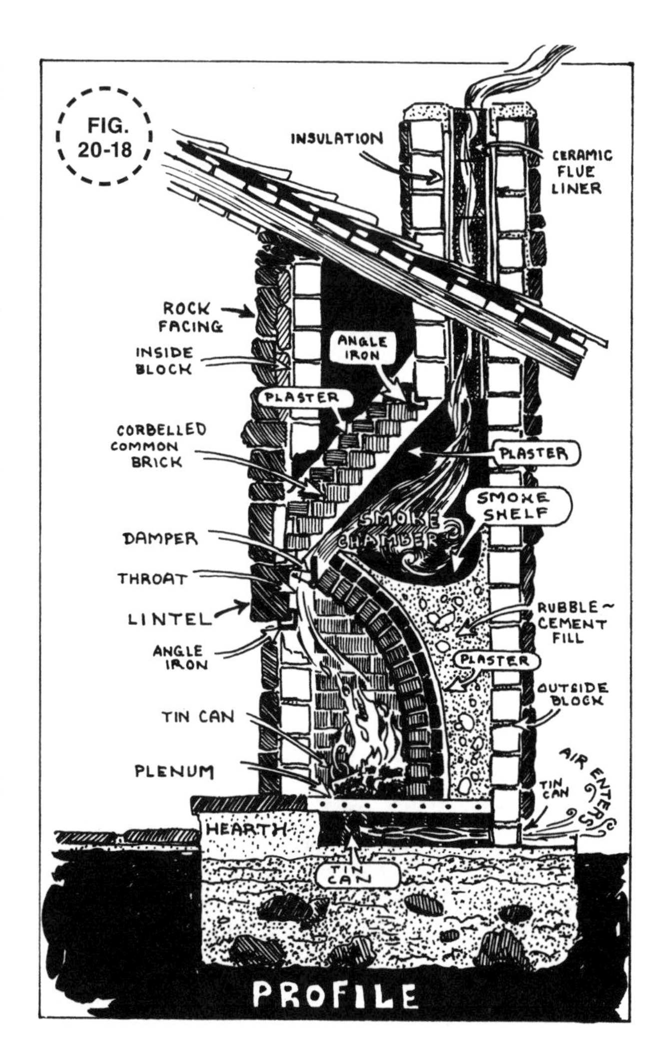

When laying the slab, be sure to take the hearth into consideration, though there is not a great deal of weight on it. I have built hearths on the floor joists and mortared the facing onto a constructed wood box frame. The frame was prepared with paper to absorb expansion and reinforced with a wire for support. This works well if the rocks are not too thick or too big. But with this method, you usually get a crack between the hearth masonry and the fireplace proper because the floor joists are bound to shrink and the whole thing is built on them. The hearth of this fireplace is built on the slab foundation.

After I finished the stone face, I cleaned off the joints with a solution of 50 percent muriatic acid,* 50 percent water. You can use a steel brush or a stiff bristle brush to take the cement stains off the face of the rock. This is a messy job. When all the unsightly cement stains are gone, wash the face down thoroughly to remove the acid residue, then let it dry. Now, I'll let you in on a secret: Take a bottle of cola, put it in a pan, and wet down a sponge with it. Then go over the rocks with the sponge. A mason told me this one. I thought the guy was off his rocker—but it really

*EDITOR'S NOTE: Take proper precautions when handling acid. DO NOT use muriatic acid in an enclosed space. Wear a respirator with acid-grade filter, heavy gloves, and face protection. Have baking soda or lime, and water, handy. Consult a professional or consider renting a sandblaster from a hardware store, alternatively.

FIG. 20-20

FIG. 20-21

works. It gives the stones a nice clean feeling (**FIG. 20-19 AND 20-20**).

As we finished the interior of the fireplace, the weather began to get real nasty, so we didn't start the exterior facing until the following spring. I used the same principles as were used in the interior face and it turned out just as beautiful, adding a rustic monument to the entrance of the house (**FIG. 20-21**).

We are indeed happy with our fireplace. It has become the central point where people gather around and communicate, especially in winter. In that season, we swing the divan to the middle of the living area to create a semicircular continuation of the hearth, making the area around the warm fireplace snug and cozy.

The fireplace is the structural, aesthetic, and functional axis of the house. It is the center of strength in its dominant form. Structurally, it is the point where the two roofs join together; aesthetically, it is where the design contrasts blend to epitomize the flow of the spaces and relationships within the house; functionally it is the gathering place that provides warmth and nourishment to all who join together around it. It joins everything together just as mortar joins stone, as nails join wood, and as real communication joins people. If care and patient awareness are utilized throughout the development of the fireplace, it will carry on the flow, doing its part in bringing it all together.

CHAPTER 21

The Monolithic Approach—A Central Fireplace

I thought a fireplace would be a bit extravagant, but my wife wanted one and I soon succumbed to her wishes. After all, we were building a house that we planned to live in for a long time, so we might as well include all the luxuries we wanted as we constructed it, instead of adding them on later as expensive afterthoughts. This is one decision we never regretted. We just love watching the open fire in winter. It keeps the house totally warm and is an excellent social center.

We planned our fireplace to be in the center of the house, dividing the kitchen and living room areas. Sure it takes up a lot of space, but it is much more efficient, being entirely indoors. It is surrounded by its own warmth instead of located on a weather-exposed outer wall. It retains its own heat, radiating it back into the house through the rock face and upstairs red brick, rather than dissipating out of the cold side of an exposed chimney (**FIG. 21-1**). This also reduces the problem of smoke condensing into creosote, because it doesn't get quickly cooled off by the outside weather.

FIG. 21-1

Since we had to build a huge chimney for the fireplace I decided to incorporate as much as possible in the one central stack so we wouldn't have to deal with several expensive and tedious chimney stacks. I designed this complex so it would include within its 2½ × 6 foot main base a chimney flue for the basement furnace, a fireplace and flue, a flue for the kitchen cookstove, and, as an afterthought, a small built-in oven. We didn't really decide on the oven until the basement block foundation was built. There was only a limited space between the heatform unit and the exterior wall. Consequently, the largest

possible oven we could have was 20" wide (**FIG. 21-2 AND 21-3**).

We used a heatform firebox for several reasons, mainly because it was easy to install and created a nice air circulation flow through its double walls. The cool air enters through two low intake vents and circulates around the fire in the space between the walls. After it heats up, it rises by convection, goes out the upper vents, and flows through the room as an auxiliary heat source. Heatforms also make it easier to construct the fireplace and chimney because they provide an inner form for the surrounding masonry (**FIG. 21-4**).

Even with all the heatform's advantages, I still sometimes have second thoughts about a firebrick firebox, because the thin sheet metal construction of the heatform cools off much faster than firebrick. So when using a heatform, the fireplace is warming up rapidly and cooling off rapidly after the fire is out, as opposed to 4" of firebrick taking awhile to heat up but retaining heat long enough after you put the last log in at night to keep the house warm until morning. And with a little effort and ingenuity, an air circulating unit can be made with firebrick, having both the advantages of retaining heat and giving off extra heat through convection. All you have to do is back the firebrick with a ⅛" thick double-walled sheet metal jacket with air ducts, so you can draw the cold air in from below and bring it out through the upper outlet vents. Then back the

outer layer of sheet metal with the necessary 4" of clay brick.

If I were to design my fireplace over, I'd figure out some way to run a duct or channel from the firebox to the outside of the house so it would not deplete the oxygen from inside the room.

We began construction by pouring a 4" thick pad that extends 6" beyond the base of the chimney complex on all sides, approximately 5 × 7 feet. A mixture of 2 parts sand, 2 parts gravel, and 1 part Portland cement was used. No large rocks were put in for volume. When you want a solid foundation under a structure that is going to be carrying a tremendous amount of weight, it is best to not break up the pad with large rocks, but to make it of continuous small aggregate. We did not put any rebar in the pad because it was set on bedrock and hard-packed gravel. It is usually wise to reinforce the pad with rebar on 12" square centers, a little below the middle of the pour.

Above the pad, we began the block enclosure around the perimeter of the base, allowing for the necessary flue and ash pit openings as the courses were raised. An 8 × 16 inch column was left hollow from the pad to the top of the base (floor joist level) to save on blocks at the right side. Then we put in a column of blocks and left another space 2' long by 16" wide in the center for an ash pit. This ash pit also extends from pad to floor level. A cleanout slot was allowed in the bottom block at the center of the ash pit at the front wall. On the left, the blocks were stacked solid to a height of approximately 5½' where the furnace's thimble pipe intersects it (see **FIG. 21-3**). At that point the 8 × 8 inch furnace flue stack was started. A hole in the bottom flue tile was cut out to accommodate the intersecting furnace pipe. The flue linings were stacked and mortared together to beyond floor level, creating an inner form for the main floor pad. So at the main floor level, just below the floor joists, two spaces were cut into the plywood bottom form, one for the furnace flue and one for the ash pit. Since the main floor had already been laid and a

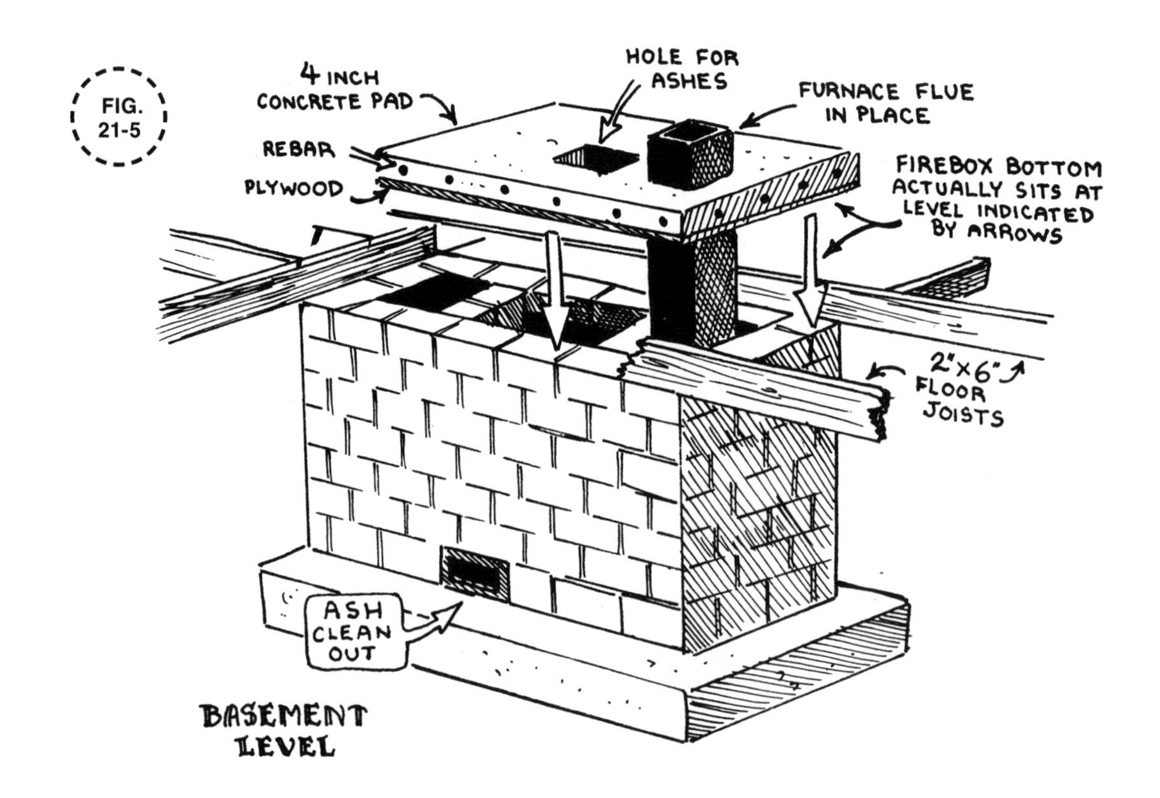

4 × 6 foot hole was framed between the floor joists, the surrounding two-by-six joists and headers were used as the side forms for the 6" pad, which we then poured. The pad was reinforced with rebar placed on 12" centers wherever possible (FIG. 21-5). The plywood was buried forever.

The blocks for the hearth were placed across the front of the already-set-up pad, allowing 4" on the three outer sides for the stone face. The top was then finished with flat sheets of commercial slate (see FIG. 21-6). If we were to rebuild the hearth, we would make it slightly wider for aesthetic reasons. This would also make it more comfortable for people to sit on. As it is now, its 16" width is too narrow for a person to sit on and be far enough from the direct fire. Also, a wider hearth would help prevent soot and debris from spilling out onto the floor. The perimeter of the block structure was laid out with a few courses of block, 4" in from the surrounding edge to allow for the rock face.

Next came the heatform unit. On top of the pad, between it and the above heatform, a layer of 4" firebrick was laid. The unit was then set on the firebrick. The heatform is a double-jacketed steel unit consisting of firebox, throat, smoke shelf, and damper assembly. Its double-walled construction allows it to take in cool air from the low air inlet vents, circulate it through the space between the walls of the firebox, and expel it as warm air out the round air flues in the throat. Once this unit was set in place, it was lined with insulation wherever it came in contact with masonry to allow for metal expansion and contraction. Since the firebox and throat area were shaped by the heatform, the smoke shelf masonry was an easy chore, using rough filler brick between the metal and the outside block work to form the curved shelf. After the insulation and necessary block were put around the heatform, a piece of angle iron was set into the block work on either side, above the throat. This angle iron supports the

12 × 12 inch flue linings and the surrounding chimney masonry.

Meanwhile, the masonry around the furnace flue was also raised and our afterthought oven was allotted for on the other side of the fireplace. Since the heatform took so much space (46" overall width in front, 33" in back), there was little room left for an oven. In fact, even to have an oven that had only 16" x 18" inside dimensions at the opening, we had to steal space from the side of the fireplace, eliminating the bricks that should be between the two structures. There is only 2" of firebrick and a thickness of fiberglass insulation between the oven and the heatform. Actually, this does provide the advantage of the oven unit being heated by the fireplace as well as by its own firebox. And there is no problem with the oven burning out, since it is constructed out of firebrick.

FIG. 21-6

The 8 × 8 inch flue opening for the oven and cookstove was started about 3' up from the floor, just level with the oven's firebox. This flue opening is located deep into the chimney structure from the kitchen side to allow the oven to have a depth of 24" before reaching it. This is a dual-purpose flue for the oven and for a cookstove. (**FIG. 21-6 AND 21-7**). Just under this oven, there are two smaller compartments, a 10 × 10 inch firebox that extends 32" in depth to the back of the flue opening and a shallower ash pit below. The doors to these three compartments are old iron furnace doors

FIG. 21-7

mounted on steel plates that were embedded in the masonry. The oven and firebox are surrounded by a lining of 4" of firebrick. The top bricks are held up with a series of 2" steel straps, which go across the top of the oven box. These straps and the ¼" rebar grate bottom of the firebox are also embedded in the surrounding masonry (**FIG. 21-8**).

The air from the oven fire comes in through the ash pit door air control and goes through the ¼" rebar grills on which the fire fuel sits. The fire then heats the oven from underneath and from behind as it goes up the flue. It burns terrifically. The long chimney pulls the smoke out just beautifully. It heats up well, except it can't be damped down to build up a good, continuous heat in the oven for proper baking. One remedy for this problem would be to put a damper inside the flue above the oven. This would keep the heat close to the oven and cut down flue drafts. One damper would regulate both the cookstove and the oven if it were placed above where the cookstove pipe intersects the flue. Right now, if we close down all the mechanisms of the cookstove and then open the fire door to the oven, the cookstove would damp down even more because the draft would be sucked into the fire door opening, reducing the amount coming down from the flue. Also, to circulate the heat around the oven to the utmost efficiency, I suggest planning the oven so it would

FIG. 21-8

include air channels around the sides. This would provide an even heat around the oven instead of only heating the bottom and back of it.

The inside block and masonry were continued up until the top of the main floor, then the rock face was started. Remember to insert quite a number of corrugated tin ties in between the concrete blocks so the stone face will hold securely to this inner concrete form. We wanted a cobblestone finish on this fireplace, so we searched the nearby creek and riverbeds for this kind of round rock. You don't have to worry about such rock exploding from heat because none of them will be in direct contact with the fire.

The mortar mixture was the standard mix, with one additional part Portland cement. A ¼" mesh screen was used to screen the sand. I prefer using coarse sand when mortaring rocks together because this material provides more adhesiveness. It has more angles that grip and interlock to each other than does the finer sand. The round stone facing went relatively fast, about 25 square feet or 2' all around the structure in a single day with two of us working steadily. Actually, you could go as high as you want with any single course until the mortar and already seated stones can't carry anymore weight without slumping or falling out. We tried to get each stone to fit so it wasn't dependent on mortar to hold it on top of the stone below it. If it was put in a precarious position where we had to hope that the mortar would hold it, it was taken off and altered by either tilting it until it was secure or trimming it with a stone mason's hammer.

The thing about stonework that makes a good piece of work is how clean you can keep the face of the rocks, rather than leaving them covered with mortar. At the end of a day or halfway through the day, depending on how fast the mortar is setting up, I clean off the freshly laid stones with a sponge, making sure to remove all traces of mortar on the faces and painting where necessary.

The stone face surrounds all the concrete block and the outer edges of the heatform. An iron lintel bar was placed across the front of the firebox, a few inches below the steel frame of the heatform. It overlaps the sides of the firebox opening by 2" on either side, so it could be embedded into the mortar face (40" long across a 36" opening). Fiberglass was put around the iron lintel where it comes in contact with the masonry, to compensate for expansion. This bar was set below the top of the firebox opening to shorten its height because there are doors on either side of the fireplace, which bring in drafts that sweep the smoke out into the room. The lower lintel keeps the smoke from rolling out of the firebox. This fireplace burns fuel beautifully even without a grate. When you don't have a grate, the coals go into the ashes, radiating their heat longer than if they were suspended over a draft. Being on the firebrick instead of on an open grate, wood burns slowly and more efficiently, saving on fuel.

The low intake openings on either side of the firebox were covered with thin layers of slate on end, spaced far enough apart to not interfere with the incoming air. The outlet vent above the box was also dealt with in the same manner, with flat stones on edge, allowing an aesthetic contrast between the round cobblestones and the flat materials (**FIG. 21-9, 21-10, 21-11**).

The concrete blocks were continued above the main floor, around the three flues, but were slightly recessed so the red brick facing would be above the main floor blocks. This portion of the stack was designed this way so the heavy red bricks would not be directly above the lower cobblestone facing, but would be raised on the more solid concrete block foundation. These red bricks were stacked almost to the ceiling where

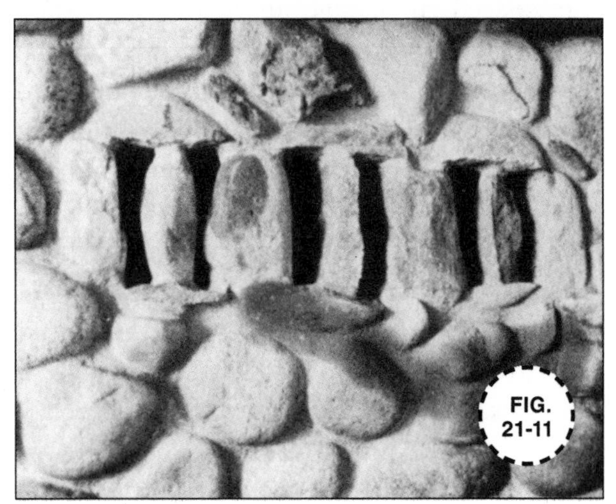

they were replaced with a couple of concrete blocks for contrast and strength. The facing was continued out beyond the roof with alternating red and beige bricks. Then the whole chimney structure was capped with a 1½" of concrete and a lip extending 2" beyond the stack all the way around it.

If this fireplace were to be reconstructed, we would make it 16" narrower at the base (one block shorter) to have more basement space and save on expensive materials and labor. The basement concrete work could then be corbelled out near the main floor to accommodate the cantilevered hearth instead of supporting it with such a massive foundation. Also as an afterthought, I would have put in bolts or hinges for shelving above the cookstove so kitchen implements would be in a handy place above it.

CHAPTER 22

Cedar Log Root Cellar

Winter has become our season of plans. As a city child, I had never bothered classifying the seasons except as they related to the school year or to the clothes I could con out of my mother and, later, to how much I was cooped up with the children. In the country, they are definitive.

It was our second winter and we were planning the work for the following building season—the seven months when there is no snow, usually referred to as summer. Our land had been bush when we—two couples and two children— arrived. We had spent the first hectic summer jointly building our water system, two cabins, an outhouse, a chicken house, a toolshed, a makeshift woodshed; clearing land for a garden, for pasture, and for orchard; and keeping our cooperative running smoothly. By fall, completely exhausted, we gratefully returned to our respective cabins. The first winter was therefore one of minimal plans, mostly for the expansion of what already existed. Our partners were designing themselves a new house to be built in slow stages and we were contemplating a wood and storage shed.

By the second winter, our most fundamental needs had been met and it was time to evaluate the accomplishments and weigh them against the dreams. We had had a basic premise when we made the move from city to country: to learn the skills for providing the essentials of human needs with our own hands and minds. The winter nights of talking accumulated as we debated the direction of our future energies, how much of ourselves we wanted to devote to the "place," and where those nonessentials that have become indispensable fit into our new lifestyle. We were learning that neither of us wanted to live at "Walden Pond," that we both wished to find a satisfactory way to combine the best of urban and rural worlds. We discussed how a homestead grows in stages, each stage being a unit unto itself, dependent for efficiency on total completion of that phase. Keeping a cow creates the need for pasture, a hay field, fencing, a barn. Truck gardening might involve further clearing, a tractor, sources of fertilizer, good transportation. We were on the last lap of stage one—shelter, water, garden, and winter fuel.

Did anything remain before we chose to tackle another stage? If we were going to the trouble of growing our own food and buying or trading for what we could not produce, were we obligated to provide these fruits of labor with adequate storage? We answered yes. Other people's basements and decrepit root cellars had proven wasteful. Storing jars in the house meant risking total loss by freezing if we happened to spend just a few nights away from home during a cold spell. The freezer, though a miraculous invention, is not useful for all foods and depends upon the benevolence of the power company. And so grew our plans for a root cellar as the building focus for the third summer.

The placement of the root cellar was a foregone conclusion. We had always assumed that one would be built sometime. So during the first summer, when we hired a backhoe for the water line trench, we had it scoop out a hole into the mountainside. We had little choice about the exact location of the hole. The flat portion of our narrow acreage butts up to the mountain slope and we merely chose a spot that was close to the two cabins, where the hillside is steepest, and where there is no spring runoff. The snugger the building fits into the mountainside, the less work there is pushing dirt up against the walls for maximum insulation. Proximity to the house is a consideration that can be overlooked only in the summer when snow is a mere memory.

Another consideration is shade and sun. Our main concern is keeping the food from freezing in the winter, since we do not store much during the two to three hot months. Therefore, we picked a site that would be shaded in summer by deciduous trees and shrubs whose leaves drop off exposing the root cellar to whatever bit of sun manages to fight through the snow clouds.

The backhoe dug out a hole whose size ultimately proved to be a lucky guess but it would have been far better to approach construction in the logical order—that is, plan the size of the root cellar first, then get a hole dug to meet the specifications. The hole should be about 3' feet larger than the outside dimensions of the building so that there is room to work, but should not be made any larger than needed. That would just add to the time or money to have the building backfilled.

Whatever books and pamphlets on food storage we could find were limited to either basement root cellars or outdoor ones, constructed of concrete. We were definitely committed to an outside root cellar. Two families would be using the structure and we felt that it would be better located on neutral ground rather than have one family disturb the other to get its food. We also considered that since fire is always a risk when heating with wood, it was advantageous not to have everything under one roof. And, of course, we had the hole.

Choosing the building material for the root cellar emphasized once again the lesson that country people usually know what resource is most suited for their area. We were all set to build with concrete even though many of the book diagrams were of root cellars so elaborate they resembled bomb shelters and even though neither of us thought concrete a particularly attractive medium. But we had decided to forgo aesthetics in favor of practicality—if the books said concrete, then it was concrete. Luckily, Jim found himself talking about root cellars to a longtime resident who warned that concrete was, in his opinion, the worst of all alternatives for this climate where there is so much rain and snow. Concrete retains moisture, doesn't breathe well, and so presents problems of ventilation. Without proper ventilation, mildew develops and food spoils. Seepage of water into the root cellar will also ruin the food there. Our friend suggested three alternative building materials: rock (being more porous than concrete), frame construction packed well with insulation, or log construction. The rock medium was intriguing but is time consuming, and we have a fairly short building season. Therefore, it was frame or logs.

Up until then all of our buildings had been built with dimension lumber utilizing some poles. We have few large trees on our land due to past logging and to a forest fire; with several mills in this area, lumber is relatively cheap. And we had always been building against time. So while we were fast becoming proficient in frame construction, we were totally ignorant of log building. But during our first summer, we had picked up a bunch of cedar logs for a sauna that wasn't getting built. The length of the logs strangely enough exactly fitted the hole. They were just lying around, soon to decay. We said, "Why not? Let's try it," and our decision was made.

Once we decided to use logs we found the whole building taking on an almost spiritual tone. Providing a space for our own food with materials from the surrounding countryside and designing the whole building to harmonize with its setting became the principal expression of our joy to be living here.

We were now ready to get down to the business of constructing. To fit the size of the hole in the mountain and to accommodate the logs already cut, we decided on a two-chamber root cellar whose outside dimensions would be 10 × 14 feet. The inner chamber would be 8 × 9 feet and would be separated from the outside room by an insulated frame wall. Using a frame wall eliminated the difficulty of interlocking logs and provided better insulation than the curved surfaces of logs. Since this wall must insulate independent of backfill dirt, no skimping should occur.

The height of the root cellar, 6' 2", was determined by the height of the tallest member of our family and by the number of logs we were able to get. It is best to keep the building as small as the need warrants. Excess space increases building time, but more importantly, proper temperature control is difficult to maintain in a larger structure. The two-room construction is essential because it allows the outside door to be closed before opening the door to the main chamber, thereby keeping the exchange of air and temperature at a minimum. The outside chamber provides an excellent place to hang meat and to store winter apples until the coldest weather (**FIG. 22-1**).

For building logs we used western red cedar, some of which were picked up at a road construction site and the rest came from an abandoned logging operation. Western red cedar is one of the most impervious woods to rot and dampness. When in the ground, it is the number one choice for this type of building (**FIG. 22-2**). If cedar is unavailable, just ask around for the best wood in your area for underground use and how it should be treated.

When choosing the trees to cut, aim for the straightest logs for uniform thickness. Uniformity of size will make notching much easier. A 6" diameter is the minimum thickness; 8"–10" would be better. But for ease of handling, the size of the building and therefore the length of the logs should be taken into consideration. Trees that have fallen naturally or have been cut and are lying on the ground must be examined carefully. If an unpeeled log has been on the ground for too long, rot may have started to work its way under the bark. If cedar logs are used, they should be well dried, because their shrinkage is tremendous. But all logs should be dry and peeled before using.

When our logs were assembled at the site, their progression in the building was planned out. Starting with the largest logs, we arranged them with an eye toward uniformity in the tiers and toward a steady succession from thickest to thinnest for the courses.

Actual construction was begun by cleaning and leveling the site. An unleveled site produces unlevel walls and therefore a totally wonky building. We went so far as to use a good level on the ground to assess our progress. Once satisfied, we built the forms for the footing, using

FIG. 22-1
BINS FOR ROOT CROPS
SIDE SHELVES ARE SUSPENDED FROM 2"×8"'s BOLTED TO THE CEILING JOISTS
APPLES
MEAT
INWOOD '75

FIG. 22-2

two-by-six boards. The forms should be strong enough or well enough supported so that they are not pushed out of line when the cement is poured. We braced our forms with outlining stakes, which were driven into the ground every few feet and nailed to their outside. To further prevent spreading, holes were drilled and wire braces were put in every 4'.

The footing itself must be solid and strong to avoid settling, to prevent shifting when the site is backfilled, and to ensure that stress is equally distributed throughout the total perimeter of the building. We decided on a concrete footing because it met these standards. Two alternatives are flat or tamped-down gravel. Neither produces the desired stability and in both cases the chances of rot are greater as the necessary vapor barrier is harder to create between a bumpy surface and the log.

The footing is 6" high and 14" wide and was poured directly on the leveled ground using a 1:5 mixture of cement to sand and gravel. Immediately following the pour, we inserted pieces of ⅝"

FIG. 22-3

reinforcing rods protruding from the center line of the footing, one in each corner and the others at approximately 4' intervals, which later served as anchors for the first course of logs. The height of these rods depended on the thickness of the first log. Steel bar or pipe can also be used as anchors; we happened to have some extra rods left over from another project (**FIG. 22-3**).

After the concrete was poured, it was dampened for three days and covered after each watering with feed bags to hold in the moisture. Since cement should cure four to five days before being built upon, we took the opportunity for a break and went camping. Once home, with renewed energy, we recommenced work by removing the forms from the now-dry cement and placed a strip of 50-pound roofing paper all along the top of the footing as a vapor barrier. This prevented the first course of logs from sitting directly on the cement, which would be constantly wet if in contact with moisture. Logs will rot faster if wet.

Our first course of logs proved our first major mistake. We took the two logs for the long walls and drilled holes to fit them over the rods sticking out of the footing. We then ripped a log in half lengthwise with the chain saw for the short walls, fitting them over their rods in the same manner. These logs were ripped to compensate for the displacement of height between the long walls and short, perpendicular walls. This half log displacement is necessary for the notching process.

The rectangle of the first course was formed by butting the ends of the short logs against the sides of the long logs at the corners. Much smarter would have been to put down the short half logs first and then notch the long logs to fit over them,

FIG. 22-4

thereby creating a more tied-together fit. These first logs are just sitting over the rods, which is sufficient anchorage as only lateral movement is possible. We did not have a drill long enough to go clear through the logs and so were unable to use thinner ¼" rods which would have been bent over the top of the log for a tighter anchor (**FIG. 22-4**).

With the first course down, we had reached the point of deciding how to make the door space. The front half log would serve as the door sill. Fitting the outside door tightly and accurately is a necessity for good temperature control and much harder to achieve if the sill is cement, wood being easier to plane. We knew of two methods of making door spaces in log walls: If enough long logs are available, a solid wall can be constructed and a door can be cut out later. Or, if there are only shorter logs, they can be notched into the structure at one end and later be tied together at the other end by a solid door frame. Since we had few long logs, we chose the latter method and stacked the logs at random lengths that were each a bit longer than the beginning of the openings. Using a chalk line as the guide, the logs were later sawed, making straight vertical lines for the door frame (**FIG. 22-5**).

The time had arrived for our first notch; we could no longer procrastinate. The challenge had to be met. Knowing ourselves to be the complete novices we were, we traded dinner for a lesson with a more experienced friend. It proved invaluable and boosted our confidence. If no one is around as an instructor, use an old log to practice on. It beats wasting a good building log.

We began each notching operation by examining the log for the two straightest, flattest, opposite surfaces and these became the top and bottom of the log as it fit into the wall. We secured the log in place for measuring the notch with a log dog at each end. Using a ruler, we took an "eyeball" measurement of the distance between the log we were working with and the parallel one below; this would determine the depth of the notch. By "eyeball," I mean that I made sure my eye was parallel with either surface to be certain I was accurately at the low point of the upper log to be placed and the high point of the parallel one below. Once the depth of the notch was determined, I followed the contour of the end of the perpendicular log below with the ruler and transcribed a series of dots ⅛" shallower than the determined depth. I marked it ⅛" shallow to allow for miscalculation and overcutting. Be sure

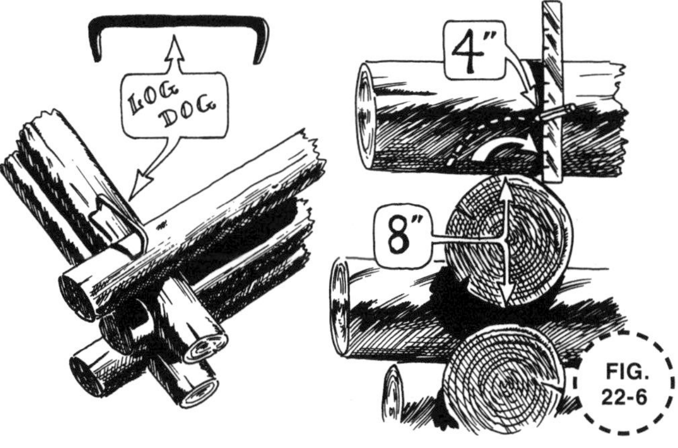

that the ruler is always straight up and down or the contour will not be accurate. Also, use a thick crayon when marking so you'll have a clear line to work with. When finished, a semicircle was drawn by connecting all the dots as in a child's game.

We later found an easier way to determine the depth of the notches. Using a scribe or an old compass, measure the distance between the log you are working on and the parallel one below. Open the scribe to that width and lock it there. Then, place the scribe an the end of the perpendicular log you want to notch to and transcribe the contour onto both sides of the end we were working on. Repeat the process on the other end—it's as simple as that. Too bad we didn't know this when we were building (**FIG. 22-6**).

Because so many of our logs were fairly narrow and we did not realize the importance of uniform thickness when they were cut, we sometimes came up against a space between the logs that was larger than half the diameter of the log at the notch. This meant that if we cut the notch large enough so that the log sat directly on the one below, we had to cut out much more than half the width of the log. This would weaken it. On the other hand, if we cut the notch to a depth that matched the radius measurement, the log did not sit on the one below, but hung suspended. To compensate, we had to shim in between some of the logs. This also occurs when the logs taper too much so that you are notching a very small top log over a rather large bottom butt. To avoid this, choose logs that are of uniform width or make a progression from widest to narrowest that is steady, and, if possible, cut your logs at the same time so they can be compared.

After we individually marked either end of the log, the log dogs were removed and the log was rolled to an upside-down position and was redogged for the chain saw kerf cuts. The kerfs were then cut to the lines of the semicircle. Remember, for safety, when the walls get too high, take your logs off and do the cutting on

FIG. 22-7

FIG. 22-8

FIG. 22-9

FIG. 22-10

the ground. Also, until you are an experienced notcher, and maybe even then, always cut out your notches smaller. It is easier to chisel out a little more to fit the curve than to deal with an oversized notch (**FIG. 22-7**).

Using a hammer, we knocked out the pieces of wood between the kerfs and chiseled a flat, accurate border on the edge of the notch as a guideline (**FIG. 22-8 AND 22-9**). With the nose of the chain saw, we smoothed out the interior of the notch (**FIG. 22-10**). It is wise to practice this step. We ruined a few good logs because of our inexperience.

Once or twice we did hit a perfect bulls-eye on the first try when the log was rolled back into place, but usually we had to chisel out several times to get a good fit.

Sometimes the notches proved satisfactory but bumps in either the top or bottom log where they met prevented a tight fit. To correct this, we wedged up the top log at each end and, using the chain-saw blade laid flat, leveled the bumps. If done right, the logs fell into place when the wedges were removed. When the fit was satisfactory, we rolled the log away once more and laid down fiberglass insulation in between the two logs and at the notches. These 2"-wide strips of

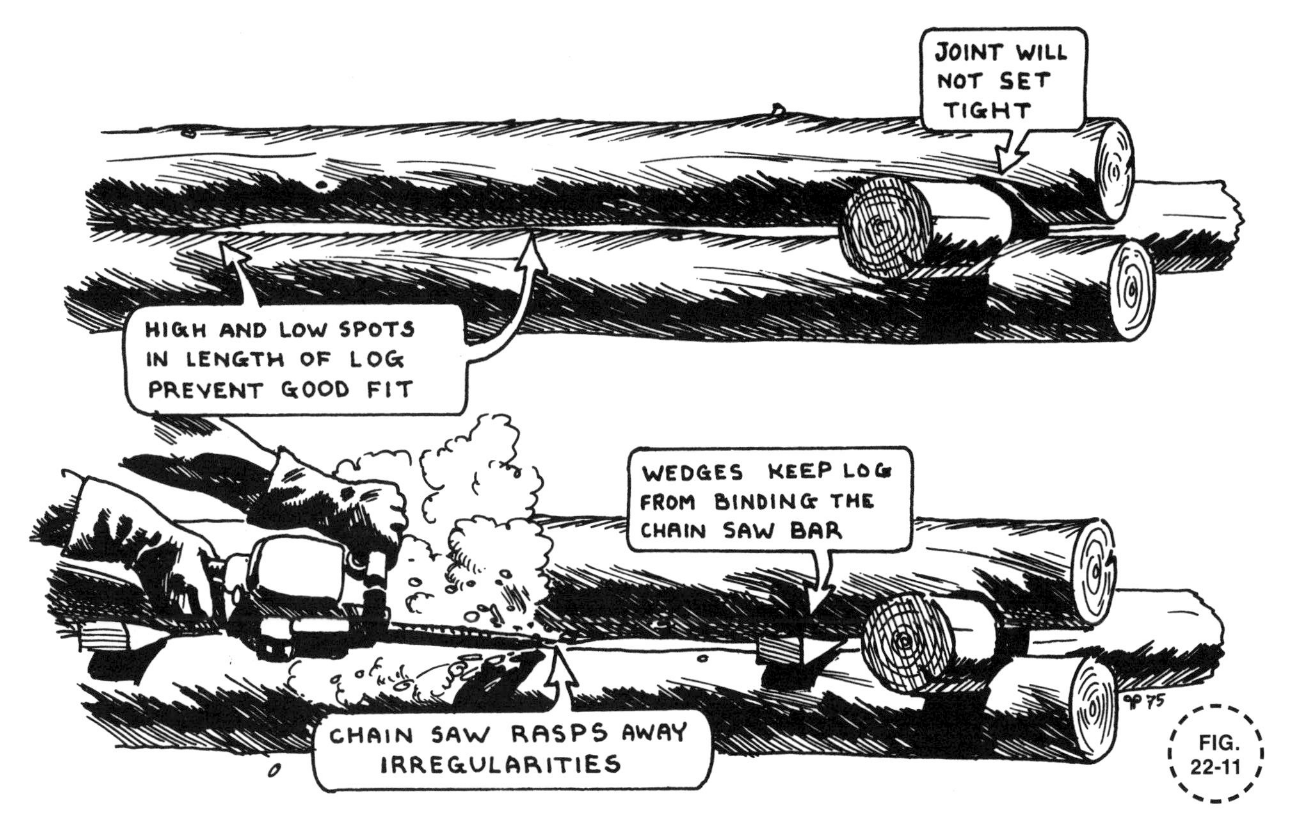

FIG. 22-11

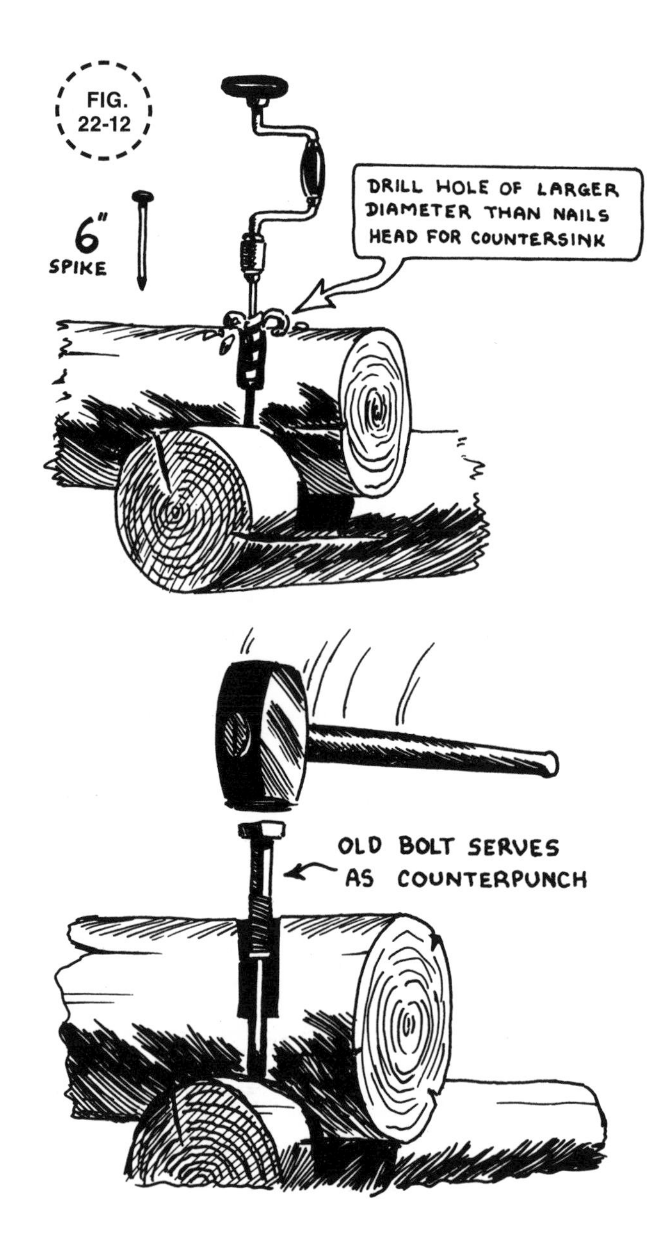

insulation had been cut with a hand saw when still tightly rolled (**FIG. 22-11**).

Fiberglass was chosen as the insulating material because it would be permanent. Since the building would be backfilled, rechinking from the outside would be impossible. Once the fiberglass was laid, we rolled the log to its final resting place and spiked it down at each notch and at several points along the log to ensure that there would be no movement when backfilling occurred. We used 6" spikes, but the size of the nails depends on the width of the logs. If you have spikes that are not long enough, a hole can be drilled partway through the top log and the spike punched down, using a long bolt as the punch (**FIG. 22-12**).

With each succeeding course, the ends of the logs were alternated so that the butt end of the log above sat on the smaller end of the one below, thereby keeping the building level.

When our courses were completed, we were left with one tier lower, or two sides lower, than the others. To compensate, we nailed plates of dimension lumber to even out the final tier. Aesthetically, logs would have been preferable, but our log supply was exhausted.

Once the log walls were up, we poured an 8" wide cement footing to accommodate the interior wall. This is not a bearing wall (one carrying structural weight), so reinforcement was not necessary. This wall was framed with two-by-six

material. Using a chain saw and chisel, a channel was notched in each log at the place where the frame wall met the logs. This provided a straight, flat surface for the end studs to be nailed to and created a snug fit for the wall with no air spaces. We stapled 3" thick fiberglass insulation to both sides of the frame so that the paper backing of the insulation formed a vapor barrier, then we covered the wall with 1" cedar boards (**FIG. 22-13**).

Planer shavings that are stuffed down between the walls are also adequate insulation. The shavings must be bone-dry and well tamped when first put in, because they settle. If the wall is not properly filled, the settled shavings will create a space at the top resulting in heat loss.

By the time we reached the construction of the roof, harvest season was approaching and as an expedient we resorted to the use of dimension lumber for ceiling joists and roof rafters. We had been wrestling with the type of roof to build throughout the summer. Originally, we had decided on a flat-log roof covered with 2' or 3' of dirt. This was cheap, easy, quick, and, in our ignorance, we believed it to be a good insulator. But the same friend that discouraged our idea of cement walls destroyed this notion, too. In our wet climate, he said, this type of roof is sure to leak and as we have rains in the fall often followed by a quick freeze before the snow, we could be left with a frozen mass of dirt covering our precious food. When it did finally fall, the snow would act as an insulator and the big "ice cube" over the root cellar would remain so until the spring thaw.

Once it was established that we would be building some sort of regular roof design, we had to decide if we wanted to make a functional space under such an exposed roof. Our neighbors had built a log root cellar with a hip roof high enough to allow for a storage area above a well-insulated ceiling. Other root cellars in the area have entire second stories built above them for use as workshops. In the end, we chose a simple peaked roof with a 45° pitch, coming down to the backfilled ground level. It was fairly economical; we didn't really need the storage or work space, and we were pressed for time. But, most important,

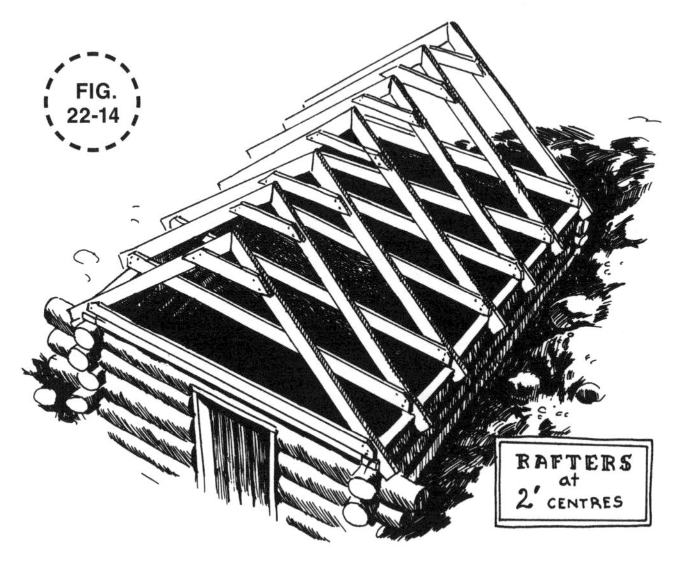

we considered our root cellar to be aesthetically pleasing as is, in its surroundings in the woods, as a space exclusively for food. This may seem extravagant, a waste of a good foundation, but for us it was appropriate and resulted in a much lovelier building.

For the ceiling we used cedar two-by-eights as joists and the random leftover 2" lumber for the ceiling boards. The two boards where the ceiling meets the roof rafters were carefully notched to attain a good fit for better insulation. The roof rafters were also two-by-eights on 2' centers (**FIG. 22-14**). The roof had to be strong enough to hold the whole winter's snow load. There would be no interior heat coming up to melt the snow.

In order to determine the pitch of the roof, we overlapped the ends of two pieces of lumber, leaving the nail loose so that the boards could move. One of us stood on the ceiling and held the

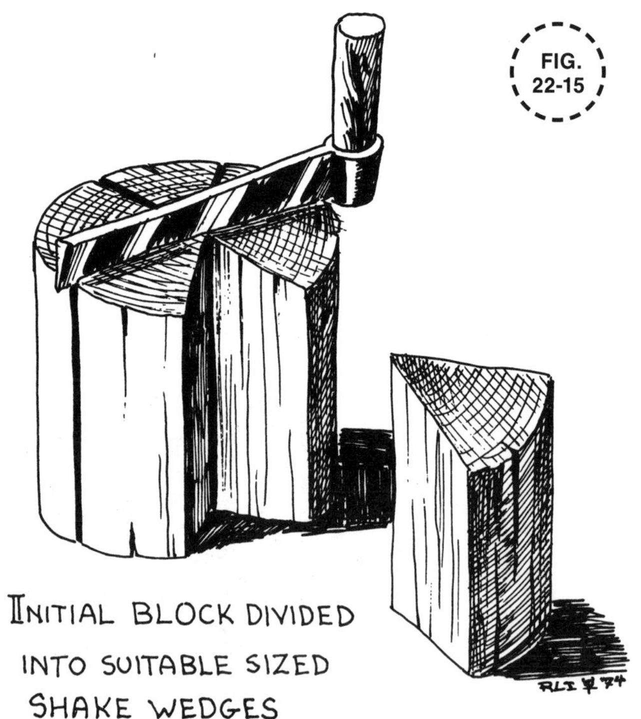

leakage from the roof. We had gathered the bolts before knowing much about bolt quality. Western red cedar provides the best shake material, but in the West, spruces, fir, and pine can also be used. We picked up our bolts on a Sunday from a logging operation site. They had been rejected because they came from decadent trees whose centers were beginning to rot out. We were most careless in our selection, lugging home many bolts good only for kindling, but we managed to get enough suitable ones to have made the hard chore of gathering them worthwhile. Bolts should be straight grained, dry, clear of knots, and, if

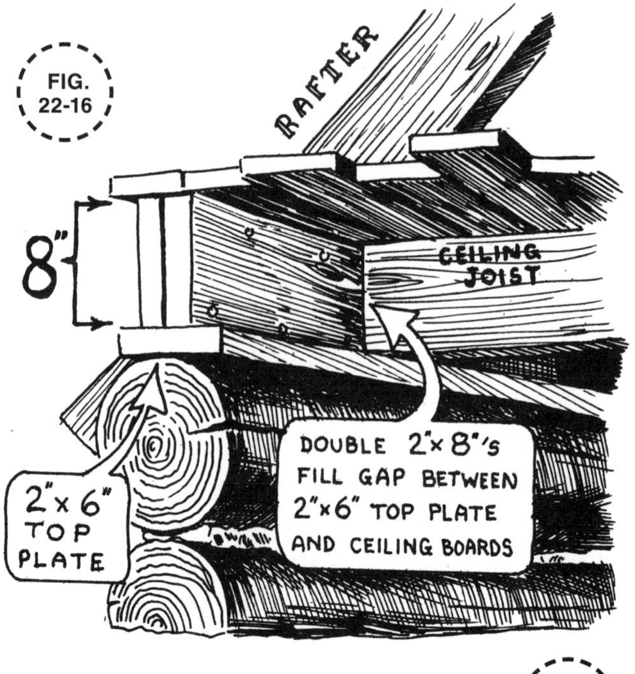

sample rafter while the other stood away from the building and studied the effect of different pitches. We were aiming for a roof that blended the building with the slope of the hillside, and had a pitch steep enough to shed some of the winter's snow. We finally chose the 45° pitch.

The rafters were cut and butted at an angle at the top, then supported there with a small two-by-eight horizontal brace. We cut two-by-eights as blocking for the space between the roof rafters and the ceiling joists and doubled them up to increase the insulative value of the wood (**FIG. 22-15**).

The roof of the root cellar is shaked. The shakes (shingles) were split intermittently during construction as a change of pace. If shake bolts can be picked up in the woods, the shakes would be a very cheap, durable, and beautiful roofing material to use, especially in a building where no fire is to be laid.

Our shakes are far from perfect, but there is no

possible, from 20" to 30" long, though shakes down to 14" are adequate. The center should not be too rotten so that the shake is of sufficient width, at least 6". The larger the shakes, the faster the roof goes on, the less nails are used, and the fewer cracks there are, minimizing the chance of leakage.

The bolts were sectioned into a few easy-to-handle chunks and the shakes were split from them with a froe, which was pounded with a birch mallet (**FIG. 22-16 AND 22-17**). The heavier the mallet, the less whacks were required to sever the shake from the bolt. But by the laws of diminishing returns, if the mallet is so heavy you cannot lift it, it is of little use.

We tried to make our shakes ⅝" thick and always turned the bolt upside down before the next split, for proper taper (**FIG. 22-18**). With cedar shakes, it is wise to cut the outer white cambium layer off with an axe. It rots much faster than the rest of the shake and will decrease the life of your roof (**FIG. 22-19**).

The shakes were attached to one-by-four boards, which had been nailed across the rafters. Because our shakes were so irregular in length, these purlins were put on separately as we were ready to add another row of shakes. The shakes were nailed in rows, thick end down, starting at the bottom of the roof. Two shakes were nailed down first, using two shake nails in the bottom and one at the top of each shake. Then a third one was nailed to cover the crack and to cover the nails of the shakes below. The distance between the first two shakes depends on the width of the one that is to be laid over the crack, but they must not be too close because shakes will swell when

FIG. 22-18

FIG. 22-19

wet. The next row of shakes was laid down in the same manner, overlapping the first row by 3" or so, and the shorter sides were butted directly against the longer one (**FIG. 22-20 AND 22-21**).

If your shakes are uniform in size, they can be nailed at regular intervals, eliminating the constant culling and juggling of shakes that we were forced to do, and the purlins can also be put on in advance. Splitting the shakes, and roofing with them, was an incredibly satisfying experience. The rhythm of splitting is akin to creating music. And there we were, laying down a lovely, brown-red, natural roof, high enough up for a good view of the fall color explosion advancing down the mountainside.

Because the roof looked so nice, and shaking

was such fun, we used the leftovers to cover the front gable, which faces west and gets rained upon. The back gable, facing the mountain, was left open for ventilation and to enable the addition of more insulation over the years as settling occurs (**FIG. 22-22**).

Once the roof was waterproof, we drove to the site of a long-abandoned mill and scraped off the top dry layer of planer shavings from several piles. We managed to accumulate just enough for a measly 6" layer of insulation before the fall rains destroyed that source for another year. We were about to abandon shavings as our roof insulation material when a neighbor arrived with a pickup load of new shavings, explaining that she had overestimated her need. With hers, we gathered enough for an insulation level of 18".

Prior to adding these shavings, we cut a hole in the middle of the ceiling to insert a square vent box the width of one-by-six cedar boards, long enough to reach above the 18" of insulation over the ceiling and extend into the root cellar 2"–20" altogether. The ends of the box were screened to deny entry to insects and small animals. This screen is a heavy-gauge stainless steel screen, which was used in a boy's detention home (**FIG. 22-23**). This vent allows air circulation inside the root cellar's main chamber to prevent mildew. During very cold weather, the vent can be closed to protect against freezing. Some vent boxes are built to almost touch the floor of the root cellar. They bring the low, cool air in. Since we have a ground-level entrance, the cool air can come through the low vents at the bottoms of the doors. Venting a root cellar is dependent upon climatic conditions and should be locally assessed.

The life expectancy of a log root cellar is about 25 years if the logs are protected from ground moisture. There are several methods of waterproofing, among which are painting the

FIG. 22-23

logs with tar, burning them on the outside with a torch, or covering them with a vapor barrier. We chose to cover the walls with building paper, stapling it around the contours of the side and back walls to avoid tearing when the building was backfilled. Once the paper was put on, we began the tedious process of bringing the dirt up to the level of the top log in back and under the roof eaves on the sides for maximum insulation. The front was left open.

If a good machine operator is available in your area with a small front-end loader, you can eliminate this tiresome task. We planned to hire this modern alternative to a shovel to push the remaining earth into a long gentle slope at the sides. We would then build a retaining wall of logs to keep the dirt in place.

To break the monotony of digging the backfill, we finished the inside root cellar chamber. First, we chinked over the fiberglass insulation with 1¾" triangular surfaced strips, which were cut on a table saw. They increased the insulative value of the walls by adding more wood to the thinnest area of them. At first we tried stuffing the insulation back into the crack before nailing on the strips, but stopped for fear that hard-packed insulation does not work as well. Instead, we just let the wood push the insulation as far back as was needed, leaving it fluffy (**FIG. 22-24**).

We were now ready to furnish the root cellar. This is an entirely personal matter and depends on what is harvested and what is put up. The number of canning jars and the kinds of vegetables and other perishables that are to be stored should determine the size during the initial planning. In our case, we resorted to doing the best with what we had to work with. We built a shelf area along the back wall, which is 18" wide for the canning jars, and divided it in half for the two families. This solid two-by-four divider acts as a brace to support the crossing shelves. Along the side walls, we bolted shelves of varying dimensions to the ceiling joists. The two-by-four shelf braces hang from the joists so the shelves can suspend from them and need not be supported from the valuable space below. The space below is a tall space for winter cabbage, which is harvested with its roots. The narrower shelves are for storing tomatoes and other wrapped vegetables. Under these shelves are racks made of two-by-fours to support bins filled with moist sand for storing root crops.

We wanted to experience a full year with the root cellar before building more permanent bins, so we temporarily stored the potatoes, carrots, and beets in the small wooden boxes that are used to transport fruit from orchards. We later decided that the permanent bins should be unattached boxes, larger than the fruit boxes but small enough to be carried out of the cellar in summer for cleaning, airing and receiving a change of sand.

The entire floor of the root cellar is covered with a sand and light gravel mixture, which can be turned over each summer when the building is aired out (a cement floor would have to be disinfected each year).

Now that we know the root cellar is satisfactory

FIG. 22-24

we will be growing more storage crops, but the size is perfectly adequate for the two families, provided that we are careful to utilize the space efficiently. One winter, we were even able to offer some space to a neighbor.

The interior work was done prior to hanging the doors so that we had light, though it was still pretty dim and insufficient. If we had used a method other than hanging for attaching the shelves and if the autumn rains had not been rapidly approaching, it would have been easier to work inside before putting on the ceiling. As it was, there was just enough time to stand back and admire the handiwork of the newly hung inside door, before we rushed to put jars and potatoes in. This inward opening door is 32 × 72 inches (2" shorter than the tallest member of the family) and is made of two layers of cedar two-by-eights nailed perpendicular with a vapor

FIG. 22-26

FIG. 22-25

barrier of 50-pound roofing paper between the layers. There are stout handles on both sides of the door, which are well secured with screws (**FIG. 22-25**).

For the outside door, a chalk line was drawn and the logs were chainsawed to create the opening. Then two-by-sixes were nailed to the top and to both sides of this opening to frame the 34 × 66 inch door (see **FIG. 22-5**). It is made of one-by-twelve cedar boards nailed perpendicularly to two-by-eight cedar with the same vapor barrier. The two-by-eight side faces out. The nails come through the one-by-twelves into the thicker outer board, hiding the nail heads from the visible surface of the door. The size of

this door was determined by the need to bring animal carcasses in and out, which would be hung in the small outer room. Beneath this door is a very shallow air space to supply the inner door air intake holes. A stone step was added in front of this door as a finishing touch (**FIG. 22-26 AND 22-27**).

Our root cellar has proven itself. The only difficulty was that the ventilation system was not properly equipped with a sufficient low cold air intake. Mold developed on jars that had not been wiped clean enough and on the boxes storing the root crops. To allow more air into the inner chamber, we drilled three holes at the bottom of the inside door (see **FIG. 22-25**). They were closed off with fiberglass insulation when the cold weather arrived. To eliminate using insulation, which can be eaten by mice, we planned to build a sliding panel over the holes.

Last winter was a mild one. An occasional nighttime temperature of -5 degrees was noted.

CHAPTER 23

Railroad Tie Root Cellar

Another simple and quite efficient root cellar design is that of our own root cellar. The first year on our place was filled with the chores of reconstructing a couple of the dilapidated, unusable buildings. There was nothing left of the old barn except what could be salvaged for dry kindling and aged paneling for our kitchen walls. The root cellar was nothing more than a concrete-front compost hole filled with well-rotted cedar. I shoveled out the old material and redug the hole because I liked this old location. The big question was whether or not to move the ugly concrete front or incorporate it in the design. Since it was already quite late in the year, with some harvests already in, I decided to keep the front and deal with it later.

It really surprised me to see how decomposed those cedar log walls were. It prevented me from using the same material. I shopped around and asked several friends and neighbors for advice. Soon a friend wanted to borrow our truck to pick up a few loads of railroad ties, which he acquired while working for the railroad. He said he could take as many of the discarded old creosoted ties as he wanted for the nominal price of 25 cents each. He asked what I wanted in trade for the use of the truck, so I asked for a load of ties. Perfect material for the surrounding walls of the root cellar. Each tie is 8 × 9 inches wide and 8' long and is thoroughly creosoted against decay. Each was still in excellent condition even after several years of contact with the ground under a railroad track.

I trucked a capacity 3-ton load (about 50 ties) to the cellar site and unloaded them. We put them in five equal stacks on three sides of the hole. One stack went along the back for the back wall and two stacks went along each side wall, making them two ties or 16' long. They were stacked 9 high to make convenient 80" tall walls; many of the ties were toenailed into each other to make the walls more secure. Upright braces spaced along the outside of these ties would have also been a good idea, but didn't really seem necessary since the surrounding soil is very porous and doesn't retain enough moisture to cause much pressure against the walls, even during winter frost.

Then, 32" in from the front concrete wall, I built a studded inner wall with a sufficient doorway opening. This wall was 8' long and the width of the rear ties. It consisted of two-by-four studs with horizontal one-by-eight cedar sheathing. It was

insulated with wood shavings that were packed well as the wall went up to insure that there would be a minimum of settling. The inner walls of the cellar were then sheathed with vertical one-by-eight rough-cut cedar. This sheathing completely blocks out any odor from the creosote and adds to the beauty of the rest of the rustic inner construction (**FIG. 23-1 AND 23-2**).

The roof is multilayered, consisting of heavy cedar logs, 90-pound roofing paper vapor barrier, two-by-six tongue-and-groove cedar, another vapor barrier of black plastic, 24" of dirt and gravel, and, finally, a gabled cedar-shake roof. The heavy cedar logs were notched and spiked across the top of the walls to tie the whole structure together and to provide a solid primary roof. It was protected by a 90-pound roofing paper vapor barrier to preserve it and keep the moisture off of it. To further seal and insulate the roof, two-by-six cedar tongue-and-groove boards were added

across the roofing paper, and black plastic was stapled over those boards to protect them from still another roof. This tertiary roof is of gravel and dirt shoveled on at least 24" thick until it filled the roof space and the surrounding sides.

For an entire winter, this was a final roof. We figured with all those layers, surely it couldn't leak. But it did. Probably through the nail holes in the roofing and plastic. The water dripped in throughout the winter, causing all the stored meat and vegetables to grow unbelievable mold because of the dampness.

Come spring, we emptied the cellar and cleaned it out. At that point, I finally realized another mistake. In my haste to put the roof on, I forgot to put in the vent system. Anybody with a little imagination knows what that meant. I got up on the roof, dug a hole in the dirt, cleaned out around the hole, and pointed my chain saw nose-down near the rear of the roof. Finally, an 8 × 6 inch hole was cut for the hot air outlet. A root cellar also needs a low air intake somewhere near the bottom to let cool air in, and a high outlet to allow the rising heat to escape. This system provides a proper air circulation, which is necessary for an efficient root cellar.

After I installed the outlet, I extended the one-by-eight cedar vent tube to well above the dirt and gravel roof so it would go out beyond the gabled cedar-shake roof, which I later put on (**FIG. 23-3**). This vent was then capped with a tiny gabled roof of its own to protect it from direct precipitation entering it and collecting in the cellar (**FIG. 23-4 AND 23-5**). The outlet vent should extend beyond the final roof to provide proper circulation. If the outlet was between two roof layers, the rising hot air would not be allowed to freely dissipate outside, but would get partially trapped within the cavity between the roofs where there is little or no circulation.

The intake vent comes in from the outer concrete front wall through an 8 × 6 inch cedar tube,

FIG. 23-3

FIG. 23-4

FIG. 23-5

which extends from that wall to inside the inner chamber. This low vent brings in the cool air (FIG. 23-6 AND 23-7). I screened this vent and the outlet vent from the outside with regular gauge screening to prevent rodents from entering but they quickly managed to nibble through the screening and made good their entry (FIG. 23-8). I suggest using a heavier gauge stainless steel screen for this purpose.

The interior is divided into shelf spaces, bins, and floor storage areas. The shelves are on either side of the aisleway in front of the bins. There are three shelves on either side, 16" apart and 24" from the floor. They are the width of two one-by-tens, approximately 19" apart, and are braced by a framework of two-by-fours, which are nailed into the sheathing in the rear and notched into upright log posts in front (FIG. 23-9). Below the shelves on the left is a storage space for sacks and boxes, on the right, a sand pit for root crops to be stored, roots-down in the sand to help them last longer. The buried root method of preserving root crops re-creates the conditions of being in the ground. This keeps them fresh much longer than if they were just placed into the root cellar on shelves or hung from the roof on hooks like the meat (FIG. 23-10).

The three covered 30 × 30 inch bins along the rear wall are for storage of apples, carrots, and onions, which should not be in contact because they easily pick up each other's taste and odors (FIG. 23-11). The middle bin is filled with sand so the carrots are preserved as if they were still in the ground. They stay crisp for a long time when

FIG. 23-6

FIG. 23-7

FIG. 23-8

FIG. 23-9

FIG. 23-10

FIG. 23-11

FIG. 23-13

FIG. 23-12

preserved in this manner. The apples should be wrapped individually when stored to prevent "one bad apple spoiling the lot" (**FIG. 23-12**).

The doors were framed, hung, and finished with ornamental handles. This cellar, now that it is properly completed, works great. It keeps a cool 50° to 55° even during the hottest days of summer (perfect place to escape the heat) and remains well above freezing even throughout the winter cold spells. I guess the several layers of roof provide a really fine insulation (**FIG. 23-13**).

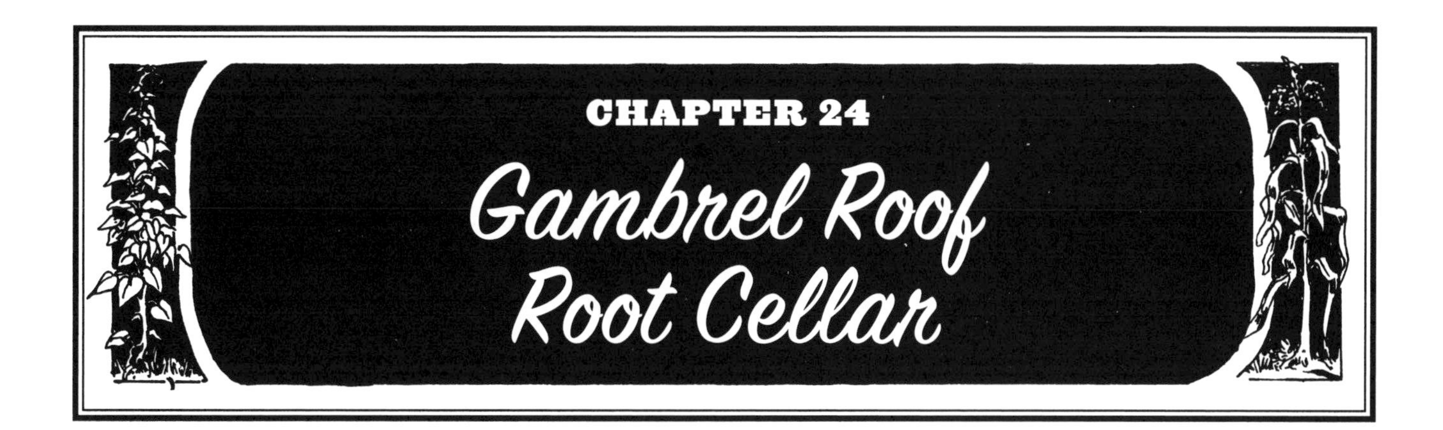

CHAPTER 24

Gambrel Roof Root Cellar

We needed a root cellar that was sufficient for two families, with plenty of roof space for an out-of-the-way storage area. We didn't want this structure to be incorporated under the same roof as our house or any other building that is heated with wood because of the chance of fire. If it were under the same roof as the wood-heated house, a fire could wipe out our home, clothing, and entire winter food supply. I would never want my food supply to be in such a vulnerable spot. Besides, to have a sufficient cellar for two families, we would need a large accessible area that could be properly protected from the extreme weather. Such a site should have good drainage and preferably be built into a hillside or mountain for protection against frost and direct exposure.

As long as we were building another space, we might as well incorporate several uses for it. We needed space for wood storage, to keep wood dry and to allow it to season. We also required additional living quarters like a space for guests to crash or a meditation area away from the general order of things (**FIG. 24-1**).

FIG. 24-1

We found the perfect site for this structure within the old rock slide to the east of our house. This slide area contained rock and gravel and didn't contain clay like the rest of our soil. Clay is not very porous, consequently a building constructed in it would probably rot out within 15 years. A large hole over 10' wide was dug out in this slide with a backhoe and the floor area was made a reasonable level. The first round of logs was then put around the 10 × 12 foot area and was lifted and leveled over hefty rock corner piers.

Because we had only a few heavy logs, all much larger than we needed, we decided to rip them at the site where they were felled. All our material was 80' long, fire-killed snags that were dry as a bone and still standing. To rip these snags, we dogged them in place on skids, raising them 16" off the ground at bottom so the chain saw wouldn't kiss the ground while cutting through. We found the top point at either end by holding up a level flush to the end. A chalk line was then snapped along the top of the log and the log was ripped down the center with a heavy-duty chain saw. No guide was used. The cuts were eyeballed, consequently they were not perfectly straight, but they were fine for our purposes. After the rip was made, each of the half log tops and bottoms were flattened out to create 1½" shelves for strips of fiberglass insulation to rest on. This insulation sealed the spaces between the wall courses (FIG. 24-2 AND 24-3).

Since we had a flat surface to work with, it was easy to draw the outline for the notch rather than having to scribe it around a contoured surface. The notch used was a triangular one with a vertical inside line to keep the flat walls plumb. The outside line closest to the end was cut at about a 55° to 60° angle, making a sort of triangular tenon for the interlocking pieces from the shoulder to the end. The half-round shoulder is jammed against the notch so it can't shift out. For added reinforcement, the ends were then spiked through the notch. A simple jig incorporated a straight vertical cut and the angle cut was made. The shape was transferred to the flat face of the log and the notch was cut out with

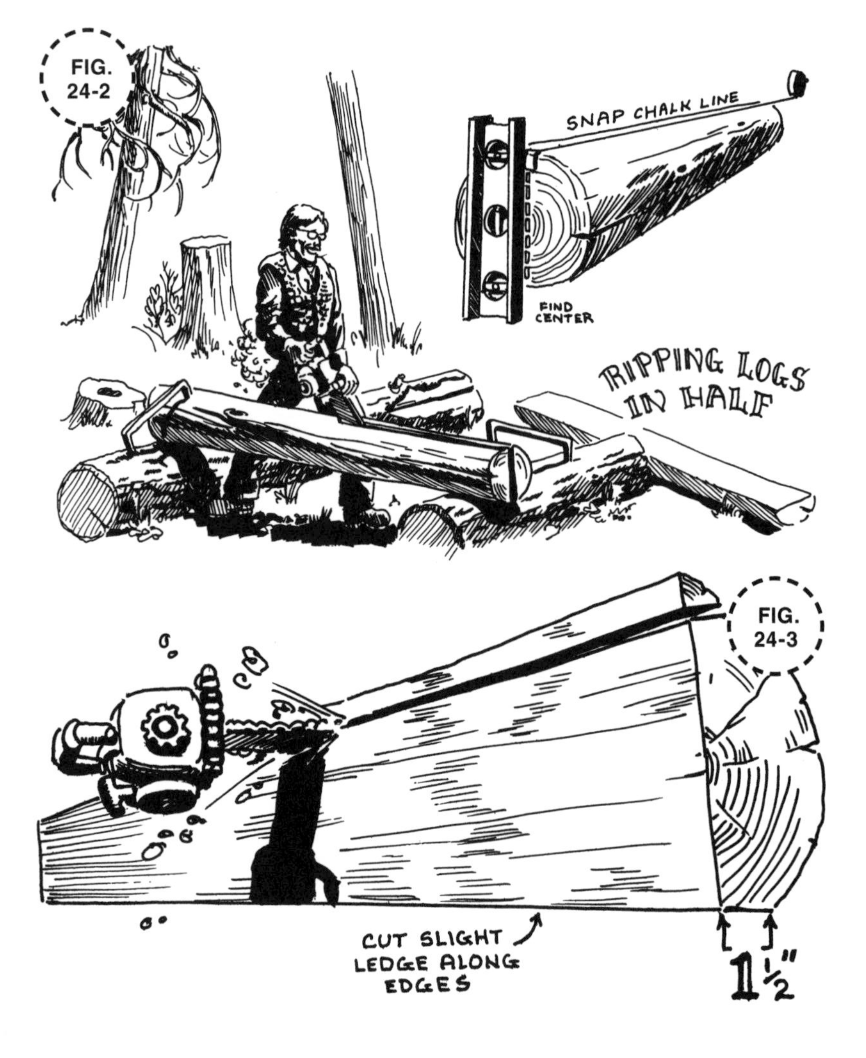

the chain saw. Two simple cuts and the notch was done. The walls went up extremely fast with this method (**FIG. 24-4**).

Working with half-round logs provided a few additional structural benefits. The facing walls went up exactly at the same height for each course because we put one face of a ripped log opposite the other. This made it easier to deal with the ceiling and roof. The faces also provided a flat surface to hang shelves and fasten bins to (**FIG. 24-5 THROUGH 24-8**).

When constructing walls with logs that are stagger-notched, you always end up with two sides higher than the other two sides. The side walls ended up 10" higher than the front and back walls. That 10" became our insulation space between the ceiling and the loft floor. Three 4 × 10 inch salvaged bridge timbers were notched into the bottom of the side wall logs as ceiling joists to support the one-by-six cedar ceiling. These ceiling joists extend 1' beyond the side walls. The ceiling was nailed on. Two 10"-high logs were added to the top of the front and back

FIG. 24-5

FIG. 24-6

FIG. 24-7

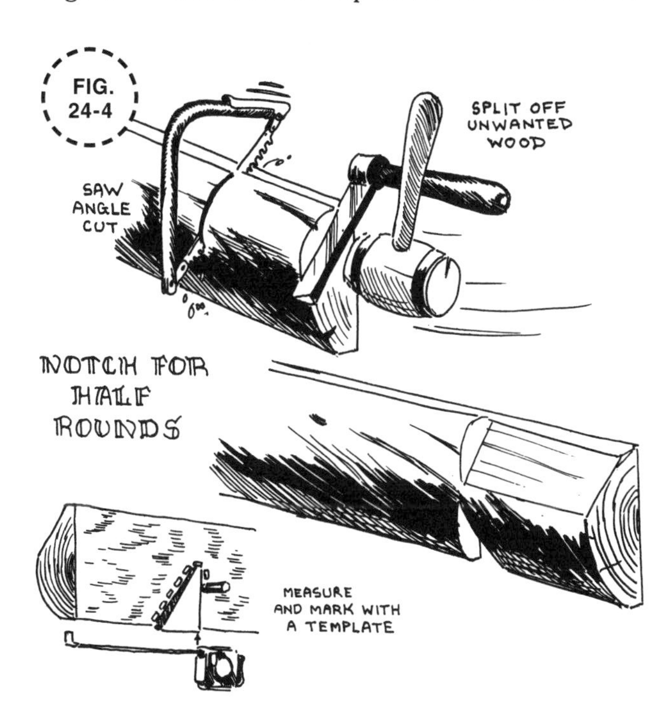

FIG. 24-4

FIG. 24-8

walls to create a box over the ceiling. This box was filled with two-by-tens on 2' centers and 10" of fiberglass insulation sprinkled heavily with lime. The lime is supposed to prevent rodents from eating through the fiberglass. It works well. Only one rodent has been able to do any damage, but he'll soon wish he never went in there.

The one-by-six cedar loft floor was then nailed in place. Once this platform was made, we built two sets of braces. These are the uprights, crossbeams, and angle braces for the front and rear walls. The uprights extend at a slight inward angle up from the four corners of the building to create a slight slant for the lower sides of the gambrel roof. This slant complements the lines of the roof. The crossing beams are notched over them and both are supported by angle braces. The side wall beams were then notched across the frameworks (**FIG. 24-9 AND 24-10**).

To cantilever the lower roof sides over the side walls, a 6" log was spiked over each of the ends of the exposed ceiling joists, which extend beyond the side walls. These two poles serve as outer rail eave extensions for the 14" overhanging gambrel roof. This roof also extends 4' beyond the front and 2' past the rear of the building to protect it from direct precipitation. This lower roof consists of 7' long rafters, two-by-four nailers for shakes, and shakes. The rafter tops were cut at an appropriate angle in a jig for uniformity and were nailed to the crossing side beams at 2' on center intervals. The bottoms were spiked to the

FIG. 24-9

FIG. 24-10

FIG. 24-11

FIG. 24-12

FIG. 24-13

eave extension rails at the same interval. The two-by-four nailers were nailed across the rafters at 1' on center for the finished roof shakes (FIG. 24-11).

On the west side of the roof, a doorway was constructed as a side entrance for wood. The gravel in front of that entrance was gradually sloped as a truck ramp so birch wood could be carried up for drying. The doorway was framed with two-by-sixs and the small gable roof extension above it was cantilevered on diagonal uprights (FIG. 24-12 AND 24-13).

The upper sections of the gambrel roof were dealt with next. First, two 18" extensions were cut and were notched in over the front and rear crossbeams. They raised the loft area to a height of 7½'. A ridgepole was spiked in above the extensions. Rafters were then placed between the side wall beams and the ridgepole. The front rafters angle beyond the ends of the beams to outline a 5' overhang at the top. Nailers were placed across the rafters and shakes were nailed above them in the same manner as the lower roof (**FIG. 24-14 AND 24-15**).

We have few buildings to store things in, so it was important that we design the root cellar to contain as much usable space as possible. The gambrel roof on our root cellar almost goes straight up for the first 6' above the walls, the result being there is only about 6" of floor space on either side that doesn't have free head clearance. The inner area is more than sufficient to store the fresh fruits and vegetables and canned goods for two families. It is comprised of two sections. The 2 × 8 foot outer chamber is an insulating corridor leading to the inner chamber. The chambers are partitioned off by a frame construction inner wall with a door. The inner area is composed of shelf space on the three walls and bins filled with sand on the floor. Some shelves are far enough apart for 2-quart jars to fit snugly on and some for 1-quart jars. The distances between shelves were figured out accurately so there wouldn't be any wasted space (**FIG. 24-16**).

The root bins are located along the bottoms of the walls. They are filled with sand so roots and root crops (e.g., carrots, beets, turnips) could be protected as if they were still in the earth (**FIG. 24-17**). None of the bins were made with lids to separate potatoes, onions, and apples. We didn't feel this precaution was necessary. Our apples, potatoes, and onions were sitting on opposing shelves and they all did well. The apples lasted

as late as July and were still crisp and juicy. They did not have even the slightest taste of potato or onion. We ate every one that we stored without having any go bad. The potatoes and onions didn't have any peculiar tastes or odors, either. They also kept very well. We attribute a lot of the success of our root cellar to proper ventilation.

A proper vent should have 1" on a side for every 1' along the wall of the root cellar. Thus an 8 × 8 foot cellar should have an 8 × 8 inch vent (**FIG. 24-18 AND 24-19**). We should have brought the vents up through the roof before putting on the shakes but we didn't do so. If the vent is extended beyond the roof, the air will circulate better, especially when the loft gets closed in and used as a room. We didn't want to have to leave a window open all year round for air circulation.

The front door construction of the cellar is unique, neat, and structurally sound. It protects against warpage and is really strong. The door is of laminated pine, planed down to a 2" thickness. It is reinforced with four 1½" deep dovetailed fir splines. To cut the grooves for the splines,

FIG. 24-19

FIG. 24-18

FIG. 24-20

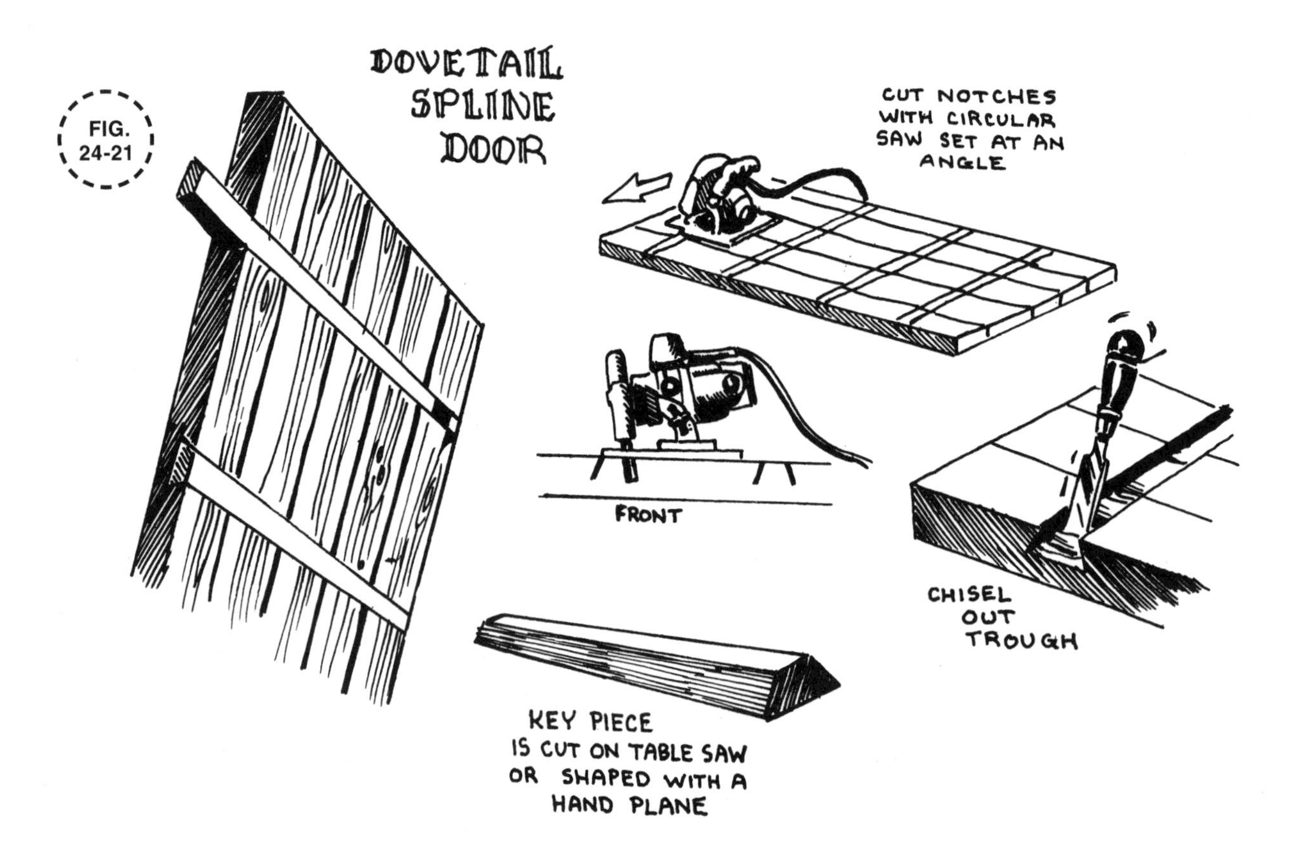

I set the blade of the saw to what I thought would be a good angle and put a jig along the door so I could run the saw against it and keep the cut really straight. After that, I cut from the other side, making the angle go the opposite direction. The piece between the cuts was chiseled out. The fir dovetailed splines were then cut out on a table saw and were driven into the slots (**FIG. 24-20 THROUGH 24-23**).

One other thing to note. Be sure to protect your root cellar from spring runoff, especially if it is built in clay or other nonporous types of soil. It should either have drain tile around its foundation or a length of drain pipe down the center to carry off the excess moisture. We have a length of 4" perforated plastic drain pipe just below ground level in the center of the gravel floor. It has helped to prevent moisture buildup and flooding.

FIG. 24-22

FIG. 24-23

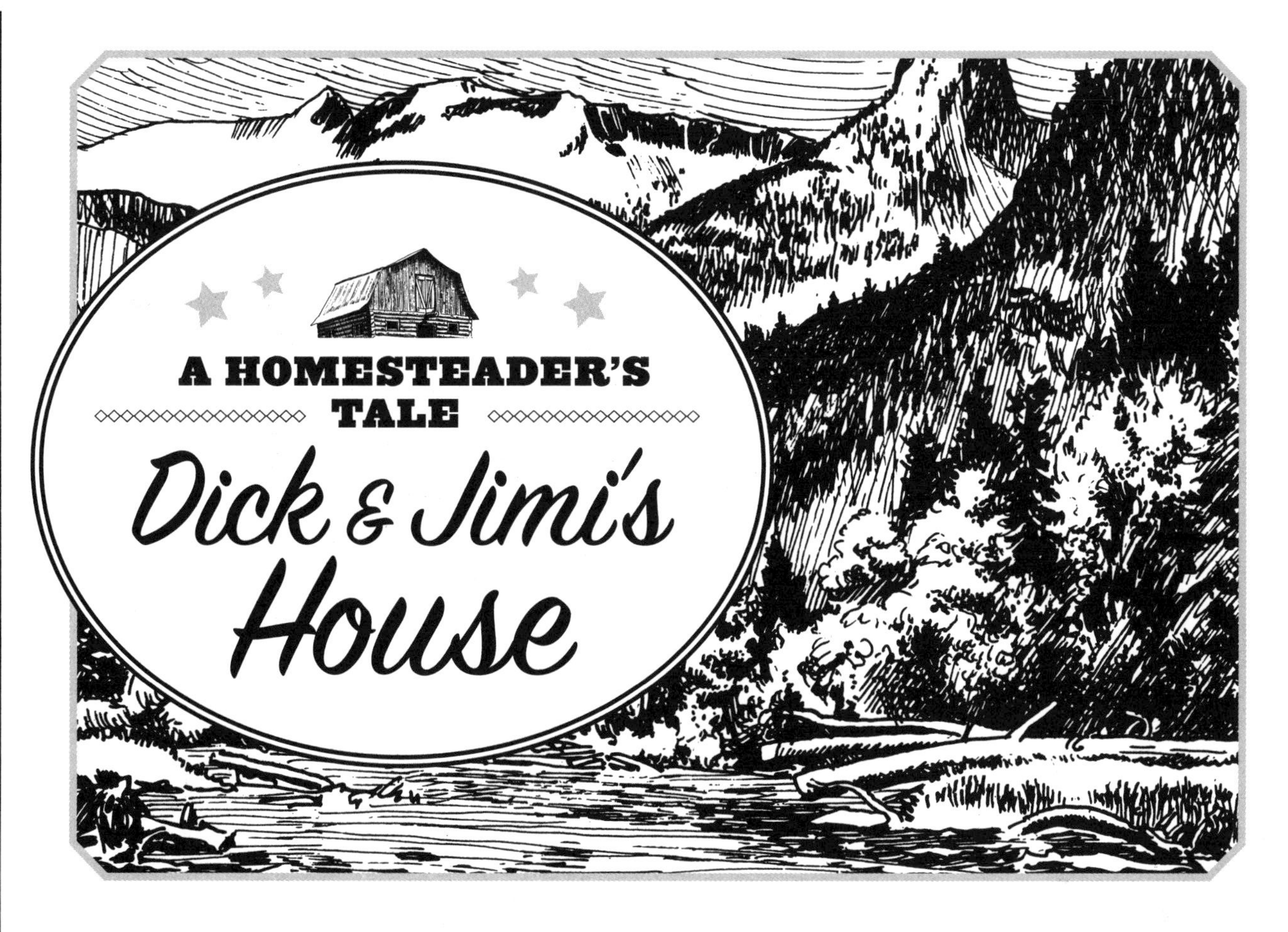

For many years, I worked at an office job in my native country, Germany. As I sat at my desk, to ease my mind from the piles of paperwork in front of me, I would sometimes look out at the harbor and wish I could trade places with the dockworkers. I wanted to feel the weight of one of their large crates on my shoulders and walk to the waiting cargo ships with the heavy container. But instead of doing something about that foolish desire, I just went to the coffee machine, refilled my cup, and forgot about such wild fantasies.

I soon became very restless and realized something was missing in my life. My body was trying to hint its needs to me. I was usually very tense and had many unexplained aches and pains. But as my body cried out, I quieted it with aspirin or alcohol. I became more sluggish and knew something was wrong. I began asking myself many questions like, "Was my body designed to sit all day at a desk and push papers around?"

I began reading about the body and how it functions. I read about germs and how they collect and reproduce in parts of the body that get very little circulation. I learned that natural resistance breaks down with lack of body movement and that a person needs plenty of exercise to maintain good health. I became aware of myself as a human animal who was meant to be outdoors, using my muscles as well as my mind to sustain my existence.

That awareness was the important factor that made me change my lifestyle. I picked up whatever literature I could find about living in the country and surviving on a homestead. My dream was to meet all my needs in such an environment, without having to waste many precious hours each day working in an office or at some other form of unsatisfying servitude.

Alpine barn built into hillside

Utilization of the hill contour for our house

My first step toward my dream was to travel around to find the place I wanted to spend the next several years developing. It had to be a place that allowed me enough personal freedom to experiment with my ideas and reap the benefits. Though there is nowhere in this world that one can find such total freedom, I am happy to say I stumbled upon an area that was a satisfactory compromise. And here is where I met the beautiful, hardworking woman who shared the same dreams and soon decided to share her home with me as well.

Our homestead is located in the foothills near a range of high mountains. It provides us with good soil, plenty of freshwater, varying terrains, and privacy. It reminds me of many areas in the Alps I enjoyed visiting. In fact, every day as I was planning our homestead, I used the information I had learned from Alpine farmers. Those farmers always built their homes and the animal quarters on the hillsides, leaving the flatland open for farming and gardening. The grazing fields were also on the slopes, so the nutrients of the animal manure would wash down onto the produce fields.

The barns were dug into those slopes to provide the animals with warm, earthen stables, which were beneath ground level on the uphill side and exposed on the downhill side. Of course, such stables need to be well ventilated. Another advantage to this design is that hay and grain can easily be unloaded into the upper loft, which is level to the ground on the uphill side.

Our house is also built into a small hillside and has very good drainage on either side because it is sheltered by the long overhanging roof, which straddles the crest of the hill. It has a full 27 × 27 foot basement, which was dug out by a Caterpillar. I wanted such a basement to use as a workshop area and as a place for a root cellar, where we could preserve some of our garden and field produce for the winter. A refrigerator just could not keep all that we needed to store. Besides, we did not want to have to depend on expensive electricity.

If there is time and means, it is a very good idea to build the root cellar under the house; then there is only one roof to worry about. There is no need for additional insulation above the cellar because very little, if any, warmth travels down to it from the living sections on the upper level. If the root cellar is built into the slope, like the animal stable, it should be deep into the ground on the uphill side and level with it on the downhill side, enabling the gardener to just wheelbarrow the produce in through the front door.

In our design, the 20 × 27 foot workshop is on the downhill side of the 7 × 27 foot root cellar, protecting it from the weather and insulating it. The long overhanging roof, which straddles the crest above, also protects the cellar because it detours the runoff, preventing it from collecting around and penetrating through the rammed-earth walls of the basement area. Even with this roof, the root cellar still retains all the natural moisture of the earth—just as an ideal garden does—because it is underground.

It is important to only build into the slope of a small hill if the hill's crest can be covered with such a straddling roof. If a structure is built into the bottom or at a lower level of a large hill or mountain, the hole for that structure will usually create a spring that will penetrate any concrete basement, unless its walls are thoroughly tarred and have a water barrier of gravel or clay tile around their footings. I have seen cellars in houses professionally built by contractors with 2' of water in them at spring breakup.

The retaining wall around a basement which is set into a slope like ours can be made of rammed earth (soil cement), since there is no danger of runoff to wash it away. Rammed earth is an inexpensive soil mixture, usually made up of 60 percent sand and 40 percent clay-type soil. It is held together with about 7 percent Portland cement. If rammed earth is used for fill between heavy post-and-beam walls, it can contain clay, straw, small twigs, even sawdust, and it will require just a small amount (no less than 7 percent) of concrete to cement it all together. A wall of this sort should be protected by a roof overhang of at least

Rammed earth forms

Sequential basement foundation wall forms for rammed earth

2½' to prevent direct moisture from eroding it.

Rammed earth is more economical and insulates much better than an equal amount of concrete. Its thermal quality lets it remain cool in summer and helps it to retain heat in the winter. A 10" thickness of rammed earth—60 percent sand and 40 percent clay with 7 percent Portland cement—as a retaining wall can easily support a single-story dwelling with a heavy snow load. We used this standard mixture for our basement walls.

We started by putting in a form for the footing around two sides of the perimeter of the basement. The form was made with one-by-six boards, two high. These were braced at the corners and at 4' intervals with two-by-four uprights that were buried 1' into the ground. They extended 8' high for the full-length walls and were held together at the top with twisted wire.

The rammed-earth mixture is a drier one than regular concrete. The mortar should be moist enough to mix properly, but be crumbly to the touch. It should not slump when it is worked a bit by hand. When the mortar was the proper consistency, it was shoveled into the footing form. It was tamped down with an eight-pound sledgehammer until it pressed into every part of the form and the excess mortar oozed out of the corners. It was then left to set, with pieces of rebar sticking out every 18" to join with the retaining walls.

I began building the forms for the north and south walls, and realized that I would have a problem securing them above the footing pads. But I figured that because of the dryness of the mortar, not much would ooze out from under the forms when they were tamped down. The one-by-six boards were raised to 36" at these walls. They, too, were braced at every 4' with two-by-fours, and this bracing was supported by the longer uprights, which framed the footing forms. To tie the walls together, I placed bent, hewn timber uprights at each corner over the footings. These uprights also acted as nailing posts for the form ends. Their bends lean in toward the building to give the walls added structural strength. I raised the forms 36" because this height is a comfortable one for the first tamping and it also is where the windows were to be started on the south and east walls. The north wall became a full wall without windows.

By the time I got around to building the form for the west retaining wall—the wall farthest into the slope behind the root cellar—I decided to try a different style. I built the forms so the inside of this cellar wall would slope like a dam. It would be 10" wide on top and slant to a width of 24" at the footing. This wedge- or dam-shaped wall is much stronger than the ordinary straight wall and is easier to make. It does not have to be supported on a footing because its thickness is relative to the amount of back pressure it might receive.

After the rammed earth was tamped in and the completed retaining wall was left to set, the eight-by-ten timber windowsills were notched into the end posts over the 3' wall sections of the south and east walls. An eight-by-ten upright support was placed between each window space to frame it in. These uprights also act as structural posts for the top wall beams, which were notched in above them. Five windows were then put into the south wall, and four windows and a door went into the east wall to bring in plenty of light to the shop area. Workbenches were built under these windows and other woodwork conveniences would soon be added.

I did not want to put a cement floor in the workshop because I do not like to walk on such floors. For the present, the floors in both the workshop and the root cellar will remain dirt with a few wood chips mixed in. I am thinking of sinking some joist beams into porous gravel and nailing a rough one-by-four floor over them. The gravel will provide drainage between the joists and prevent them from prematurely rotting.

Southwest corner

Inside the workshop

The two root-cellar sections are separated from the workshop area by an 8"-thick, insulated double wall. The insulation used between the walls is oat straw, which has been bathed in an asphalt emulsion. This emulsion is a very effective and long-lasting mixture of three fungus-retarding agents: asphalt, lime, and water-glass silicate. I submerged the straw in the emulsion, left it in the sun to dry, then pressed the treated

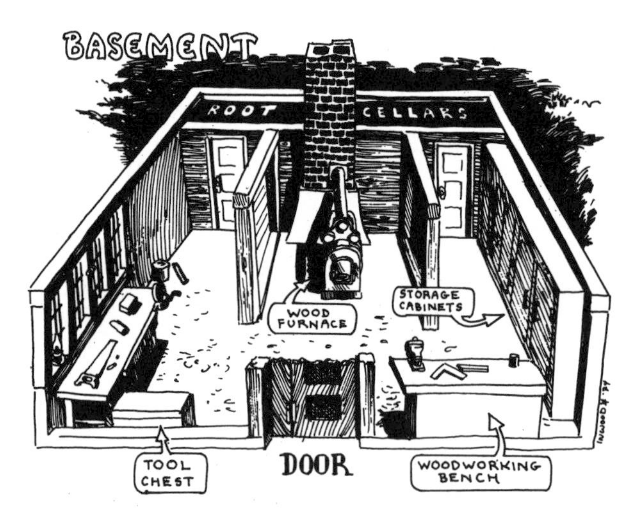

Natural curve used for corner brace

straw in between the walls with my fists.

The root-cellar sections are divided by a rammed-earth wall. On the north side of the wall there is a shelved area for the storage of canning jars, crocks of sauerkraut, sacks of potatoes, a barrel of eggs, and more delicate produce like apples. Apples are not supposed to be mixed with odorous vegetables like cabbage and onions, because they end up tasting like them.

At that time, we preserved the eggs by mixing a solution of water-glass silicate into the water in the barrel. This nontoxic chemical compound turns the water into a gelatinous mass that reduces the capillary action of the liquid, preventing air from entering the barrel and spoiling the eggs. An old, nearby farmer told us of a simpler method that we now use. He mixes lime with salt and puts the two ingredients into the barrel of liquid. The mixture produces a diluted, mortarlike substance that coats and seals the eggs, keeping them fresh for several weeks at a time.

The south area of the root cellar has a 2½'-high bin that is filled with sand. We bury carrots and beets in between the layers of sand. This preserves them throughout the winter as well as if they were still in the ground. The various crops like cabbage and brussels sprouts were harvested with their root systems attached and were replanted into the moist, earthen floor. This method of retaining freshness works well because the plants' roots think they are still in the soil and continue to supply the plants with life-sustaining nutrients.

There are four holes, 14" deep, in this south cellar, for milk. We placed old flue-linings in these

Dual root-cellar layout

holes and put in the containers of milk. The milk keeps cold enough in these linings to stay fresh for several days. The milk even feels cold after it has been in a hole for many hot summer days.*

Though the walls of these cellars remain dry, they receive enough moisture from their soil floors to keep them cool. For proper circulation, they have air vents near their ceilings to allow the warm, rising air to escape. These air vents—8 × 8 × 16 inch cement blocks—are built into the north and west walls. The cool air comes in from the workshop and pushes the old, musty air out through the high vents. Root cellars need this type of vent system to prevent mold and fungi from forming in the otherwise stagnant dampness.

*EDITOR'S NOTE: Not recommended. Please refrigerate eggs, as well as milk and all other dairy products.

After the basement walls were finished, 28'-long fir joists were put in over them. They

Floor joists set into foundation wall

spanned between the east and west walls. The joist ends that rested above the west wall were tarred and cemented onto that wall with the rammed-earth mixture. This provided a flat surface for the living-area wall plate and floor to rest on. It also gave the building added sheer strength. Many of the rough one-by-six boards that were used as the retaining wall forms were nailed over the joists as a subfloor for the upper area.

On the uphill side of the slope, we can walk from ground level through the woodshed and into the living section. Here again, I used some techniques I learned from the Alpine farmers. Our walls were post and beam with modified rammed-earth mixture as the filler. The triangular bracing for these walls is a variation of the styles used in this type of construction in Germany.

The posts were made from cedar logs that I hewed to 8" square at the site where they were cut, so the horse would not have to drag any excess weight. I used cedar—it is the most porous of the local woods. Being porous, it is rot resistant, light in weight, and also very insulative. I hewed these timbers by first chopping a score cut against the grain every 1' or so with an axe. If the wood had a very straight grain, which most large cedar logs do, the score cuts could be spaced farther apart. I used a coarse adze or a broadaxe and made them roughly square. At the building site, I finished the 8" squares with a shipbuilder's adze.

The posts were spaced along the south wall according to the width of the windows we had. We bought ninety-four old 4 × 2 foot and 3½ × 2 foot framed windows for $2 each. We wanted at least four of the 4 × 2 foot windows along the south wall to bring in the sunlight throughout the year. There is something very romantic about waking up on a cold winter morning to the bright sunshine reflecting off the white snow-covered landscape. It really helps to start the day off right. And to me, light is worth more than warmth. I spend a lot of time outdoors and enjoy it when the outside can come inside by way of large window spaces. The Japanese say that light is necessary for comfort. Dark spaces are

Typical exposed timber configuration

Hand-hewn timber

Exposed timber and rammed-earth-form setup

depressing; light spaces bring out inner happiness.

After the posts were toenailed onto the bottom wall plates, they were braced with slightly curved diagonal timbers. These braces were hewed from bent cedar pieces to add an aesthetic touch to the walls. In European post-and-beam architecture, there are sometimes many bracing designs used to break up the monotony of similar diagonal patterns. Three common patterns are Der Mann, Der Womann, and Der Wildemann. My design was much simpler, the aesthetic beauty being in the curved pieces that I hand-hewed.

The windowsills and the top beam were not put in place until after the rammed earth was tamped in. They would otherwise have been obstacles during that process. To prepare the posts for the rammed earth and to prevent air spaces from forming when the mortar dried and contracted, I attached one-by-one strips of board to the outside and inside edges of the posts. An inside wall form of plywood the height of the windowsill was then nailed to the posts, and two one-by-six boards were attached to the outside.

A rammed-earth mortar,

consisting of 5 parts sawdust, 5 parts sand, and 1 part Portland cement, was mixed very dry (moist to the touch but crumbly) for the walls. Some said this mixture wouldn't last because the sawdust would soon rot out, but I remembered reading about sawdust-and-cement floors in 19th-century Victory-type ships. It worked for those ships, why couldn't it work for my walls? It in fact does work very well, probably because the Portland cement contains lime, which prevents fungus from forming on the sawdust. This mixture was pushed between the forms and tamped down with the sledgehammer.

Eight inches of a rammed-earth mixture of this type gives more insulation than an equal thickness of regular cement. It is more porous and contains many tiny dead-air spaces like a hollow cinder block. It is durable enough for filling in between posts and beams, because it does not have to carry any structural weight, and will last quite a long time if a certain precaution is taken—this being that the roof directly above the wall have an overhang of at least 2½' to prevent direct precipitation from eroding it.

We were very happy with the way the south wall came out. It is aesthetically pleasing and it gives us a very fine view. There is, of course, plenty of light coming in from its 16' of window space. In fact, just below the windows inside the house, we have a wide sill for starting plants or sprouting grain and seeds. We even have a continuous oat sprouter, made from a hollowed-out log on a slight slant. It sprouts oats by the bucketful for chicken feed, so we have eggs with dark orange yolks in the middle of the winter. This sprouter is warmed by a couple of

South wall exterior

South wall interior

Entrance on west wall

hot bricks, exchanged twice daily and reheated on the wood stove.

The southern half of the west wall is also rammed earth–filled post and beam. This wall is level with the crest of the hill and is protected by a long overhanging roof that covers the woodshed in front of it. The entrance to the living area is located here, under the 14' overhangs.

The northern half of the west wall, and the north and the east walls are of frame construction with two-by-four studding and cedar one-by-four sheathing. They were insulated with asphalt-emulsion-treated straw as was the inner wall to the root cellar. This treated straw is proving to be a very good insulator. It provides the walls with a lightweight, porous sealer that contains many dead-air spaces, which effectively keep out the cold and retain the heat.

For our chimney, we made a concrete pad above the rammed-earth wall, which divides the root cellar compartments. We used two commercial 8 × 8 inch flue linings and mortared in red bricks around them, filling in any spaces that were between the bricks and linings. Near the bottom of the chimney, we left one brick loose in front of the lining opening to clean out the creosote. To make sure we were safe from chimney fires, we wanted a solid brick chimney instead of just having stovepipe going out through the wall.

Many people I know of have lost their homes due to creosote buildup that finally caught

Basic post-and-beam layout

Basic post-and-beam layout

fire in a flimsy stovepipe, and then ignited their cedar-shake roof or their framed ceiling.

Throughout the open living area, there are a few six-by-six upright posts, which brace the horizontal rafter supports and hold up the roof. These uprights relieve the roof pressure on the walls and prevent them from spreading under the weight of a heavy snow load. In the center, there is also a 22'-long center post which spans from a pier on the shop floor to the peak of the roof. The upright supports are spaced approximately

"Wind bracing" of curved timbers

"Wind bracing" of curved timbers

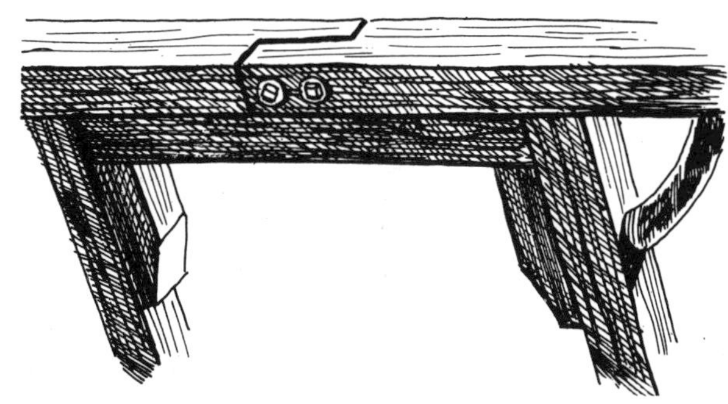

Lap joint on center beam

Beam support in kitchen area

9' on center, bracing the peak and going north and south across the east floor. Since the roof peak is off-center, there is no need for upright supports on the west side because the rafters span less than 11' between the peak and the west wall.

Notched in above the posts, going north and south, are the six-by-six horizontal rafter supports. They are braced to the uprights by diagonal wind stiffeners. These wind stiffeners are curved pieces of log that add another aesthetic touch to the interior of the house. Their structural purpose is to prevent the walls from parallelogramming or collapsing due to unexpected heavy winds. Most natural forces that can destroy a structure do so mainly by damaging the roof—with the exception of wind, which can actually compromise the integrity of the overall structure. Wind has been known to destroy many buildings that would otherwise be sound but lack bracing against its unexpected force.

The center ridge beam is not one continuous timber, but rather two separate pieces. These pieces were spliced with a side-lap joint and were bolted together. They were then supported from underneath by a short beam, which spans between the two nearest uprights. This support beam is braced by two diagonals, securing it to the posts. The ridge beam spans 3' beyond the north and south walls, providing them with plenty of overhang for protection.

Because I was working with poles 5" in diameter, and some even larger, and because I did not plan on using store-bought insulation for the roof, I spaced my rafters roughly 28" on center. This rafter placement is sufficient to carry

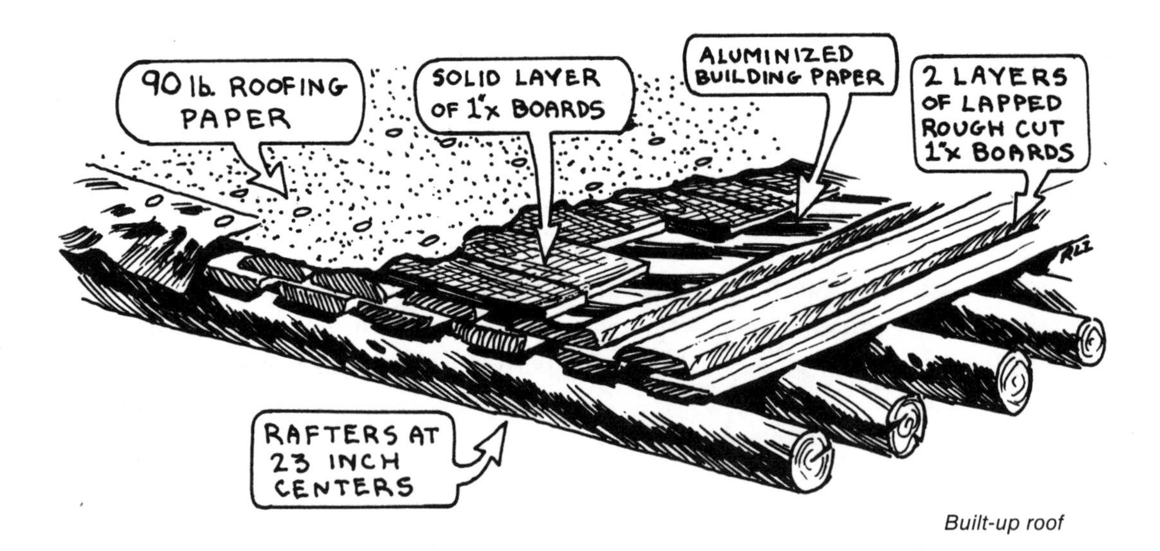

Built-up roof

a heavy snow load since the rafters are braced in the center of their span by the crossing support beams. The rafters on the east side of the building are pine and larch because these types of wood have very little taper. They span from above the center crossbeam to 5' over the east wall. This overhang protects the entranceway below and detours the runoff so it won't come near the basement retaining wall.

The rafters on the west side are cedar because they only span a short distance from the off-center peak to the west wall. Since these rafters overhang 14' beyond that wall to cover the outside woodshed and shield the crest of the hill from runoff, they have to be more weather resistant than the others. Cedar, though it usually tapers drastically, is the most weather resistant of all the local species of wood. Since these

rafters only span a short distance inside the house, their tapers are not very noticeable.

Our subroof is a double thickness of slab boards, which are on the average about 6" wide. The second layer laps over the sides of the first the way roof shakes overlap. The boards of these two layers are spaced half their width apart to create dead-air corridors between them. When these corridors are properly sealed off, they effectively insulate the roof because they are not exposed to outside weather. And since the air is trapped in these areas, much of the rising heat from the house cannot pass through them. It is locked, instead, between the boards, and is not lost through the roof. Of course, there is some heat loss, but not enough to prevent our house from staying toasty when our wood stove is going.

Skylight over kitchen

Above the two layers of slab boards, there is a sheet of aluminum builder's foil, placed shiny side down, to reflect the ultraviolet heat rays that have come through the air spaces. This foil barrier holds in much of the otherwise escaping heat and is well worth its nominal price. A continuous layer of one-by-four boards are nailed over the aluminum foil to provide a flat surface for the finish roofing of rolled 90-weight asphalt paper.

This style of subroof gives us about the same insulation as would 3½" of fiberglass and is far less expensive. But even with 3½" of fiberglass insulation, there would still be some heat loss,

Brace supporting end of rain trough

Brace supporting end of rain trough

resulting in the snow melting on the roof and icing up at the eaves. When the snow melts and collects on the eaves, the mass of ice will travel upward by capillary action and back up under the eaves, thus entering into the house and dripping through the subroof. To prevent this ice buildup, we covered our eaves with an 18"-wide flashing of aluminum. The ice slides off this flashing instead of building up over it.

We have a skylight built into the roof over the kitchen area. This skylight was easy to install. It was framed much like a window casing, except more care was taken to seal this frame so no water vapor would seep through. A very good way to seal such a frame is to cover all the joints with tar. Also, make sure that the window you use is of a substance that will not shatter or break under a heavy snow load. We used tinted, corrugated plastic. This works fine and is much cheaper than a safe thickness of glass.

As a finishing touch to the roof, my wife hollowed out a thin cedar pole to be used as a trough at the edge of the north eave. This ingenious device keeps water from randomly dripping off that eave and splashing on whoever is walking under it. It also controls the flow of the melting snow and allows us to collect it in a barrel for domestic use when our water system freezes, which it occasionally does in the dead of winter.

We have been very comfortable in our little house and are especially pleased with the light-giving south wall. In fact, we still have not curtained it off with the woven draperies my wife has been producing on her loom.

If we have any regret over the design of our house, it would be about the roof. If we had it to do over again, and had the necessary amount of money, of course, we would have put in 6" of fiberglass or its equivalent. This would better control the heat loss and keep our house even toastier throughout the winter with the minimum expenditure of cordwood-cutting energy. But then again, cutting wood keeps one healthy during the long dormant winter.

We have found, through the experience of our last few years here, that trying to maintain an almost self-sufficient homestead is a full-time job and sometimes necessitates more physical, mental, and emotional energy than we wanted to put out. But I must also add, the rewards are many and the faith we have gained in our personal resourcefulness in solving life problems is well worth the energy we have given. Energy given is energy received.

A Homesteader's Tale: Chris & Margee's House

I originally researched the material for this book in order to design and construct my own home. Once the book was done, and after several years of living in a shoddy but cozy old log house, I finally moved my family into the new log manor we worked so hard to build over so many years. This house was designed for my wife and our son and me. It provided us with inspiration and space to grow in the ensuing years.

What we wanted was a good-size log house that could be built by two people. Margee and I figured out various designs, but each was either too small or too impractical. The logs we would be using were from trees that a friend asked us to clear for him. They would have to be trucked in our three-ton, staked-bed truck. The longest logs we could carry on its 12'-long bed would be twenty footers, with butts toward the cab and a heavy-duty chain binder around them. Even that was kind of pushing it, not to mention being illegal. In fact, one day I almost flipped the truck, driving a load up the hill to the building site. The front tires came off the ground and the front end hung in midair until I got some of the load off the truck.

Keeping in mind this limitation on the length of the log, we figured out a design that would give us ample room, but be no easy task to build. The design called for eleven logs per tier in a modified, double-interlocking-square floor plan, with a basement under the 20 × 20 foot southeast square. Sound complicated? You're lucky you didn't have to build this monster!

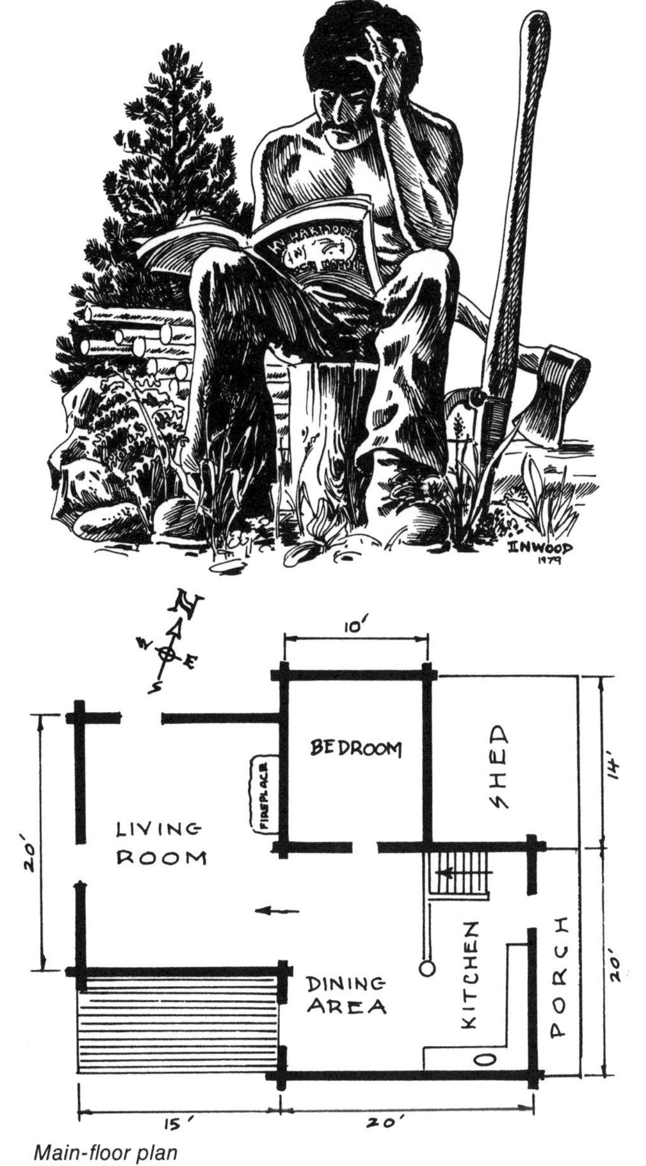

Main-floor plan

Being in a northern climate, with long, cold winters, we wanted a basement for the appliances and the wood furnace. A lot of builders prefer to pass on this convenience because of the cost and difficulty, but we felt we needed this subterranean space and are glad we expended the energy. Being underground, it is one place that won't freeze if we leave during the winter.

Actually, the basement was not that hard to construct. A friend of mine with masonry experience was between jobs and agreed to trade his labor for a few days in return for our two billy goats and my help on his barn. The deal was made. I hired a Cat to come in and dig out an ample hole, and we immediately began forming a 12"-wide by 8"-high perimeter footing around the 20 × 20 foot floor space. We poured the footing, reinforcing it well with continuous horizontal lengths of rebar tied in at each corner.

Then we poured the 4"-thick floor, setting in four two-by-fours lengthwise, as nailers, in case we wanted a wooden floor later. These on-edge two-by-fours also acted as guides for tamping the concrete.

To smooth the floor, we used a straight board and straddled it between two nailers, sliding it back and forth over the concrete to work it in. Be sure you don't forget to form a drain hole at this point because it's a hell of a job to do as an afterthought.

Using this method, the floor was poured in three stages. Each pouring filled an area between two sunken two-by-fours.

As I was tamping the floor, my friend began laying the concrete blocks. He suggested I use this material instead of forming up walls and pouring cement, because the blocks were cheaper (at that time), and easier (for him, anyway).

I wanted to learn his trade for future reference, but as we went along, I personally could only lay a few inside blocks. Most of my time was being spent mixing his "mud" and keeping him supplied with blocks.

The basement walls went up extremely fast. In less than a week, we were 12 tiers up, or 8' high. I highly recommend at least this height for proper headroom.

After the block laying was completed, my friend told me how to form up the cap and instructed me on the filling procedure of the blocks. Since I was forming up a cap for horizontal logs, I had to modify the method he explained to compensate for the stagger, making a 4"-high cap on the east and west walls and an 8" cap on the north. The south wall is 28" lower to allow for windows.

As a makeshift form for these caps, I used scrap cedar one-by-twelves straddling either side of the top rows of blocks with these boards. To hold them in position above the blocks, holes were drilled at the designated points, 8" down from the top of the boards for the north and south sides, and 4" down at the east and west sides. A doubled-up piece of wire was placed through one hole, with a nail in the loop, braced against the inside board. The loose ends of this wire were put through the outside board and held by another nail, which was twisted until the boards squeezed tightly into the straddled blocks. The tops of the forms were held apart the width of the blocks by scrap nailers, nailed into the upper ends of the form.

Constructing the basement

After the forms were in position, I crumpled some heavy paper into the tops of every other hole to keep the concrete out, then placed a length of rebar down each of the uncovered holes. This rebar should be long enough to go to the bottom of the hole, with its top sticking out far enough to be wired onto a crossing round of horizontal rebar which circles around the perimeter.

Concrete cap corner prepared for stagger

Floor joists

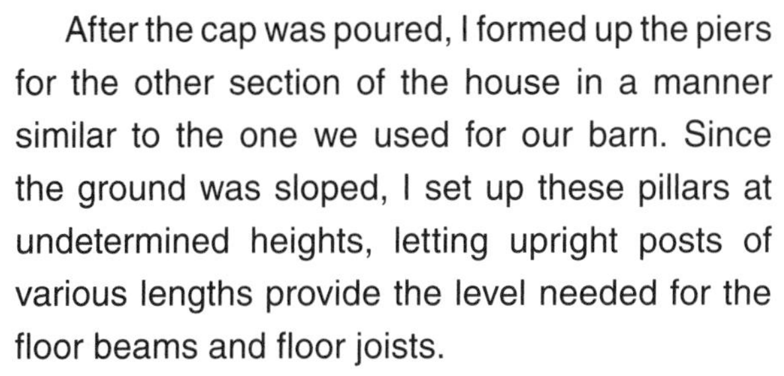

After the cap was poured, I formed up the piers for the other section of the house in a manner similar to the one we used for our barn. Since the ground was sloped, I set up these pillars at undetermined heights, letting upright posts of various lengths provide the level needed for the floor beams and floor joists.

The floor beams were set over the posts at approximately 6' intervals, going east and west. Because I wanted the floor to be as low as I could get away with, I used two-by-four materials for the joists, spacing them across the beams at 12" intervals. This is sufficient when using a two-by-four fir subfloor—the floor is quite solid.

We wanted our son's bedroom, the dining area, and the kitchen to be a step up from the living room for a multilevel effect. We also wanted to keep from cluttering the basement with several upright braces for the floor above. With these purposes in mind, we used one-by-twleve floor joists for this area. These joists span over 9' before crossing the floor beam above the

Joist placement over basement floor beam and post

basement. The floor is solid and the joists are well within the 1" thickness to 1' of span formula.

The floor beams under our son's room go north and south, and the two-by-four joists span across them to the top of the basement wall on the south and to a beam on the north.

And next came the log work: The sill logs were either butted or notched together, depending on how the intersecting walls met. We used round or saddle notches for the outside corners and hidden-shoulder dovetails for the inside corners or junctions (south living-room wall to west dining-room, east and west bedroom to south bedroom, and north living room to west bedroom).

We did not want to deal with chinking, so we decided to take the extra time to lay the logs in the Swedish, chinkless method, but little did we know what we were getting ourselves into. Using this method, I could lay two full logs a day; some of the professionals can lay three or four. Because we had eleven full sides per tier, we did not see much progress for some time.

To begin the "simple" outside-to-outside cornered logs—those with one saddle notch on either end—you begin by setting your log across the logs you will be joining it to. For interior finishing and aesthetic appeal, we, as most builders do, plumb the inside wall, and let the outside taper in or out, depending on the log. To plumb the log, take your level and hold it upright against the log, parallel to the log you are notching. Shift the log until it comes up against the plumb level and dog it in place with a log dog.

Next, take your scribe, making certain that its handle remains at all times exactly horizontal, and scribe the shape of the intersecting ends of

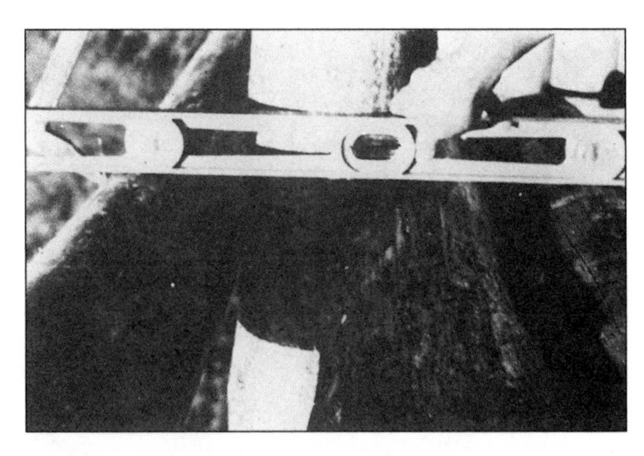

Initial plumbing of log

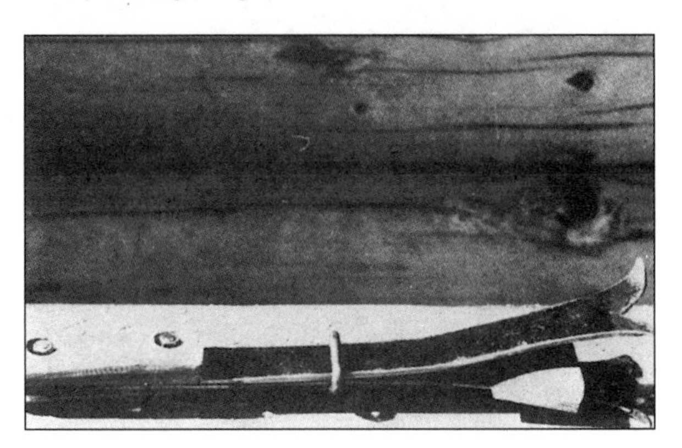

Homemade scribe with block

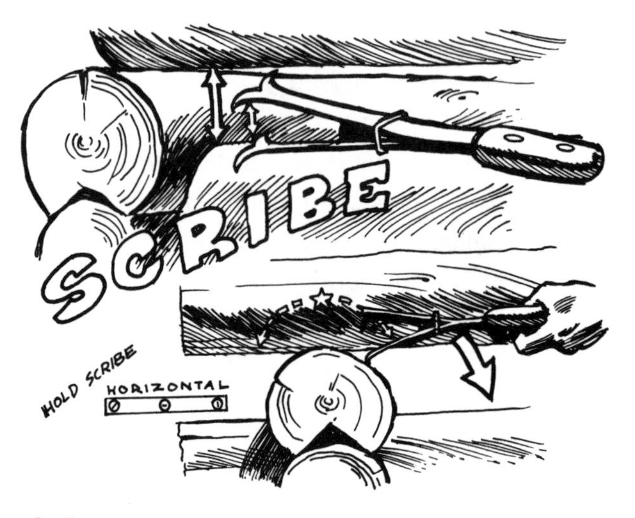

Scribe gauge

Author lining notch with homemade gouge, also shown at right close-up

the log lying beneath onto the ends of the log you are notching. To figure out how wide the tips should be spread for the preliminary notch, find the narrowest distance between your log and the log below. Spread the tips and block them at a width ¼" shallower than this space. Be sure to start your scribe mark at the apex of each log end and go down from that point, making the lines on the sides meet under the log. If the scribe is not held perfectly horizontal, the ends of the notch will either be too tight or too loose.

After making the scribe mark, turn your log over, marked side up, and again dog it in place. Now outline the scribe mark with a shallow gouge, making sure to cut in deep enough to prevent splitting when chainsawing. A friend of mine made me an invaluable tool for this purpose from an old leaf spring.

To cut out the notch, first chainsaw several vertical kerfs into it, then knock out the cut pieces and clean the seat in swift, steady motions with the tip of the chain saw.

After this primary notch is cleaned out, roll the log over and place it in the previous position, plumb to the inside wall. If the log does not seat on the notch, turn it over and try again. If it does, notch it in place. Now take the scribe, again being careful to hold the handle horizontally, and trace along the length of the bottom log, carefully scribing its contour on the log to be notched. To figure out the depth necessary for this scribe, examine the bows and dips in the upper and lower logs, and adjust the tips so they will cut into the top log enough to provide at least a 3" seat along its course. With bumpy logs, this seat can vary from 3" to more than 5".

This final scribe line should begin at one end of the top log, go around the saddle notch, go the length between that and the saddle on the other end, over this notch, and then run back to the original end.

Once the scribing is carefully completed, roll the log over again and begin the cutting. Your success with these cuts depends on how well

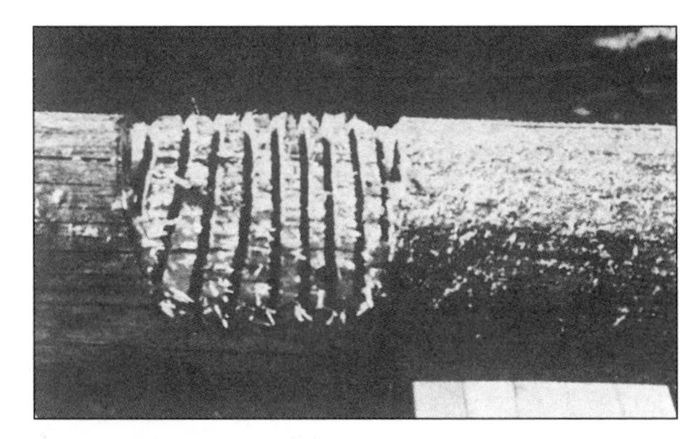

Notch with kerfs

Knocking out kerfs

Cleaning out notch seat with chain saw

you scribed and how well you follow your lines. It might help to darken the line with a pencil.

I usually begin with a long, lengthwise cut from notch to notch. The area beyond the notches should be trimmed with a hand tool because this area won't be filled with insulation. To make the first cut, turn the log about 22° to the left or right and rip vertically along that line just deep enough so your cut can be joined by the next cut, which will be at right angles to this first one. You need to cut at this acute angle to prevent the log from riding on the inside shoulders, rather than on the scribed line. You should by no means make the cuts deeper than half the log's diameter, because this would weaken its strength. This rip should be made at high revs, with your saw held at approximately a 45° angle. Never attempt to saw exactly on your line; most of the time it deviates too much from the log's contour. Trim the line later with a chisel or sharp hatchet. When you finish with the first length cut, turn your log about 45° and cut vertically down the opposite side, meeting the cuts in the middle.

Next, make the final notch cuts. The notch is again outlined with the gouge in the same manner as before, but this time going as deep as possible to save on "chain-saw jitters"; the chain saw is used enough as it is. Then, cut some small kerfs, going deeper in the center than at the edges, which gives the notch a concave seat for the fiberglass insulation that will fill it. These seats are also cleaned out in the same manner.

After this and the end trimming is completed, your log is ready to be rolled over to see if it fits snugly in place. This trial fit should

Scribing for long cut

Scribing and cutting

be done before tacking on the fiberglass-and-oakum sealer, because getting a perfect fit the first time is rarer than getting a hole in one on a tough golf course. Not even a pro can boast of more than a perfect fit once a week. Often I have to turn the log as many as five or six times before the fit is exact. The major problem is usually that the log is riding on the notches or on an inner ledge. To find the problem points (riding points), hit the log in several places along its length and over the notched areas, then turn the log over. Wherever there is a crushed point or new discoloration is usually the place where your log is contacting the lower log. Either chainsaw or hew these spots down with an axe. If the log is riding on the notch, carefully mark the spots and trim them with the gouge. Then try to seat the log again and again, until you achieve the perfect fit.

These round notches were a breeze compared to the inside, hidden-shoulder dovetails. This is a tricky notch to execute, but is quite aesthetically pleasing. To begin the notch, cut the log to be joined to the proper length. This length should be long enough to notch one end and

Making the length cut

Outlining final saddle-notch cut

Trimming scribe line

Length cuts and trimming completed

Final saddle kerfing and cleaning

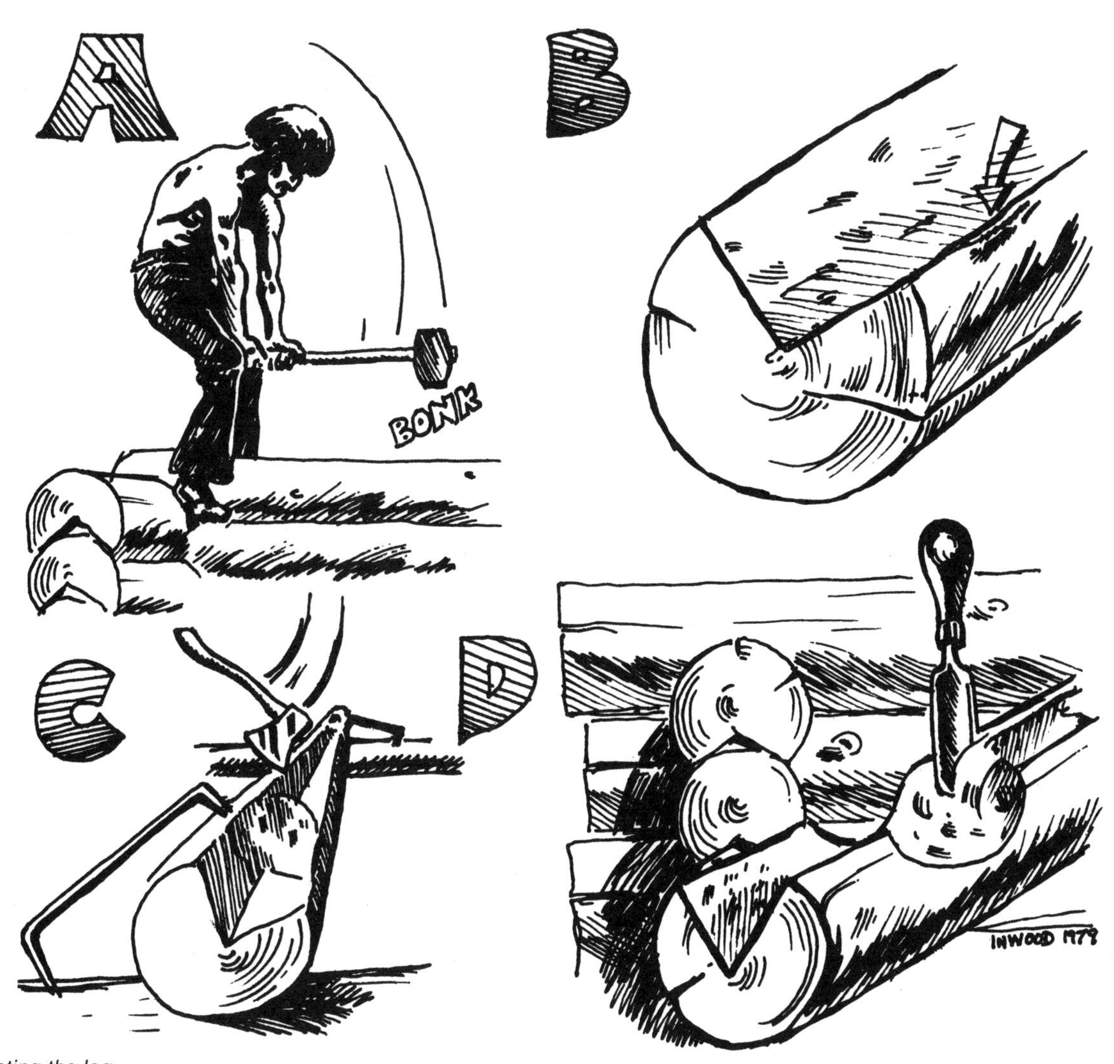

Seating the log

to bring the other end half into the width of the joining log. Scribe your primary notch on the other end. Measure the diameter of this end, find the center point, and with a level, mark a plumb line from top to bottom. Measure 2½" in either direction from this line and mark two other plumb lines, 5" apart, centered on this end. Now measure, at the top of the log, a line 3" in from the end. Eyeball straight behind the center end line, measure 2" in either direction at this point, then draw another line. This line should be 4" long and centered. Make a straight, diagonal line on either side, joining the 5" lines with the 4" lines.

Now adjust your log so the end is hanging free of any obstacles and make your primary cuts. Being careful that your chain saw is held in

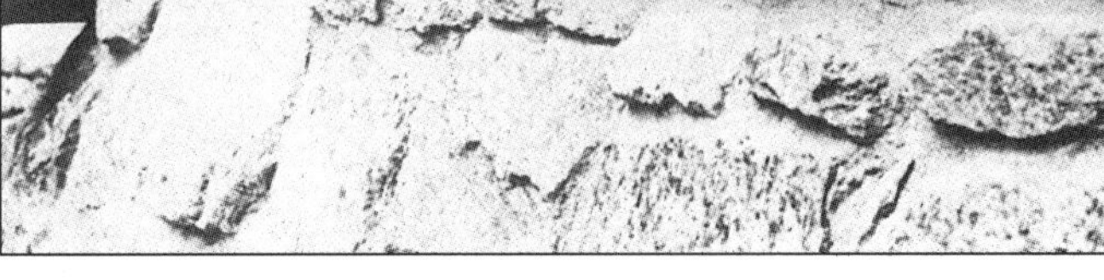

Well-fitted wall logs

Close-up of well-fitting notches

exact plumb and you are cutting straight up and down, guide the saw across the log on either side of the intersection of the diagonal and crossing lines (the 4" lines), then rip down the diagonals until you meet these cuts. The end piece that is left should be a vertical, dovetail shape.

Make your round notch on the other end. This will allow the log a seat and eliminate the need for the dogs. Where the dovetail end intersects the crossing log, draw a line on the lower log, then trace the body of the dovetail. With the exception of the bottom wall logs, this primary dovetail cut will already be in the lower log.

After an inset is made into the lower log to receive the male dovetail end, place it into the slot. If it is not a snug fit, don't worry about it. It will not be seen. If the male end does not fit into the receiving female end, trim it down.

Now, as before, adjust your scribe to the desired width for the length of the log and the saddle-notched end. Make your scribe lines, then carefully scribe the contour of the upper log's shoulder on the facing sides of the lower crossing log, making certain that the scribe handle remains exactly horizontal at all times. Now remove the top log and gouge out the meat along the scribed line to the depth of the male end. When the log is ready to be set into place, it should fit snugly into this notch, with its shoulders hidden within it.

Once this log is in place, it is ready to receive the lower half of the next crossing log. First plumb and position this next log over the male dovetail end, and trace that piece on the underside of this log to be notched. Turn this log over, making certain it is exactly upside down. Check the plumb

Marking plumb lines

Male dovetail

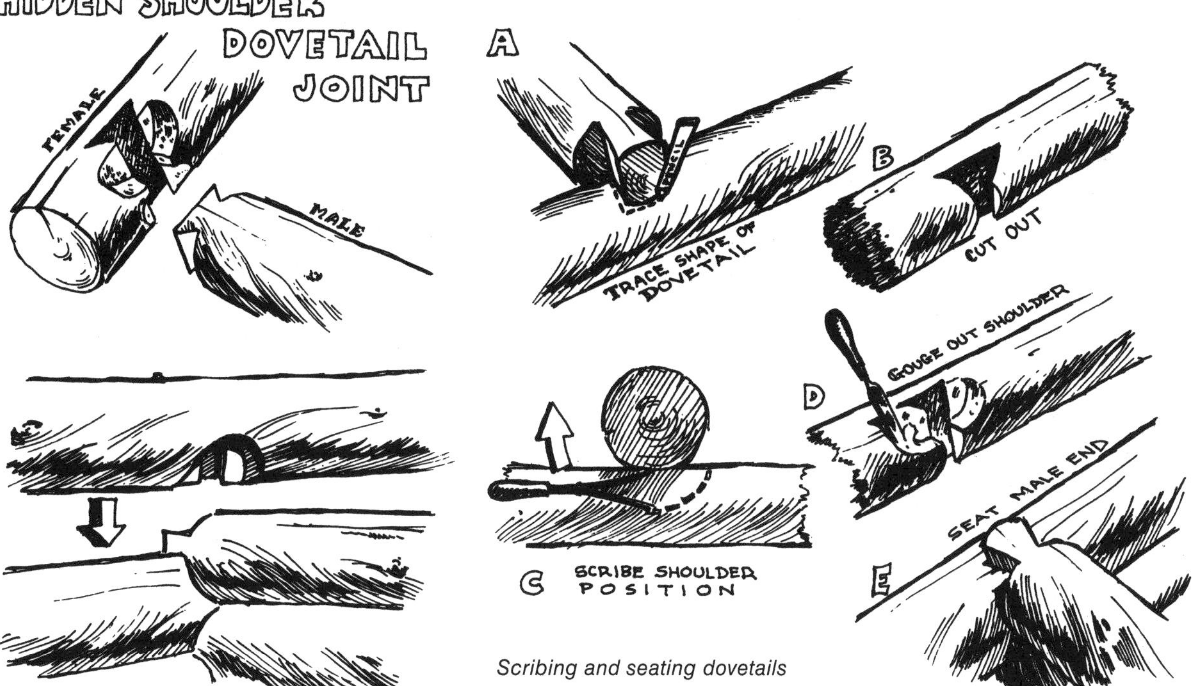

Scribing and seating dovetails

Top female log seating

line on the end with your level, then carefully cut out the female.

Now, scribe the remainder of the log. Using the same adjustment on the scribe, trace the intersecting shoulder of the dovetailed end on either side of the facing female slot. Gouge along these lines and trim out the notches. Try not to get frustrated if it takes you half a day to complete a log with one of these notches on it. That's how long it took me. (There are easier ways, but none as aesthetically pleasing.)

After a log is ready for fitting, tack shipbuilder's oakum along the inside edge, completely around the grooved inside area, including the end notches. This substance waterproofs and seals the joint. Inside the grooves and notches, a sufficient strip of fiberglass is set for insulative purposes.

Our house has eleven logs per tier and thirteen tiers (143 logs). Using the method just described, I can lay two logs a day. Alone, that works out to about seventy-five days of labor, not counting the felling, peeling, and hauling. We had a particularly rainy year and could only work, on the average, three days a week. You don't want to work on a log building in the rain; I tried it once and fell off. Because of these difficulties, I decided to hire a couple of my friends, professional log-builders, to assist me on the house. I hired myself out on a lucrative tree-planting contract and made enough per day to pay both of their wages. Though I felt this was

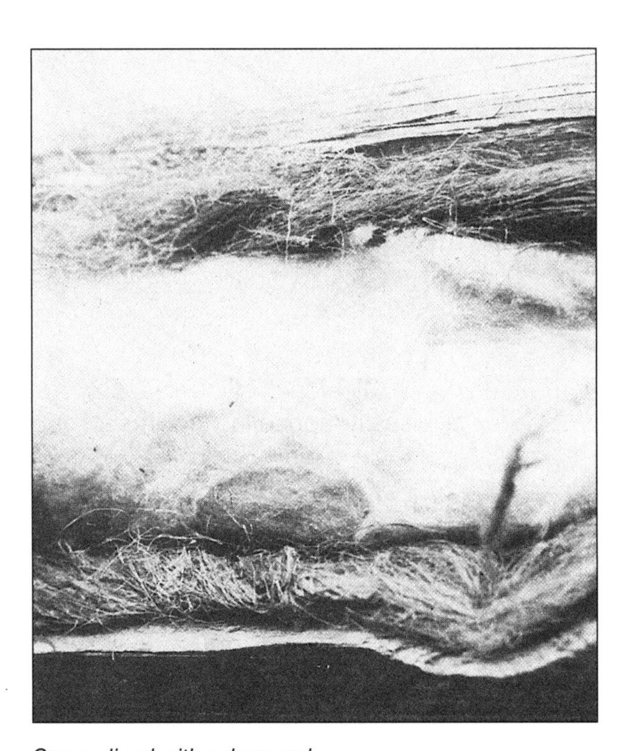

Groove lined with oakum and filled with fiberglass insulation

Finished hidden-shoulder notches

Log ascending purlins

Log after being notched

Ceiling and joists over living room

a cop-out, now I know it was the best move I could have made under the circumstances.

After my contract was completed, I returned and joined in, then finished the last few tiers alone, or with Margee, or with the help of occasional visitors. One such visitor, an old friend from the city, came by and was immediately indoctrinated into the lifestyle. We purlined up a few top logs and I cut in the notches to seat them in place. The illustration shows one such top log which fits over the ceiling joists of the living-room north wall.

The ceiling joists were round-notched over the walls and their tops were hewn for the two-by-six tongue-and-groove cedar subfloor and ceiling. This material was used because it provides a visually appealing ceiling for the lower areas.

The tops of the top plate logs were then hewn to provide a flat surface for the two-by-six framed pony walls of the second story. We decided to use frames to save time and energy. We had to have a roof on by winter and we were becoming exhausted. These walls were sheathed on the inside with spruce tongue and groove because this is a light wood and would contrast with the

Main bedroom

Bathroom

Southwestern view

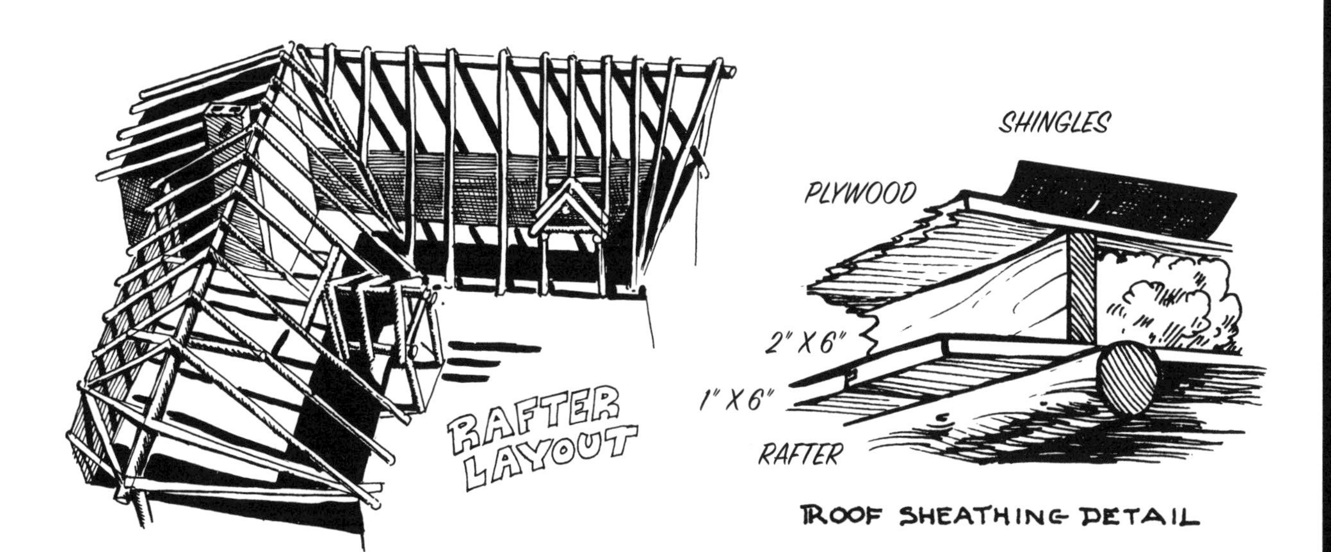

exposed log rafters. The bathroom was sheathed in cedar and Gyproc.

On the outside, the walls were braced with plywood. The plywood was given a vapor barrier of black paper and was exterior-sheathed with one-by-four cedar in a lap siding style. We didn't buy actual lap siding because the cost was prohibitive. We chose this style because we thought that the lap siding would blend well with the logs.

We wanted the exposed, log-rafter ceiling over the bedroom, studio, and bathroom, and were willing to do the extra building to have it. A sheathing of one-by-four hemlock, another light wood, was nailed over the rafters, and a network of two-by-six boxes for the insulation was built above it, crossing the rafters. The box section was covered with a layer of ⅝" plywood and the roof was finished with two-tone green asphalt shingles. (I am not a cedar-shake fan and would never roof a house with such an inflammable material.)

Studio bay window and bedroom dormer

We deviated from the normal wall construction to put in a dormer and a bay window in the bedroom and studio. This also changed the basic rafter design in the interior. The inside framing of these special openings was constructed out of log so it would blend with the rafters.

After the roof was completed, we escaped south for the winter and returned in the spring for tree planting. After tree planting, I went to

Log framework of studio bay window

Log framework of bedroom dormer

Unfinished fireplace

Stonework chimney in studio

work on the stone fireplace, again trading labor with the mason friend who had helped with the basement. We built the pad and set in the heat form, then built the necessary block work around the flumes. With this core completed, I went north to a quarry of blasted granite and brought back a few truckloads of angular stone. Since what is beautiful in stonework is a matter of personal taste, I built most of the fireplace alone. The stones go beyond the ceiling, into Margee's studio.

Later, after the fireplace was completed, I began cutting out the door and window openings. Before making these cuts, I measured the openings and framed them with plumbed and leveled one-by-fours, then carefully chainsawed the vertical cuts, cutting flush against the guide boards. The most accurate way to cut is to start near the top and cut with the tip up, higher than the body, at about a 45° angle. This way the blade is cutting down into a few logs at a time, guiding itself down a straight line. Always use a safety chain and cut at a fast speed to help prevent the saw from jumping back at you. Always saw downward, trying not to bind the tip more than necessary.

After the vertical cuts, I set up a horizontal guide on the outside of the bottom jam, or sill log, under the opening. This guide was level with the inside board. I made several kerf cuts down to the guides and broke off the excess pieces with a mallet, then hewed the sill flat with an axe. When cutting this sill, you should not go any deeper than a third of the way into the log because of

Vertical sawing through logs

Horizontal cuts

Door spline

the previous V-groove along its length.

Now measure the height of your opening and follow the same procedure at the top.

After the openings were cut, I splined them with two-by-two splines to hold the log ends in place during the settling process. When making these cuts for the spline, you should gouge a bit deeper so you can back the boards with insulation to fill the uneven spaces. The length of these splines should be 2" shorter than the opening, to allow for settling. Do not nail these splines in.

When setting in a window or door frame, you should be careful not to affix it to the actual

Window spline

walls, but only to the spline; otherwise, the wall's settling could break or contort the frame.

We looked forward to living in our house by winter. We are happy we built it the way we did, and appreciate the rustic beauty that only natural building materials can provide.

Chris & Margee's Barn

We were living in a $225 a month two-story, termite-ridden, cess-pool-stinking, rural rat-shed rental with bees buzzing in the chimney, gas leaks in the kitchen, and water so soapy it sudsed by itself in the sink. But we were satisfied there; that is, until the nearby city's smog penetrated through the mountains and putrified the once-pure air. With it came the tourists and the traffic checks and the tickets and the helicopter harassment. It was time for us to move on. We had a little money, so we considered losing ourselves in the high plateaus of Peru and ski the Andes until the money ran out. First, though, we wanted to visit our friends.

When we got to their homestead, we expected them to relate their hardships in trade for ours. But, though they were living out of a temporary one-week wonder, which they built out of scrap lumber, and though the temperature was just above freezing, they had nothing but good things to say about their new home. All was going well for them. They were working together for something they believed in. They loved the peacefulness of their riverfront forest and would not trade it for anything.

Instead of sinking under the weight of our complaints, they began teaching us new things. They showed us edible mushrooms and plants that grew wild, which were there for the taking. They showed us the forest and told us their plans for a new log house. Before long, we helped them cut and peel logs and prepare firewood.

Our friends, being former city dwellers like us, did not have any knowledge of building with logs or any other material. But they were determined to build their log house. Within hours after the project had started, many neighbors, both old-timers and new "textbook pioneers," offered their assistance. Some of them helped physically and some expounded theories over homemade elderberry wine. We couldn't believe the cooperation and helpfulness of these people. They were true neighbors.

After working with our friends for a while and learning from some of their mistakes and experiences, we decided that their area would be a fine place in which to settle. The folks were really friendly and the valley, surrounded by large mountains, was very beautiful and protected.

We searched for a short while and soon located an old homestead complete with house, barn, root cellar, and outbuildings. Our vacation money went for the down payment on the place—so much for South America.

Soon after we moved in, the roofs of the old barn and storehouse collapsed. We examined the remains of the barn and found that most of the massive 12"- to 16"-diameter cedar logs which comprised its walls had decayed into rotted pulp, good only for composting. We invited a few neighbors to help demolish the structure and shared whatever firewood we could salvage.

The root cellar was the next to go. The only thing usable from that mess was the concrete front. The rest went to the dump.

At least we could live in the house for a while. Even now, it is still sound. It has no actual foundation, just a few rock piles it rests upon. Its walls and floors slant in various directions, but it is warm and cozy—and the price was right.

We really got into the rural frame of mind fast. We wanted to immediately become self-sufficient and raise many useful animals. We bought a horse for riding and farm work. We purchased a Black Angus heifer for milking and breeding, and goats for immediate milk.

Since the winters were so cold there, the first construction project was to build a log barn to house this menagerie. I wanted to try building with log, firstly, because of its rustic beauty, and, secondly, because it is plentiful in our area and there were many sections of land that could have used a little selective thinning out.

We have an adequate woodlot on our land, yet we wanted to get the logs for the structure elsewhere for the time being. A neighbor from across the river needed some land cleared. He made me a proposition I couldn't refuse: If we would clear the land for him and raise his pig along with ours (yes, we had pigs, too!) until it was ready to be slaughtered, he would let us take all the tamarack we needed for the barn. He suggested we use tamarack because of its durability and straightness. We found that it was also a very easy wood to work with because it had few limbs or knots except near the top.

We trucked the logs to the barn site and immediately began peeling off the bark. We did not have a peeling spud, so we improvised with an old shovel that a neighbor lent us. Its round tip was cut off an inch from the end and the new edge was filed sharp to cut more easily. This proved to be an excellent tool for the purpose.

The bark stripped off easily because the logs were cut in the early spring when the sap began running. But as the weather became warmer, we found it increasingly harder to peel the logs, not only because the sap was drying and we had no shelter from the hot sun, but also because we made the mistake of having our work area downwind from the pigpen.

As we cut and peeled the logs, many neighbors came to give us a helping hand and offer us advice. Some of the advice, I must admit, was rather unsound, but most of it was necessary and useful.

I started trading labor with some of the locals. One neighbor helped me set up batter boards and showed me how to rough out a square 22 × 22 feet. He checked the square by measuring a line 6' on one side and 8' on its perpendicular. In order for the corner to be an accurate 90° angle, the diagonal of the lines had to be 10' long. We measured each corner, then double-checked the crossing diagonals. They had less than 1" difference in length.

In return for this help, I worked on his place with him. Though I didn't know much about construction or farming, I was eager to learn and he was happy to have my help. It was all a valuable experience for which I was grateful.

We decided to use ten 24" footings as the primary foundation for the barn. We dug the holes: one at each corner, one at the center of the sides without doorways, and one on each side of the two doorways. Each hole was 12" deep and more than 24" wide. The footings were 24"-high pyramids, 12" square on top graduating to 24"-square bottoms for maximum support. We made rough one-by-six forms and braced them with thin strips of one-by-four on the outer side. Each one was set in a hole. I checked the levelness of the forms by filling a hose with water and stretching it from corner to corner on one side. The ends of the hose were bent up to just above the tops of the forms. If the forms were level, the water levels within each hose would be horizontal with the tops of the forms. If that wasn't the case, we would adjust the depths of the forms. If the water level was too high, the hole was filled until the form was the height of the others.

The cement, gravel, and sand were then measured on a platform, mixed with water, and poured into the forms. We used a mixture of 1 part Portland cement, 2 parts gravel, and 3 parts fine sand. We also added quite a few large rocks for volume. This saved on cement and didn't seem to take away from the strength of the footings.

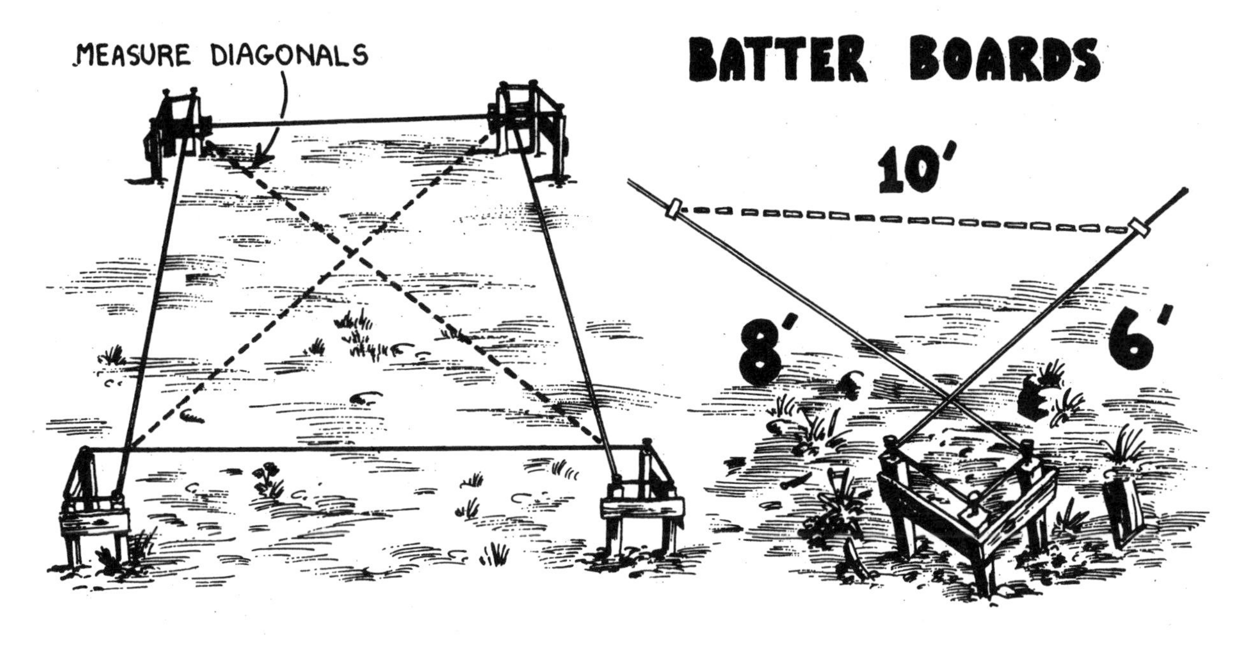

Luckily the area for the barn was relatively level, so each footing stood between 8" to 12" above ground. The sill logs would rest on top of these footings.

We figured that the best floor for a barn is a natural one of porous soil. Urine seeps through it and after a season or two, the decayed bedding and uncollected manure break down on it and produce an excellent humus topsoil, which can later be put into the garden.

There is one problem which arises, though, when building a log structure without a floor. All four of the bottom logs will be sill logs of the same level. When using any notching method to join logs, the logs must be staggered. Two parallel logs should be set on the pillars and two crossing logs should be notched a half log above them. The space between the crossing logs and the center pillars can be filled by making these pillars half a log higher than the others. The higher center pillars will then support the raised end logs. The space that is left between those logs and the ground can later be filled with stonework or other foundation material.

Another way to rectify this problem is to make notches in the underside of the two lower logs. These notches should be at least one third of the log's diameter in depth, and should be long enough to cover the top of the pillars they will be resting on. Notch the crossing logs to these with dovetail notches or any other notch you prefer.

Dovetail notches are widely used when working with milled or hand-hewn logs because of their ability to interlock. Their design prevents them from coming apart no matter in which direction stress is applied. They pull each wall log toward the center of the building, making it impossible for the logs to fall outward.

Interlocking dovetail notches

The same principle holds true when using round logs. Because of this interlocking ability, we found it safe to use this notch without having to spike the corners. We saved ourselves money on materials and prevented premature decay in the notches. Nails and spikes contract and expand at different rates than wood. After a period of time, the moisture and oxidation that collects between the two surfaces will rot the wood and weaken the corners.

Dovetail notches are simple to make. First, I used a level to find two parallel lines (north to south) at each end of the log. The lines were about 1" to 1½" in from the outer circumference of the log end. Then I measured the diameters of the crossing logs to be notched above. I marked these measurements on the side of the prospective ends of the lower sill logs.

A sharp pruning saw was used to cut in against the grain to meet the depth of the parallel

Establishing a vertical line for flat sides

Removing side slab

lines. I knocked each of the pieces out with an old axe and a hammer. Then I placed the log in position.

I rested a square on the end of this log and slanted it at a slight (30°) angle, then marked a line across the log end. A saw cut was made at the rear of the notch to the same depth as the slanted line. This piece was then knocked out and a wedge-shaped log end was formed to interlock with a coupling notch.

A crossing log was put in position above the first logs. The level was again used to find two parallel lines, and the appropriate pieces were sawed and knocked out. As the top log was held in position, I rested a square on the slant of the bottom-log notch and measured the distance between the high and low points. The distance and slant were duplicated on the side of the upper-log notch with the high point being at about half of the log's diameter. I drew a line, sawed a cut against the grain, and chopped the piece out with a hatchet.

The log was then fitted into place, which usually meant hewing its bottom until it fit snugly. The amount of hewing needed with a relatively straight log depended on how carefully the slants and distances of the log below were measured. If care was taken, it would fit tightly without any hewing. That was seldom the case. If the notch was cut too deeply, the log bottoms and tops would have to be hewed accordingly until

Making saw cut

Forming the dovetail notch

Transferring the angle to the top log

the notch interlocked properly. If the notch was not cut deep enough and the gap between the logs was too wide, then the bottom notch would have to be hatcheted or axed out until it was wide enough for a proper fit. This process was repeated until the walls were finished.

Corner of finished walls

Since we were lucky enough to have plenty of good logs to choose from, we built the barn walls out of solid logs wherever possible. But I miscounted the number we needed, and we ended up not having enough long logs for the whole structure. Consequently, the back wall was constructed of 8½'-long pieces. An 8½' piece on either side left room for a 5' doorway in the center.

Though the logs were heavy tamarack, which had dried for less than two months, they were not very difficult to lift. Using a simple parbuckle, my wife and I managed to hoist most of them into place, and waited for visitors to assist us with the others.

This parbuckle consisted of a long section of rope with both ends tied around the sill or a low log. The center of the rope was brought to the inside of the building, thrown over the wall, and looped around a log that was placed at the foot of the two skids. As the rope was again pulled to the inside, it raised the log on the skids and hoisted it to the top.

We spiked the fitted logs on either side of the proposed doorways to keep them in place after the openings were sawed through. Some builders spike either side of the window areas, but we felt that would limit us in window design.

We notched in a crossbeam above the walls before we placed the top wall logs in. The crossbeam is important structurally because it supports the ceiling joists. These joists have the dual purpose of holding up the loft floor and of making an additional tie-in for the walls.

Cutaway of interior stall layout

Each joist was placed at 19" on center except for the one at the rear. There we allowed a 30" space between that joist and the rear wall, for a trapdoor. This trapdoor is an inside entrance to the hayloft. It is large enough to easily allow hay bales to be thrown through it for the hungry occupants waiting below.

At this point we located a portable cement mixer. A fellow up the road, who eventually became a good friend, needed to use my truck, so in trade, he loaned me his mixer for a day. That entire day, and into the late evening, my wife and I mixed many loads of cement and gathered many large rocks for the stone retaining wall. By lamplight, before the night ended, a very solid and beautiful secondary foundation was completed. This retaining wall prevented logs from sagging and sealed in the sections between the footings.

Next came the roof. We chose to use a typical gambrel roof, because this style provides the most usable storage space for hay and grain.

The first frame was constructed 14' wide, 7' high, and 22' long from front to rear. This space proved to be more than adequate for at least four tons of hay and many sacks of grain. We used 5"- to 6"-diameter poles for the frame, and reinforced it with center posts, side braces, and double crisscrossed poles at front and rear. I found, from seeing many old barns with collapsed roofs and still-strong walls, that you cannot overbuild the roof section. It needs all the support you can give it.

After the frame was secured, a ridgepole complete with notches for a second set of braces was placed above it. The braces and a top ridgepole were then spiked in.

We cut twenty-four poles, which were 4" to 5" in diameter, to a length of 8' for the lower set of rafters. The bottoms of these poles were notched and spiked into either end of the twelve crossbeams. The tops were attached to the top of the frame. After these rafters were in place, twenty-four more were cut. Three

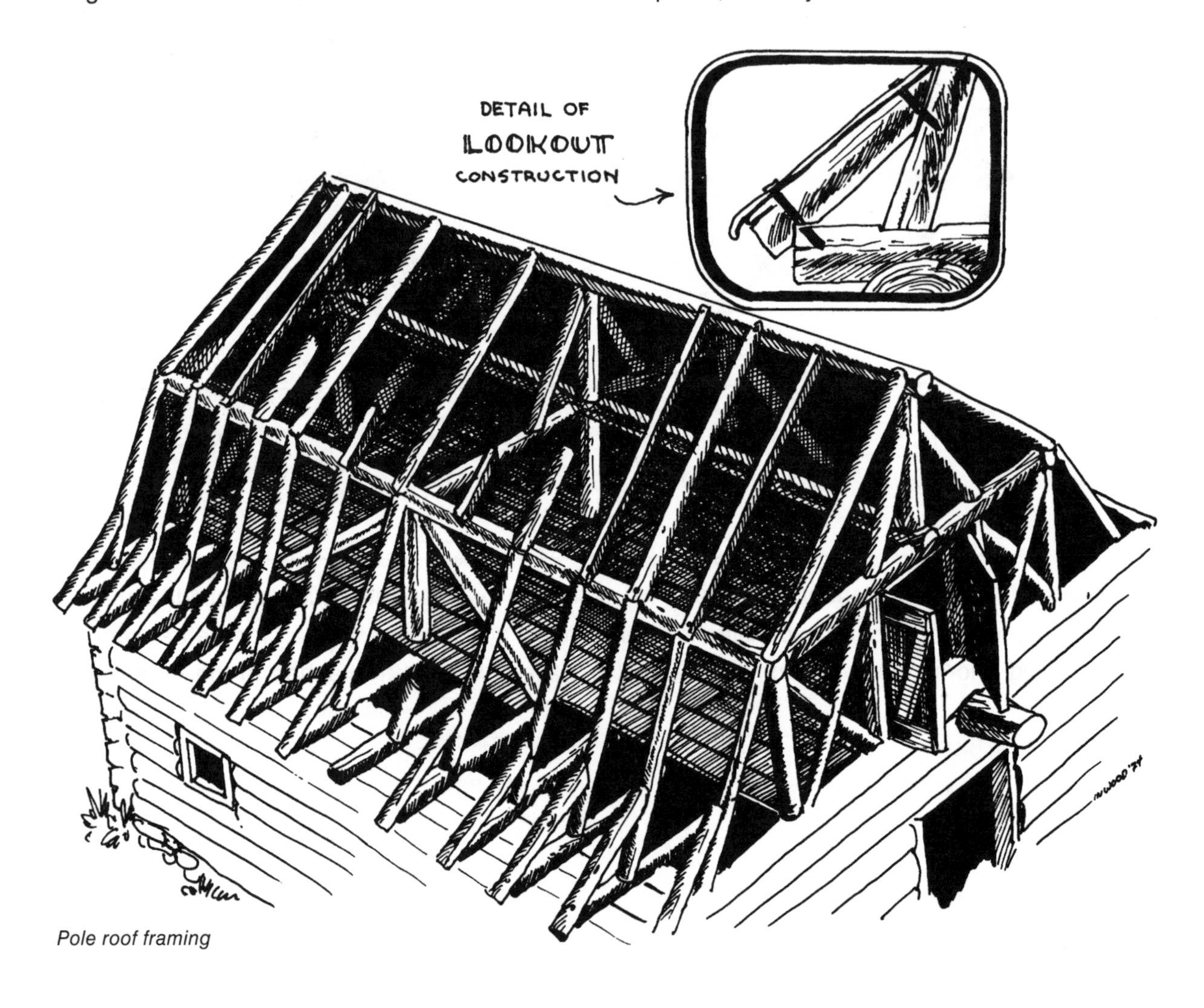

Pole roof framing

vertical sections, each 4' in height, were set in the notches at the front, center, and rear of the ridgepole. After we braced them, we wondered how to hoist the final ridgepole. My wife was pregnant, and we thought it unwise for her to chance the chore. Before long, though, her cousin and his massive friend came for a visit. Without even being asked, the friend carried the ridgepole to the roof and hoisted it up. He then slammed the spikes in place and helped us connect the other rafters.

I examined the framework and decided that the pitch of the top rafters looked too awkward because of their height. We removed the few connecting rafters, chainsawed 1' from each of the three extensions, and sledgehammered the excess pieces from the ridgepole. We later replaced it and followed the friend's suggestion of enjoying a few beers for the remainder of the afternoon. A few inches were later cut off the upper rafters to compensate for the shortened extensions, and they were again spiked into place.

Thus far, building expenses had been nominal. We had spent less than $80, including the cost of feed for the neighbor's pig. But at that point, milled lumber was needed.

Shopping for usable but inexpensive lumber can be a big chore. We went from mill to mill to find enough economy-grade material to floor the hayloft and use as nailers over the rafters. Lumber prices are constantly on the increase. Even the lowest grade one-by-fours we found were over our budget. We finally located a private party who was selling his boards. He included enough one-by-eight and one-by-ten cedar, to

Ridgepole, brace, and upper rafters

do our flooring and enough other 1" lumber to use for nailers and gable ends.

The tops of the ceiling joists had to be hewn before the loft floor could be laid. After the hewing, cedar boards were quickly nailed across the joists and a space was left for the trapdoor. The nailers were then attached to the lower rafters. After they were attached, we found out it was not necessary to czonstruct a solid covering. We then spaced the upper nailers 6" apart to save labor and materials.

We considered many different roof coverings. Although it was decided that a cedar-shake roof would have been the most aesthetically pleasing and least expensive, we chose to use an aluminum roof. The shake roof would have been impractical because of snow buildup on the upper sections, and because of the difficulty of climbing on it in the middle of winter to clean it off.

A freak mid-September snowfall reinforced our decision. The barn needed a roof immediately. The cost of the aluminum roofing was $100, but it was worth it. Though it lacks natural beauty, it is the most durable and maintenance free of possible materials. It also sheds snow and is easy to install. Two of us put most of it up in a day. It was late November before it snowed again.

Now, it was time for the interior work to begin and it was time for the barn to have doors and windows. I measured where I wanted to have the front door and nailed up two-by-fours as vertical guides for my chain-saw cuts. Then I cut through the logs to make the opening. That opening was framed with a double thickness of two-by-six, which acted as hinge supports and door frame.

It took me awhile to figure out what design of door I wanted. I needed a solid door for weather, but I did not want a massive door that would be too heavy for the hinges. I looked through the scrap pile of log pieces that I had left over and found some short sections of 4"- to 5"-diameter cedar and tamarack. I set up a guide, using a two-by-eight platform with two nails sticking out of it about 3" apart.

The log pieces were each cut to 38" because I wanted to have Dutch doors for ventilation and light. I hammered a log section down onto the nails of the platform and made it stand upright as I ripped it down the middle with the chain saw.

½ ROUND SLABS OF ALTERNATING CEDAR AND LARCH WOOD

This process was quick and easy.

After the half-rounds were cut, they were laid out in a line. I alternated the tamarack and cedar for appearance and structural reasons; nails hold better in tamarack. The cedar backing was then nailed on and one door was ready for hanging.

The second door was done in the same manner and applewood handles were added as a finishing touch. We are very satisfied with the way the doors came out.

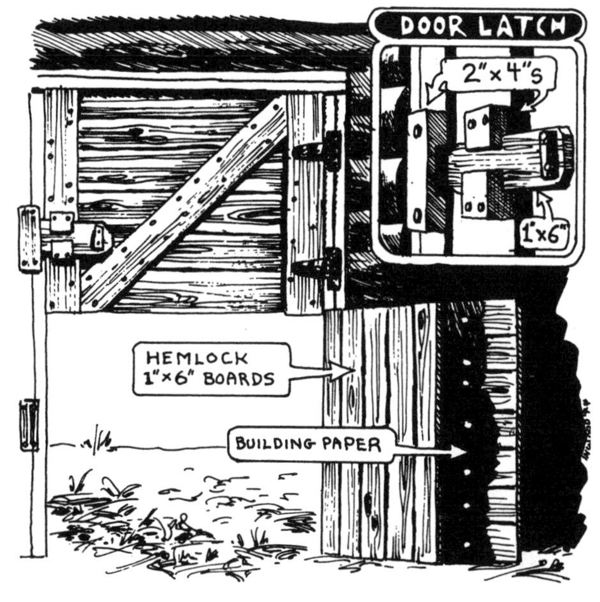

Dutch-style front door

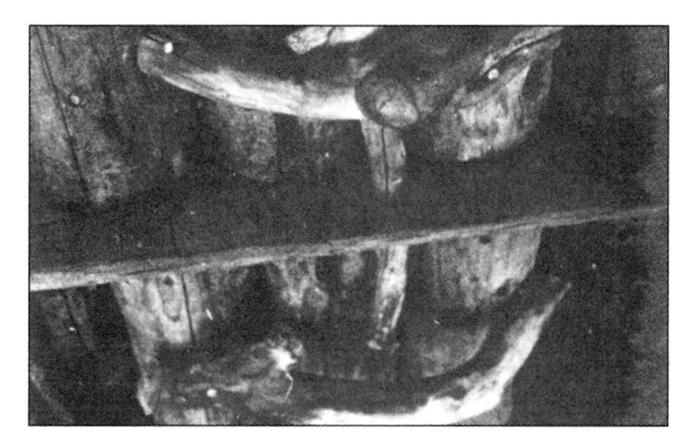
Applewood door latch

Rear door

Rear door latch

Recycled auto windows

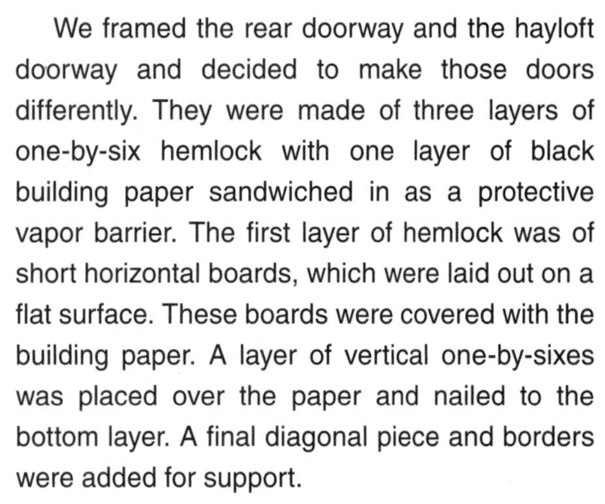

We framed the rear doorway and the hayloft doorway and decided to make those doors differently. They were made of three layers of one-by-six hemlock with one layer of black building paper sandwiched in as a protective vapor barrier. The first layer of hemlock was of short horizontal boards, which were laid out on a flat surface. These boards were covered with the building paper. A layer of vertical one-by-sixes was placed over the paper and nailed to the bottom layer. A final diagonal piece and borders were added for support.

The window spaces were next. I wanted to make sure the occupants of this building had plenty of light, but I knew I had to safeguard against window breakage, especially with our obstreperous animals. After searching through many of the junked cars in the area, I found some nearly rectangular glass in a few early model Dodges and DeSotos. This safety glass was easy to install and is very difficult to shatter.

We paneled the hayloft gables with the cedar one-by-sixes, and then we wanted to start on the interior. We were none too soon—the weather was getting colder and the animals were still without shelter. But instead of starting on the barn's interior, we were forced to construct a root cellar because the frost was already destroying our pumpkins and winter squash. The root cellar did not take long to build.

On completion of that project, a friend of ours saw the barn, and, after praising it, asked us if we wanted to take care of his milking cow while he and his family went away for the winter. Fresh milk and butter, homemade cheeses, ice cream, and yogurt—what homesteader could refuse such treats? He and his family planned to leave in two weeks. Though our Black Angus was not yet giving milk, she would eventually, so we designed one of the stalls to house his cow now, and be used by ours later.

After figuring out the various stall sizes, we dug holes in the ground a foot deep to bury the bottoms of the upright supports. Their tops were notched and spiked into the ceiling joists above. The size of the stall depended on the animal

Heifer stall

it housed. The two identical stalls at the right front were built to house the heifers or steer for the year. Those stalls are about 5' wide and 8' long. Each contains a feeding manger and an area for a water bucket. They have crisscrossed log gates which hinge and latch to the upright supports.

On the other side of the barn, we built two more stalls. The 9 × 5 foot one in front is for the horse. It does not have a gate. As long as there is an adequate place in the manger to tie a horse, there is little chance the animal will leave. Cows, on the other hand, should be hooked to a ring, and, if possible, have gates on their stalls. Our Black Angus heifer, Beulah, once unhooked herself, got out of her stall, ripped open the milk cow's dairy mash, ate what she wanted, and manured on the rest. When I came in to milk, she ran to the Dutch door as the bottom partition closed. She took a leap, came down on that section, and flattened it.

The 12 × 6 foot area to the rear is the milking stall. It has a manger and a water-bucket holder.

Cow-milking stall

It was built large to give the milker and cow plenty of room during milking.

Behind the milk cow's stall is a space about 3½ × 8 feet for storing a couple of bales of hay and a few sacks of grain. A ladder goes up to a trapdoor, which serves as the inside entrance to the hayloft. A simple rope-and-pulley system was designed to lift and lower the trapdoor.

Next came chinking. It has been proven that the easiest method of chinking a structure is to have a chinking party. Invite all your friends and

Hatchway to upstairs

Trapdoor in hayloft

neighbors to help you and you will complete the chore in one day. We set up a cement mixer and used a mixture of 1 part lime, 1 part masonry cement, and 5 parts clean, fine sand. To rechink the walls after they have settled, just use masonry cement and sand (1 part to 3 parts), because too much lime causes the mixture to be less durable and masonry cement already has lime in it.

We made palettes out of pieces of plywood and supplied each helper with a palette, an inexpensive pair of rubber gloves (to prevent lime burns), all the food they could eat, and all the wine punch they could drink. First we hammered in many nails along the top of the logs, spacing them about 2" apart. They were bent in toward the building to hold in the chinking.

I controlled the mixer, but could never keep up with the enthusiastic workers. The work went well, but I noticed that as the huge punch bowl emptied, the chinking got sloppier. The following day, after sleeping off our hangovers, my wife, a friend, and I went over the walls with a wire brush and a whisk broom to level the sections that protruded. We then smoothed out the rough surfaces. Chinking should be concave and should blend into the wood. This prevents water from collecting on it and working its way between it and the log, causing premature decay of the walls.

After completing the chinking, the barn was almost finished. All that was left was repairing Beulah's damage.

It has been over a year now since we completed the barn. We are very happy with it and we are proud of our achievement. I sometimes think back to the pleasure I had when building it. Each problem and each hassle that was overcome gave me a little more confidence and reinforced my belief in this type of life. And now, every time a neighbor or a friend wanders up to offer help, and every time someone makes a useful suggestion to us, my newborn faith in

Wall chinking

my fellow human grows stronger and becomes more a part of me. Every time I work with someone, whether it be on their project or mine, I learn a little more about communicating my own needs—I become more in touch with them as I experience them.

More and more, I'm realizing that the only way to be successfully self-sufficient, without the need of an established order, is to cooperate and communicate with those around me who also want to live the good life. It's surprising how many of us there are.

Index